PLANT SPECIES AND PLANT COMMUNITIES

PLANT SPECIES
AND
PLANT COMMUNITIES

Proceedings of the International Symposium held at Nijmegen, November 11–12, 1976
in honour of

Professor Dr. Victor Westhoff

on the occasion of his sixtieth birthday

edited by

Eddy van der Maarel & Marinus J. A. Werger

Dr. W. Junk bv Publishers   The Hague – Boston – London 1978

Additional material to this book can be downloaded from http://extras.springer.com.

ISBN-13: 978-90-6193-591-9   e-ISBN-13: 978-94-009-9987-9
DOI: 10.1007/978-94-009-9987-9

This Symposium was sponsored by the Faculty of Science of the University of Nijmegen.

The articles in this book have also been published in our periodical *Vegetatio* (see page 177).

Prof. Dr. Victor Westhoff

CONTENTS

Editorial note

This volume is not just a liber amicorum, a Festschrift. It is a report on a Symposium, a being together of friends, in this case friends of Professor Dr. Victor Westhoff. We thought that such an event would do justice to the extraordinary personality of Westhoff. And so we not only discussed scientific matters, on a theme that has enjoyed Westhoff's interest for a long time, but we also had food together, made fun, made music and read poetry. The special atmosphere of these few days cannot be described appropriately in this book.

We still have some things to say. First of all we were very much pleased, almost embarrassed by the tremendous response we received from almost all Dutch friends, pupils and colleagues and from the selection of foreign friends we approached with the idea of the symposium. This response made it necessary to ask some of the potential contributors to content themselves with poster stands or with the presidency of one of the four scientific sessions. We are very grateful indeed to all these colleagues and friends for their enthusiasm and readiness to contribute to the symposium in whatever way.

Secondly we enjoyed the effort of our publishers. Most contributions presented at the symposium or displayed at the poster stands have been published in VEGETATIO in the meantime; they are collected in this volume. Only a few contributions have not been included according to the authors' wishes; these were either summaries of previously published work, or preliminary reports considered not yet suitable for publication.

Dr. W. Junk Publishers are gratefully acknowledged for their rapid cooperation in collating the available papers in this volume.

Thirdly, we have to thank the Director of the Faculty of Science, Dr. C. J. M. Aarts, for his immediate agreement and support, Prof. dr. ir. J. J. Steggerda, chairman of the Faculty's Board, who kindly addressed Prof. Westhoff at the symposium dinner, and Prof. dr. H. F. Linskens, Director of the Botanical Laboratory, for his help with the initiation of the programme. A very special word of thanks we would like to say to Peter Toll, LL.D., of the Faculty's Directorate who took care of the general organisation of the symposium and the festivities involved in a both cheerful and very effective way! Through him we also thank the many Faculty employees who contributed in various ways to the success of the event. Without their help the symposium would not have been possible at all!

Finally we wish to express our feelings of gratitude to staff and students of our own department of Geobotany, especially to Mrs. A. van de Zand-Barten. Without their continuous help the symposium would not have become what it still is in our own memory: an unforgettable feast!

The editors

Eddy van der Maarel
Marinus J. A. Werger

OPENING ADDRESS

Ladies and Gentlemen,

Professor and Mrs. Westhoff: Dear Victor and Nettie: I welcome all of you at this special symposium in honour of a colleague who is to celebrate his 60th birthday and who occupies a very special place in our life as a plant ecologist, vegetation scientist or (Westhoff's preference!) geobotanist. At the same time we have come together in a relatively quite atmosphere (what we really do too seldom) to listen to lectures and to discuss, both publicly and privately, a theme that is of special concern to our 'hero of the feast' Victor Westhoff: Plant species and plant communities.

Dear Victor, this symposium was planned as a surprise. However, one of the personal characteristics you excel in is your natural curiosity. Therefore, I presume that you have been more or less aware of what is going to happen here! We, your colleagues and friends, wish to express our acknowledgements for your achievement as an eminent geobotanist and a remarkable human being. We do this by organising this symposium for you.

Our feelings towards you are clearly demonstrated by the number of participants, which is well over 150. It would go too far to mention all of them by name, but you will agree with my wish to make some exceptions, especially for the colleagues from abroad.

First of all, I welcome Prof. J. Lebrun, Université de Louvain-La-Neuve, president of our International Society for Vegetation Science. Monsieur Lebrun: soyez le bienvenu ici!

Then I would like to welcome Dr. E. Duffey from the Institute of Terrestrial Ecology (Nature Conservancy) at Monks Wood, one of our outstanding colleagues in nature conservation and vice-president of the British Ecological Society, that old and famous organisation, which nominated Victor Westhoff an honorary member earlier this year.

From the wide circle of University colleagues, professors of Geobotany or its equivalents, I may welcome the following: Prof. J. M. Géhu, Lille, Prof. C. H. Gimingham, Aberdeen, Prof. R. Hundt, Halle, Prof. G. Jahn, Göttingen, Prof. E. Landolt, Zürich, Prof. N. Malmer, Lund, Prof. J.J. Moore S.J., Dublin, Profs. S. and E. Pignatti, Trieste,

Prof. P. van der Veken, Ghent, and Prof. G. Wendelberger, Vienna. We are happy indeed that most of them are on the list of speakers, demonstrators and symposium presidents.

Further, I may welcome two colleagues from abroad who have become more or less 'nijmegenised', Dr. Mike Dale, St. Lucia/Brisbane and Prof. Wolfgang Holzner, Vienna, both guest lecturers at the Department of Geobotany here.

I would like to add those Dutch friends who are on the list of speakers: Prof. Jan Barkman, Dr. Wim Beeftink, Dr. Geerd Sissingh and last, but not least, Dr. Marcus Adriani, our nestor and talented speaker, who will present the introductory lecture.*

Finally, I must mention some distinguished foreign colleagues who would have liked to come but were unable to attend the symposium: Dr. D. Ranwell, Norwich, Prof. S. Rivas-Martinez, Madrid, Prof. H. Sukopp, Berlin and Prof. R.H. Whittaker, Cornell University.

At the end of my opening speech I cannot resist taking the opportunity to say some personal words to you, Victor, as a token of our twenty years of friendship. I do this 'short and snappy' by connecting your name with that of your main teacher and great example: Josias Braun-Blanquet. He is still active, as you know, and I am happy to announce a small paper of his for this symposium. In a recent paper in Vegetatio on the Braun-Blanquet approach in perspective I made a comparison between the history of vegetation science and the history of music and there I dared to compare Braun-Blanquet's significance for phytosociology with that of Johann Sebastian Bach for music. Well, Victor, I feel it appropriate now to continue along these lines and to connect your name with that of the musical genious you admire so much and even rank above all others: Wolfgang Amadeus Mozart.

With the wish that it will be both serious and cheerful,

---

* The scientific part of this lecture will be placed at the beginning of the series of contributions, whilst the personal words Adriani added will follow after these opening words.

both harmonious and melodious, I open this symposium!

Eddy van der Maarel

Veuillez me permettre, Mme et M. Westhoff, mesdames et messieurs, d' ajouter encore deux mots en français. Tout d' abord pour vous féliciter, tous les deux, à l'occasion de votre anniversaire remarquable: voyez cequi se passe au jour d'hui: regardons cequi se passera demain! Nous vous admirons, M. Westhoff, vous, l' homme de la science géobotanique, l' homme qui a demérites. Il y a une grande reconnaissance de la part de vos amis et de vos élèves: c' est beaucoup et c' est très sèrieux. Il y a presque trop: les deux jours du symposion ne suffisent pas. Mais quand-même: In der Beschränkung zeigt sich der Meister!

Votre amitié m' est très précieuse à cause de votre inspiration continue pour les conceptions scientifiques, par votre inspiration pour les grandes responsabilités des chercheurs.

Quel délice de regarder ensemble des tableaux, des dessins, des gravures, d' écouter ensemble la musique d' un Scarlatti, d'un Franz Schubert, d' un César Frank, de se dire qu' un Paul Verlaine, 'comme poète a touché aux profondeurs de la joie et de la misère de l' homme. Et ce tout, cette totalité, Victor, pour vous impregné par votre sens inné d' un holisme. Mes félicitations cordiales pour vous deux, Victor et Nettie!

Marcus J. Adriani

PLANT SPECIES AND PLANT COMMUNITIES: AN INTRODUCTION*

M. J. ADRIANI[1] & E. VAN DER MAAREL[2]

[1] Van Itersonlaan 50, Oostvoorne, The Netherlands

[2] Division of Geobotany, Toernooiveld, Nijmegen, The Netherlands

This symposium is dealing with various relations between plant species as basic systematic units and plant communities as basic vegetation units. It will thereby touch upon relations between the respective ecological disciplines as well, disciplines we used to call autecology and synecology. This twofold theme has been a favourite one for a long time to the colleague and friend whose 60th birthday we are celebrating: Victor Westhoff. When we take his thesis in 1947, on the vegetation of dunes and salt marshes on the Dutch islands of Terschelling, Vlieland and Texel, as a starting point, Westhoff's ecological endeavour spans a period of 30 years.

Throughout this period he has approached plants and plant communities in his very personal way: both causal and final. His 'Leitmotiv' can be expressed by Goethe's words 'Wie fass ich Dich, unendliche Natur'', which we find as a motto for a section of the book 'Inleiding tot de Plantensociologie', 'Introduction to Plant Sociology' by Meltzer & Westhoff (1942). With this early work (but how mature it was already!) we may link Westhoff's oeuvre with that of Josias Braun-Blanquet, his great example and teacher: the 1942 book was in fact a thorough introduction to the Dutch reader of Braun-Blanquet's (1928) 'Pflanzensoziologie'.

As Westhoff & van der Maarel (1973) pointed out once again in the Handbook of Vegetation Science, the essence of the Braun-Blanquet approach can be stated in three ideas:

(i) Plant communities are conceived as types of vegetation, recognized by their floristic composition. The full species compositions of communities better express their relationships to one another and environment than any other characteristic.

* Nomenclature of plant species follows Heukels-van Oost-stroom. 1975. Flora van Nederland. 18e druk. Noordhoff, Groningen.

(ii) Amongst the species that make up the floristic composition of a community, some are more sensitive expressions of a given relationship than others. For practical classification (and indication of environment) the approach seeks to use those species whose ecological relationships make them most effective indicators; these are diagnostic species (Character-species, differential-species, and constant companions).

(iii) Diagnostic species are used to organize communities into a hierarchical classification of which the association is the basic unit. The vast information with which phytosociologists deal must, of necessity, be thus organized; and the hierarchy is not merely necessary but invaluable for the understanding and communication of community relationships that it makes possible.

Although it is believed here and there that this threefold essence may lead to a sterile system of abstract community types, Braun-Blanquet (e.g. 1959, 1964, 1968) has made it clear all over again that this approach is leading to the diagnosis of a real integrated plant community. This integration is based on the structure of the plant community, that is its organisation in space. With Mueller-Dombois & Ellenberg (1974) we may consider the biomass structure including the stratification of vegetation layers, as a most important aspect. Besides we may distinguish a horizontal, a pattern structure.

A second essential element of structure is the set of functional relations between plants. To Braun-Blanquet this element was even of primary importance: he treated 'the social life of plants' in the very first chapter of his book. Westhoff has given many a new impulse to the study of what we may call the integration, identity and stability in the plant community (Langford & Buell 1969). We may mention here his Dutch contribution on biotic factors (Westhoff 1958), and particularly the papers on the use of structural characters in the classification of vegetation

3

(Westhoff 1967, 1968). His ideas have led to an attempt to arrange syntaxonomical units in larger structural-physiognomic units, i.e. formations. And so in the book 'Plantengemeenschappen in Nederland,' 'Plant communities in the Netherlands', the syntaxa are described in 13 formations (Westhoff & den Held 1969).

The involvement of structural characters in the description of plant communities is not only found in Westhoff's own work, but also in that of many of his collaborators and pupils (e.g. Doing 1963, 1966, van der Maarel 1966, Segal 1970, Londo 1971, Werger 1973).

Despite the voluminous oeuvre on vegetation and vegetation science Westhoff produced, he has never considered himself a real phytosociologist, but rather an autecologist. Indeed Victor Westhoff has a passion for the plant, the 'sensitive plant' as Shelley titled one of his great poems and Westhoff himself described in a section of poems in his first and only published collection of poems called 'Levend barnsteen', 'Living amber'. This parallel interest in plant species and plant communities led to a sort of cross-fertilisation in the late nineteenfifties when Westhoff published his first accounts on what he called the phytosociological position of plant species. The elegance of this approach was demonstrated with *Carex buxbaumii*, *Listera cordata* and *Scirpus planifolius* as examples in the international literature (Segal & Westhoff 1959, Westhoff 1959, Westhoff & van Leeuwen 1966) and many other species in Dutch journals (e.g. Westhoff & Ketner 1967, Westhoff & van Leeuwen 1962, Westhoff & Passchier 1958) and numerous reports from his students at Nijmegen.

The approach consists of a careful phytosociological description of the vegetation in which a particular species occurs optimally or just marginally. Through tabular arrangement of the relevés and a detailed phytosociological interpretation, and with the help of a complete spectrum of syntaxonomical species groups, the phytosociological position of that species can be understood. The environment was not so much described directly through an analysis of soil and microclimate but rather interpreted by means of the powerful diagnostic value of the entire species combination of the vegetation to which the species under investigation belongs. In this respect Westhoff's work comes very close to the approach developed by Ellenberg, culminating in the latter's book 'Zeigerwerte der Gefässpflanzen Mitteleuropas', 'Indicator values of the vascular plants of Central Europe' (Ellenberg 1974).

Unlike Ellenberg and other outstanding ecologists Westhoff has not paid much attention to soil analysis, neither to experiments with plants. Apart from the more

personal reason that he has little of an 'apparatnik' Westhoff holds a number of more general considerations. One of them is the typically Braun-Blanquetian awareness of the historical factor. This temporal dimension of a species' environment cannot be understood through a snapshot of the present physico-chemical complex.

Another consideration which is concerned with the experimental approach is the one put forward repeatedly by Heinrich Walter, notably in the preface of his book 'Die Vegetation der Erde' (Walter 1973, 1973a): by far the largest experimentator in nature is Nature itself. If we only carefully describe the changes in plant communities caused by changes in the environment and describe the latter changes as well we have results as obtained from experiments available.

Along similar lines is a remark by Walter in his recent book 'Die ökologischen Systeme der Kontinente' (Walter 1976). Continuous and careful observations together with relatively few but effective measurements often allow a relatively rapid (and cheap!) insight in the functioning of ecosystems. After having expressed some doubts on the rapidly increasing investments in ecophysiological equipment and automisation of registration Walter ends with words which could almost be Westhoff's. Its translation reads: 'The ecologist must live together with his plants at the site as continuously as possible and while observing them experience all taht happens. We are dealing with plant behaviour research. Only in this way the essential can be understood'.

We would do unjustice to Westhoff, however, when we would not directly add how interested he is in more experimental ecophysiological work as soon and as long as it has an immediate bearing on the real field situation. (e.g. Steubing & Westhoff 1966, cf. Adriani 1945). This appears also from present research by his pupils (e.g. Blom 1976).

Another link with Walter's work is Westhoff's interest in the socalled Gesetz der relativen Standortskonstanz, the law of relative habitat constancy. This idea, which comes rather close to Vinogradov's idea of ecological compensation is concerned with the changes in the constellation of habitat factors governing the occurrence of a species towards the boundary of its distribution area (see Werger & van der Maarel, this volume, for references).

We now come to another important aspect of Westhoff's work on the relation between plant species and plant communities, that is the human impact on nature as it is reflected in the occurrence as well as in the disappearance. of particular species and the accompanying changes in plant community composition. He integrated van Leeuwen's (e.g. 1966) Relation Theory in his views on the relation between anthropogenic and natural environments and between the species characterising them. This resulted in many papers by Westhoff and his students on the behaviour of rare plant species characteristic for apparently anthropogenic or semi-natural conditions, however, with a far from trivial natural background. Needless to say how such work has been of constant significance for nature conservation and nature management, the attitude and approach of which are very much Westhoff's concern. He realised more than other people that in relation to nature man has come to the point of no return and that the last remains of the biosphere should be carefully acknowledged and managed (cf. Westhoff 1952, 1969, 1969a, 1971, 1971a).

The study of plants and plant communities as a scientific enterprise, too, has received much attention from Westhoff. In a number of papers he explained (e.g. Westhoff 1951, 1970) the logical and methodological basis for the division of the entire approach, which he prefers to indicate as geobotany, 'the study of plants and plant communities in their situation in the biosphere' (Westhoff 1969). Within this conception, which, as you know, is emphasized in the name of his University Department, there is a place for the major disciplines he believes in: plant geography, plant autecology or rather: plant species ecology and plant synecology, or rather: vegetation ecology.

Although we have now mentioned over 20 publications by Westhoff, this introduction is by no means a bibliography of his work. For a more complete survey we may refer to two real bibliographies concerning phytosociology in the Netherlands, the first by Westhoff (1961), the second by his collaborator Schenk (1978), which contain many more references. Still, one magnum opus has to be mentioned, which is both the ultimate demonstration of Westhoff's abilities and a good ending for this introduction: it is 'Wilde planten: flora en vegetatie in onze natuurgebieden', 'Native plants: flora and vegetation in our nature areas', written by Westhoff with three of his friends (Westhoff, Bakker, van Leeuwen & van der Voo 1970-1973).

References

Adriani, M. J. 1945. Sur la phytosociologie, la synécologie et le bilan d'eau de halophytes de la région néerlandaise méridionale ainsi que de la Méditerranée francaise. Thesis A'dam. 217 pp. Comm. S.I.G.M.A. 88.a J.B. Wolters Uitg. Mij. Groningen.

Blom, C.W.P.M. 1976. Effects of trampling and soil compaction on the occurrence of some Plantago species in coastal sand dunes. I. Oecol. Plant. 11: 225–241.

Braun-Blanquet, J. 1928. Pflanzensoziologie. Grundzüge der Vegetationskunde. Biologische Studienbücher. 7.1. Ed. Berlin. x + 330 pp.

Braun-Blanquet, J. 1959. Grundfragen und Aufgaben der Pflanzensoziologie. Vistas in Botany 1: 145–171.

Braun-Blanquet, J. 1964. Pflanzensoziologie, Grundzüge der Vegetationskunde, 3rd ed. Springer, Wien-New York, 865 pp.

Braun-Blanquet, J. 1968. L'école phytosociologique Zuricho-Montpelliéraine et la S.I.G.M.A. Vegetatio 16: 1–78.

Doing, H. 1962. Systematische Ordnung und floristische Zusammensetzung niederländischer Wald- und Gebüschgesellschaften. Diss. Wageningen (Wentia 8; Belmontia II (8), 1962).

Doing, H. 1963. Over de oecologie der inheemse berken en de systematische indeling der berkenbossen (mit Zusamm., with summary). Jaarb. Ned. Dendr. Ver. 22: 97–125. (Belmontia II (9), 1964).

Doing, H. 1966. Enkele opmerkingen over het begrip „hoofdformatie" (with summary). Gorteria 3: 5–11.

Ellenberg, H. 1974. Zeigerwerte der Gefässpflanzen Mitteleuropas, Scripta Geobotanica 9: 1–98.

Langford, A.N. & M.F. Buell. 1969. Integration, identity and stability in the plant association. Adv. Ecol. Res. 6: 83–135.

Leeuwen, C.G. van 1966. A relation theoretical approach to pattern and process in vegetation. Wentia 15: 25–46.

Londo, G. 1971. Patroon en proces in duinvalleivegetaties langs een gegraven meer in de Kennemerduinen. (with summary). Thesis Nijmegen, 279 pp.

Maarel, E. van der. 1966. On vegetational structures, relation and systems, with special reference to the dune grasslands of Voorne. The Netherlands. Thesis Utrecht 170 pp.

Mueller-Dombois, D. & H. Ellenberg. 1974. Aims and methods of vegetation ecology. Wiley, New York, 547 pp.

Schenk, W.A. 1978. Bibliographia phytosociologica: Neerlandia 1960–1975. Excerpta botanica Sectio B sociologica. (in press).

Segal, S. 1970. Strukturen und Wasserpflanzen. In R. Tüxen (ed.). Gesellschaftsmorphologie, p. 157–171, Junk, Den Haag.

Segal, S. & V. Westhoff. 1959. Die Vegetationskundliche Stellung von Carex buxbaumii Wahlenb. in Europa, besonders in den Niederlanden. Acta Bot. Neerl. 8: 304–329.

Steubing, L. & V. Westhoff. 1966. Kationenaustausch-Kapazität der Wurzeln und Nährstoffpotential des Bodems in psammophilen und halophilen Pflanzengesellschaften der niederländischen Meeresküste. Vegetatio 13: 293–301.

Walter, H. 1977. Die vegetation der Erde in Bd. I, 3. Aufl. Fischer, Jena/Stuttgart, 743 pp.

Walter, H. 1973. Vegetation of the earth in relation to climate and the ecophysiological conditions. London, 237 pp.

Walter, H. 1976. Die ökologische Systeme der Kontinente (Biogeosphäre). Fischer, Stuttgart, 131 pp.

Werger, M.J.A. 1973. Phytosociology of the upper Orange River Valley, South Africa. A syntaxonomical and synecological study. Thesis Nijmegen. Pretoria, 222 pp.

Westhoff, V. 1947. The vegetation of dunes and salt marshes on the Dutch islands of Terschelling, Vlieland and Texel. Thesis Utrecht, 131 pp.

Westhoff, V. 1951. An analysis of some concepts and terms in vegetation study or phytocenology. Synthese 8: 194–206.

Westhoff, V. 1952. De betekenis van natuurgebieden voor wetenschap en praktijk. Broch. Contactcie. Natuur- Landschapsbescherming, Amsterdam, 36 pp.

Westhoff, V. 1958. Biologische factoren. In: Het milieu van onze gewassen, p. 82–103. Staatsuitgeverij, Den Haag.

Westhoff, V. 1959. The vegetation of Scottish pine woodlands and Dutch artificial coastal pine forests; with some remarks on the ecology of Listera cordata. Acta Bot. Neerl. 8: 422–448.

Terminologie und Methodik, insbesondere zu der Struktur

Westhoff, V. 1961. Bibliographia phytosociologica: Neerlandia 1935–1959, nonnullis operibus ex annis 1870–1934 addatis. Excerpta botanica sectio B sociologica. Band 3: 81–220. Stuttgart.

Westhoff, V. 1967. Problems and use of structure in the classification of vegetation. The diagnostic evaluation of structure in the Braun-Blanquet system. Act. Bot. Neerl. 15: 495–511.

Westhoff, V. 1968. Einige Bemerkungen zur syntaxonomischen Terminologie und Methodik, insbedondere zu der Struktur als diagnostisches Merkmal. Ber. Int. Symp. Pflanzensoz. Syst., Stolzenau/Weser 1964: 54–70.

Westhoff, V. 1969. Die Reste der Naturlandschaft und ihre Pflege. In: Handbuch für Landschaftspflege und Naturschutz. Band 3, p. 251–265.

Westhoff, V. 1969a. Verandering en duur. Oratie Katholieke Universiteit, Nijmegen. Junk, Den Haag, 19 pp.

Westhoff, V. 1970. Vegetation study as a branch of biological science. Misc. Papers. Landbouwhogeschool Wageningen 5: 11–30.

Westhoff, V. 1971. The dynamic structure of plant communities in relation to the objectives of conservation. In: E. Duffey & A.S. Watt (ed.). The scientific management of animal and communities for conservation. Blackwell, Oxford, p. 3–14.

Westhoff, V. 1971a. Choice and management of nature reserves in the Netherlands Bull. Jard. Bot. Nat. Belgique 41 (1): 231–245.

Westhoff, V., P.A. Bakker, C.G. van Leeuwen & E.E. van der Voo. 1970. Wilde planten. Flora en vegetatie van onze natuurgebieden, deel I. Ver. t. Behoud v. Natuurmonumenten, Amsterdam, 320 pp.

Westhoff, V., P. Bakker, C.G. van Leeuwen et al 1971. Wilde Planten. Flora en vegetatie in onze natuurgebieden. Deel 2. Ver. t. Behoud van Natuurmonumenten, Amsterdam, 303 pp.

Westhoff, V., P.A. Bakker, C.G. van Leeuwen et al 1973. Wilde Planten. Flora en vegetatie in onze natuurgebieden. Deel 3. Ver. t. Behoud v. Natuurmonumenten, Amsterdam, 359 pp.

Westhoff, V. & A.J. den Held. 1969. Plantengemeenschappen in Nederland. Zutphen Thieme, 324 pp.

Westhoff, V. & P. Ketner. 1967. Milieu en vegetatie van Carex hartmanii Caj. op Terschelling, in het kader van een oecologische vergelijking tussen deze soort en Carex buxbaumii Wahlenb. Gorteria 3: 119–126.

Westhoff, V. & C.G. van Leeuwen. 1962b. Catapodium marinum (L.) Hubb., Scirpus planofolius Grimm en Trifolium micranthum Viv. op Goeree. Gorteria 1: 33–38.

Westhoff, V. & C.G. van Leeuwen. 1962a. Ökologische und systematische Beziehungen zwischen natürlicher und anthropogener Vegetation. Ber. Int. Symp. Antropogene Vegetation, Stolzenau/Weser 1961: 156–172.

Westhoff, V. & E. van der Maarel. 1973. The Braun-Blanquet approach. In Handbook of Vegetation Science (Ed. R. Tüxen)

Part V. Ordination and Classification of Communities (ed. R.H. Whittaker), p. 617.726. Junk, The Hague.

Westhoff, V. & H. Passchier. 1958. Verspreiding en oecologie van Scheuchzeria palustris in Nederland, in het bijzonder in het Besthmerven bij Ommen. Levende Natuur. 61: 59.67.

DIE QUELLFLUR- GESELLSCHAFT DES CRATONEURO-ARABIDETUM BELLIDIFOLIAE (Koch 1928) IN DER SUBALPINEN STUFE GRAUBÜNDENS

J. BRAUN-BLANQUET* **

Station Internationale de Géobotanique Méditerranéene et Alpine, Rue du Pioch du Boutonnet, Montpellier, France***

Keywords: *Cratoneuro-Arabidetum bellidifoliae*, Graubünden, Subalpine

In seiner Arbeit über 'Die höhere Vegetation der subalpinen Seen und Moorgebiete des Val Piora' beschreibt W. Koch 1928 erstmals einen an frisches kalkreiches Wasser gebundenen Quellflur-Verband unter dem Namen *Cratoneurion commutati* und behandelt weiter eine hierher gehörige Assoziation, das *Cratoneuro-Arabidetum bellidifoliae*.

Diese Assoziation ist bisher als einzige des Verbandes auch aus den Bündneralpen bekannt geworden. In der Uebersicht der Pflanzengesellschaften Rätiens (J. Br.-Bl. 1948, S.286) haben wir die Gesellschaft aus den Bündneralpen skizziert. Diese moosreiche, durch die subalpine Stufe Graubündens zwischen 1280 und 2000 m allgemein verbreitete Quellflur reicht ausnahmsweise im Unterengadiner Val Minschun bis 2520 m (Aufn. 33 unserer Tab.), bleibt aber streng auf kalte, karbonatreiche Quellen beschränkt.

Nach dem Vorherrschen der einen oder anderen der zwei dominierenden Moosarten können zwei Subassoziationen unterschieden werden:

1. Subass. *cratoneuretosum commutati* (Aufn. 1–12) subass. nov.
2. Subass. *cratoneuretosum falcati* (Aufn. 13–35) subass. nov.

Beide Subassoziationen benötigen eine Wassertemperatur von rund 6,5 bis 7 Grad Celsius. Nadig hat Wasserverhältnisse, Fauna und Flora eines *Cratoneuro-Arabidetum* – Quellbaches am Fuorn (Unterengadin) bei ca. 1800 m sehr sorgfältig untersucht. Er fand das Wasser alkalisch, kalkreich, mittelhart, sauerstoffgesättigt, sehr

rein; Ammoniak und Nitrite fehlen, Nitrate waren nur in Spuren vorhanden (Nadig 1942, S.295, 343 u.f.).

Wie wir am Fuorn zu beobachten Gelegenheit hatten, werden diese Quellen gern von den Gemsen besucht.

Die Aufnahmen unserer Tabelle des *Cratoneuro-Arabidetum bellidifoliae* stammen von folgenden Lokalitäten:

1. Las Maisas, Val Lavinuoz. – 2. Alp Varusch bei Scanfs. – 3. Il Fuorn, am Ofenpass (1937). – 4. Val Brüna am Ofenpass (1936). – 5. Juppa im Avers (1952). – 6. Val Sesvenna (Scarl). – 7. Oberhalb des Schwellisees bei Arosa (1937). – 8. Am Fuss des Piz Alv, Oberengadin (1941). – 9. Alp Astras im Val S'charl. – 10. Val Tschitta ob Bergün (1952). – 11. Am Aufstieg zu den Flühseen im Avers (1952). – 12. Oberhalb Langwies (Schanfigg). – 13. Val dal Fain am Bernina. – 14. Ob Madulain. – 15. Ova dils Pugls (Unterengadin) (1937). – 16. Platta – Fex im Oberengadin (1952). – 17. Buffalora am Ofenpass (1952). – 18. u. 19. Alp di Plaun ob Feldis (1949). – 20. Buffalora (1924). – 21. u. 22. La Schera im Nationalpark (1925). – 23. Val Tuoi ob Guarda (1947). – 24. Val Marmoré (1952). – 25. u. 26. Buffalora (1951 u. 1948). – 27. Oberhalb Parpan (1927). – 28. u. 32. La Schera im Nationalpark (1919 u. 1929). – 29. Albanahang oberhalb Silvaplana (1929). – 30. Giufplan (Jufplaun) am Ofenpass (1941). – 31. Julierpass (1950). – 33. Val Minschun oberhalb F'tan (1954). – 34. Alp Flix oberhalb Sur im Oberhalbstein (1953). – 35. Nordfuss des Castellins im Oberhalbstein (1953).

Der Artenliste der Tabelle wären noch eine Reihe weiterer mehr oder weniger zufälliger Begleiter, die nur in ein bis drei Aufnahmen notiert wurden, anzufügen. Als unwichtig sind sie hier weggelassen.

Das *Cratoneuro-Arabidetum bellidifoliae* ist bei ungestörten edaphischen Verhältnissen am Quellrand Dauergesellschaft. Meist grenzt es an das ebenfalls basiphile *Caricetum davallianae*, das nach Koch gleichfalls an fliessendes Grundwasser gebunden bleibt. Die Gesellschaft

---

* Herrn Prof. Dr. Victor Westhoff zu seinem 60. Geburtstag gewidmet.
** Herr Dr. R. Sutter (Bern) war bei der Ergänzung des Manuskript freundlichst behilflich.
*** Communication Nr. 219.

7

CRATONEURO - ARABIDETUM

	1	2	3	4	5	6	7	8	9	10	11	12	13	14	15	16	17	18	19	20	21	22	23	24	25	26	27	28	29	30	31	32	33	34	35	Stetigkeit
Nummer der Aufnahmen	1	2	3	4	5	6	7	8	9	10	11	12	13	14	15	16	17	18	19	20	21	22	23	24	25	26	27	28	29	30	31	32	33	34	35	
Höhe (m.ü.M.)	2350	1850	1780	1950	1950	2000	2060	2150	2180	2300	2460	1280	2300	1680	1900	1900	1950	1950	1985	2000	2000	2000	2010	2020	2050	2080	2100	2130	2150	2150	2150	2110	2520		2440	
Exposition	N	N	N	N	N	S	NW	SW	S	E	N	NE	S	E	W	S	S	SW	W	NE	W	W	W	W	W	W	W	N	N	N	flach	SW	S	NW	N	
Neigung (%)	5	5	15	10	10	5	20-25	5-10	5	20	5	5-10	15	10	10	15	2	10	10	15	15	20	7	10	10-15	10		10	25	10	10		10	80	5	
Wassertemperatur (°C)	5	5	6		6	5		8	5	5	5																		c.4							
Deckungsgrad (%)	80	90	70	80	90	95	90	80	90	80	80	90	90	90	90	100	95	95	95	90	85	80	90	95	95	80	90	90	95	80	80	90	80	80	80	
Aufnahmefläche (m2)	20	4	20	4	20	20	10	4	16	10	4	10	10	3	15	10	4	4	4	10	c.10	20	8	10	4	c.10	c.10	10	c.4	20	4	4	4	4	20	
Artenzahl	5	8	8	14	10	14	10	16	15	10	11	13	13	11	9	13	14	17	14	14	17	17	13	18	11	14	15	17	17	14	10	16	11	13	7	

Assoziations- und Verbands- Kennarten
(Cratoneurion commutati)

Arabis bellidifolia Jacq.	+		1.1-2	1.1	+.2	1.1	2.2	1.2		+.2	1.2	1.2	1.2	1.1	1.1	1.2	2.2		2.2		1.1	1.1	2.2	2.2	2.1	1.2				2.1	2.1	1.2	1.2	1.2	1.2	31
Philonotis calcarea Schimp.		2.2	2.3	+.2		1.2	2.2-3	+.2	2.1	+.2	+	1.2		2.3		3.3	+.2	1.2	2.2-3	1.1-2	1.2-3	+.2	2.2	1.1	1.2	2.2	2.2	2.2	1.2	2.1	2.2-3	1.2	1.2	1.2	3.3	28
Cratoneuron falcatum (Brid.)		4.4	2.2	5.5	3.3	5.5	2.3-4	1.2	2.3	4.4	3.3	4.2	5.4	2.3	5.5	3.3	4.3	3.3	5.5	1-2-3	1.1	1.1	1.3	5.4	5.5	4.4	5.5	5.5	5.5	4.3	4.3	2.3	3.3	1.2	3.3	24
Cratoneuron commutatum (Hedw.) Roth	5.5	4.5	5.5	5.5	5.5	5.5	4.4	4.4	4.4	4.4	3.3	3.3	1.2	4.3		2.1	4.4	4.4	5.5	3-4.3	3-4-5		4.4										3.3	3.4		13
Aneura pinguis Dum.					4.4		1.2															+			2.3											3
Orthothecium rufescens (Dicks.) Br. eur.													1.2	1.2				+.2							(2.3)											1

Ordnungs- und Klassen-Kennarten
(Montio-Cardaminetalia, Montio-Cardaminetea)

Deschampsia caespitosa (L.) P.B. var. alpina	(+)	2.2	+		1.2	+	+.2	+.2	+	+	(+)		+.2			+	+.2	(+)	(+)	+.2	1.1	1.1	1.2	+	+	+.2	(+)	+	+	1.2	+.2	2.2		1.2	+.2	29
Saxifraga stellaris L.	+.2	2.2				2.2	2.2-3	2.2	1.2	1.2	1.2	2.2	+		2.2	+.2	1.2	1.2	2.2-3	1.2	1.2	+.2	3.4	+	2.2	2.2	2.2	2.2		2.1	2.2	2.2	2.2	1.2	1.1	28
Epilobium alsinifolium Vill.	1.2	2.1-2	3.3	1.1		1.2	+		2.3	(+)	(+)						+.2				1.1	1.1		2.1		+.2	+.2	+		2.1	2.1	1.1			+	16
Bryum ventricosum Dicks.	1.1	2.3	+.2					1.1				1.2			2.2			2-3.1			1.1 2-3.1							r	1.2							14
Cardamine amara L.					+.2	1.2		+.2		1.2		+					+.2	+.2		1.2				+.2								+.2				10
Mniobryum albicans L. var. minor (Miller) Beck							2.2	2.2						1.2				2-3.2																		6
Caltha palustris L.													+.2	+.2				1.2	2.2					1.2				(+)								6
Alchemilla vulgaris L. ssp. coriacea (Buser) Camus						1.2											1.2	1.2	1.4	1.4												1.3				6
Bryum schleicheri Schwägr.							+.2		+.2										1.2				2.3	+.2			(+)			+.2		1.4				4
Cratoneuron decipiens (De Not.)																												(+)	+							1
Cratoneuron irrigatum Zett.																							1.2				+	+	1.3	5.5	5.5					1
Brachythecium rivulare Br. eur.												3.3																								1
Epilobium palustre L.																		+										+	+							1

Begleiter

Agrostis alba L.	1.2	+	2.2	2.1		2.3	+		1.1	1.2	2.1	+	3.3		1.1	1.1	2.2	1.2	2.2		1.1	1.1	1.2	1.1	1.1	2.2	(+)	2.1	+	1.1	1.1	1.1	1.1	2.1-2 2.2		31
Polygonum viviparum L.	1.2	+	+	+		+.2	+.2	+.2	2.1	+	+	+	+	1.1	+	+.2	+.2	+	+	2-2-3 (+)	1.1	+.2	+	1.1	1.1	1.1	1.1	+	1.2	+	+.2	+.2	3.2 2.3	2.1-2		21
Saxifraga aizoides Schleicher			3.3		3.3	2.3	+.2 3.3	2.3	2.3	4.3	1.2				3.3	2.3		1.1			1.1	1.1	1.2	2.1	2.2	1.2					2.1	1.1			+	19
Equisetum variegatum Schleicher			1.1	1.1				1.1									+.2	1.1		+.2	1.1	+.2	+.3	2.1		+.2	(+)	1.2		+	+			+.o		16
Bellidiastrum Michelii Cass.						+	+.2	+.2	+.2								+.2		+.2	+.2	+.2		+.2	1.2		+.2	(+)		+.2	+.2	+.2		+.2	+.2		16
Juncus triglumis L.					1.2		+										2.2							+.2		1.1		1.1								13
Salix arbuscula L. ssp. Waldsteiniana (Willd.) Br.-Bl.		+			1.2	+	1.1		1.1	1.1														1.2												10
Carex frigida All.		1.1		1.1		+	+.2		2.1	2.1	1.2					1.1								+.2					1.1							8
Tussilago farfara L.																		+						+.2									r			8
Soldanella alpina L. Ard.								1.2								1.2		+						+.2			(+)									7
Sesleria coerulea (L.) Ard.																						+.2		1.2		(+)	(+)	1.1								7
Carex fusca L.																								+.2			(+)		1.1							6
Carex davalliana Sm.																												(+)								5
Poa alpina L.																												(+)	1.1							5
Carex panicea L.																									1.1					1.2						5
Carex ferruginea Scop.																		(+)				+.2						(+)								5
Juncus alpinus Vill.							1.1																					(+)				1.1				4
Eriophorum angustifolium Honckeny										(+)						1.1		(+)						+.2						1.2						4
Triglochin palustris L.						1.1										2.2												(+)								4
Equisetum palustre L.																+								+.2				(+)								3
Allium schoenoprasum L.					1.2																									+						3
Ligusticum mutellina (L.) Crantz.																								+.2						+						3
Viola biflora L.		(+)		1.1				1.2																+.2												3
Pinguicula alpina L.																	+.2							+.2												3
Campanula cochlearifolia Lam.																								+.2					1.1			1.1				3
Crepis paludosa (L.) Moench		2.1																							1.1										+	3
Bryum sp.																						+												1.1		3

kommt in nahezu übereinstimmender Zusammensetzung auch in den Pyrenäen vor (Br.-Bl. 1948, S.119).

Nahe verwandt mit dem *Cratoneuro-Arabidetum bellidifoliae* erscheint eine *Cortusa Matthioli-Crepis paludosa*-Assoziation, die in der Schweiz als Endemismus auf das Unterengadin und das angrenzende Val Müsteir beschränkt ist. *Cortusa Matthioli* in den Savoyeralpen, bei Tignes (Rouy) angegeben, taucht wieder auf im hohen Norden des arktischen Russland.

Summary

Two subassociations of the *Cratoneuro-Arabidetum bellidifoliae* from the subalpine belt in Graubünden (Switzerland) are described for the first time. The association occurs along springs with running, base-rich water and is usually bordering the equally basiphilous *Caricetum davallianae*. The association also occurs in the Pyrenees. Synsystematically it is related to a *Cortusa Matthioli – Crepis paludosa* association, occurring as an endemic community in the Lower Engadin.

The two subassociations are *cratoneuretosum commutati* and *cratoneuretosum falcati*, characterized by the dominance of *Cratoneurum commutatum* and *C. falcatum*, respectively.

Literatur

Braun-Blanquet,J. 1948. Uebersicht der Pflanzengesellschaften Rätiens. Vegetatio 1: 285–286.
Braun-Blanquet,J. 1948. La Végétation alpine des Pyrénées orientales. Monograf. de la Estacion de estudios pirenaicos, Barcelona. Station Internat. de Géobot. Méditerr. et Alpine Comm. No. 98.
Koch,W. 1928. Die höhere Vegetation der subalpinen Seen und Moorgebiete des Val Piora (St. Gotthard-Massiv). Zeitschr. Hydrologie 4, 3/4: 131–175.
Nadig,A. 1942. Hydrobiologische Untersuchungen der Quellen des schweizerischen Nationalparks im Engadin. Ergeb. Wiss. Unters. Nat.-park 1, 9: 272–432.

OBSERVATIONS ON NORTH-WEST EUROPEAN LIMESTONE GRASSLAND COMMUNITIES: PHYTOSOCIOLOGICAL AND ECOLOGICAL NOTES ON CHALK GRASSLANDS OF SOUTHERN ENGLAND*

J.H. WILLEMS**

Dept. of Vegetation Science and Botanical Ecology, State University, Utrecht, The Netherlands

Keywords:
Braun-Blanquet approach, Calcareous grassland, *Cirsio-Brometum*, Southern England, *Ulex europaeus* scrub

Introduction

During the summer of 1970, calcareous grasslands were studied at several localities in the southern part of the British Isles, as part of a survey of this type of vegetation in North-West Europe (Willems 1973, Willems & Blanckenborg 1975).

Calcareous grasslands in Britain have been the subject of many studies over several decades (a.o. Tansley 1911, 1922, Tansley & Adamson 1925, Lousley 1950, Wells 1975). However, detailed phytosociological studies have been few, the most notable are those carried out by Shimwell (1971a, b). His study areas were mostly situated on the harder limestone of upland Britain, except for the few areas on the chalk of Salisbury Plain.

In this paper particular attention will be given to the synsystematical and ecological aspects of calcareous grasslands and allied vegetation types. Also much attention is paid to the cryptogams occurring in these grasslands. Relations of the British types of calcareous grasslands with those on the western part of the European continent will be discussed.

Methods

Vegetation relevés were made and afterwards classified on the basis of the principles of the French-Swiss School of Phytosociology (Braun-Blanquet 1964). Methods used for measuring pH and calciumcarbonate content of the soil, are given in Willems (1973) and Willems & Blanckenborg (1975).

Abundance/dominance values of cryptogams are given for those relevés in which cryptogams were identified in the field (Table I — relevés 49–60). In those relevés in which a representative sample of cryptogams were taken and later sorted and identified, presence in a relevé is denoted by an X in the tables.

Festuca rubra s.l. and *F. ovina* s.l. could not always be distinguished from each other in the field with certainty, for instance when sheep had grazed the vegetation shortly before the relevés were made. Following Wells (1975) a.o., these taxa are listed together in the tables.

The synsystematic place of the communities

The first attempt to classify limestone grassland in Britain was made by Moss in 1911 who described a "limestone grassland association (*Festucetum ovinae*)", while Tansley & Rankin (1911) described a "chalk grassland association (*Festucetum ovinae calcareum*)" from southern England (Tansley 1911). Attention was given not only to the floristic composition of these communities but also to the soil and subsoil on which they occurred.

* Plant nomenclature follows Heukels & van Ooststroom (1970) or, for taxa not mentioned by these, Clapham et al. (1968) for phanerogams and Margadant (1959) and van der Wijk et al. (1969) for bryophytes.
** Contribution to the Symposium on Plant species and Plant communities, held at Nijmegen, 11–12 November 1976, on the occasion of the 60th birthday of Professor Victor Westhoff.

The author would like to express his appreciation and gratitude to Messrs. F. van der Meulen and J. Wiegers, in 1970 students in Biology at the Utrecht University, for their great and pleasant assistance during the fieldwork. He is also much indebted to Drs. J. Wiegers of the Hugo de Vries Laboratorium, University of Amsterdam, and T.C.E. Wells of the Institute of Terrestrial Ecology, Huntingdon, U.K., for the useful discussions and critical comment on the manuscript.

11

Braun-Blanquet & Moor (1938) using the lists of species given by Tansley (1911) and Tansley & Adamson (1925) from calcareous grassland near the Sussex-Hampshire border, placed these grasslands in the *Xerobrometum brittanicum*. The authors noted the mesophilous aspects of this association and suggested a close relationship with the *Mesobrometum*. They also emphasized the richness of cryptogams, especially Musci, in the *Xerobrometum brittanicum*. This remark on the cryptogams is misleading insofar as British investigators of calcareous grasslands included the cryptogams in their accounts from the very beginning, whereas their colleagues on the continent followed them in this practice only many years later, if at all (a.o. Barkman 1953, Bornkamm 1960, Willems 1973). Comparing the number of cryptogams in chalk grasslands in both the British Isles and the western part of the European continent, it can now be concluded that there is not much difference in the number of taxa; calcareous grasslands are rich in cryptogams in North-West Europe.

Braun-Blanquet & Moor (l.c.) listed the following mosses as the most important cryptogams occurring in the *Xerobrometum brittanicum*: *Pseudoscleropodium purum, Camptothecium lutescens, Fissidens taxifolius, Rhytidiadelphus squarrosus, Dicranum scoparium*, and *Hylocomium splendens*. However, according to Barkman (1966) these mosses are not "good" xerophytic taxa.

The *Xerobrometum brittanicum* prov. Br.-Bl. & Moor 1938 clearly belongs to the *Mesobromion*, not to the *Xerobromion*. This was already stated by Tüxen (1928). Tansley (1939) supported this opinion in his classical work "The British Isles and their Vegetation", as becomes clear from a footnote on page 535.

In 1971 Shimwell published the results of a detailed investigation on the calcareous grasslands of the British Isles belonging to the class *Festuco-Brometea*. Shimwells' investigation included the *Xerobromion* as well as the *Mesobromion* from eastern and central England, Scotland, Ireland, and Wales. He described a.o. a new association, the *Cirsio-Brometum*, belonging to the *Mesobromion*. The calcareous grasslands discussed in the present paper belong to the *Cirsio-Brometum*.

Brachypodium pinnatum, Bromus erectus, Cirsium acaulon, and *Asperula cynanchica* are regarded by Shimwell (l.c.) as differential species for the *Cirsio-Brometum*. These species are already considered to be character species for the alliance *Mesobromion* and the order *Brometalia* within the whole area. They usually occur together in the calcareous grasslands belonging to the *Mesobromion* in the whole north-western European area (a.o. Scherrer 1925, Tüxen

1937, Oberdorfer 1957, Bornkamm 1960, Willems 1973). Therefore the conclusion has to be made that the species mentioned by Shimwell are not truly differential species to characterize a new association within the alliance *Mesobromion*.

Yet, Shimwells' investigation as well as the one presented in this paper indicate that the combination of species in the calcareous grasslands from southern England is sufficiently characteristic to bring this vegetation into a separate association. This can be based on the occurrence of the combination of the differential species within the alliance as such are *Thymus druceï, Filipendula vulgaris, Viola hirta*, and *Pseudoscleropodium purum*.

In view of the above, it seems to be better to unite the subassociation *typicum* and *brachypodietosum* published by Shimwell in Table V, and to name this community *Cirsio-Brometum typicum*. Only *Arrhenatherum elatius* can be understood as a true differential species for the sub-association *brachypodietosum* distinguished by Shimwell, which must be considered to be insufficient to distinguish a subassociation (Korneck 1974, Willems & Blanckenborg 1975). The *Arrhenatherum elatius* community, described by Shimwell, can be seen as a variant within the *Cirsio-Brometum typicum* in the circumscription here proposed. Following the code of phytosociological nomenclature (Barkman et al. 1976) the present author designates relevé JL 29, Table V (Shimwell 1971), as the type relevé for this *Cirsio-Brometum typicum*.

The sub-association *Cirsio-Brometum astragaletosum danici* Shimwell 1971, can be maintained based upon the differential species *Astragalus danicus, Pulsatilla vulgaris, Genista tinctoria*, and *Serratula tinctoria*. Relevé JL 49, Table V, in Shimwells' paper is the type relevé for this community.

The plant communities

The chalk grassland communities mentioned in this paper are all assignable to the class *Festuco-Brometea* Br.-Bl. & R. Tüx. emend. R. Tüx. 1961, order *Brometalia* (W. Koch 1926 n.n) Br.-Bl. 1936, alliance *Mesobromion* (Br.-Bl. & Moor 1938) Oberd. 1957, and the association *Cirsio-Brometum* Shimwell 1971.

The following types of vegetation can be distinguished:

Anthoxanthetosum odorati subass. nov. (Table 1, relevés 1 42: type relevé 25).

Within the association *Cirsio-Brometum*, the subassociation *anthoxanthetosum odorati* is characterized by the differential species *Anthoxantum odoratum*, *Achillea millefolium*, *Agrostis stolonifera*, and *Hippocrepis comosa*.

Phytocoenoses belonging to this subassociation are grazed by sheep, cattle, and horses. The pH of the soil fluctuates between 7 and 8. The altitude varies from about 30 m to about 145 m above sea level. The slope varies from gentle (about 15) to rather steep (about 45). The subsoil consists of chalk from Upper-Senonian age. This community is met with in the South Downs in East Sussex (Fig. 1).

Two variants can be distinguished:

1. *inops* variant (relevés 1–18)

This variant is characterized by the absence of differential species with a high frequency. Only *Crepis capillaris*, *Senecio integrifolius*, and *Rhynchostegium confertum* are restricted to this community but these occur with a low frequency. The number of phanerogams varies between 20 and 34, with an average of 28, species per relevé; the number of cryptogams varies from 0 to 8, with an average of 3 species per relevé.

This community was encountered on a steep east-facing slope named 'High and Over' on the western bank of the River Cuckmere between Seaford and Alfriston, Sussex, National Grid Reference (N.G.R.) TQ 511013.

The vegetation was formerly sheep-grazed but more recently cattle have replaced the sheep.

The slopes are not grazed evenly, the upper slopes being less intensively used. Once a year, usually in early spring, the vegetation is burnt to prevent seedlings of shrubs, especially *Crataegus monogyna*, from growing and spreading.

Two relevés (1 and 2) are characterized by the dominance of *Brachypodium pinnatum*. The vegetation in which these relevés were made occurs at the foot of the slope. The soil at these sites is deeper than elsewhere on 'High and Over' due to an accumulation of material originating uphill.

2. *Phyteuma tenerum* variant (relevés 22–42)

Differential species for this community are *Phyteuma tenerum*, *Trifolium pratense*, *Rhytidiadelphus triquetrus*, *Rhinanthus minor* (high frequency of occurrence), and *Tragopogon pratensis* (low frequency of occurrence). The number of phanerogams varies from 26 to 45, with an average of 34 per relevé. The number of cryptogams fluctuates between 1 and 9, with an average of 5 per relevé.

This community is described from a slope 'The Long Man' of Wilmington, North-West of Eastborne, Sussex, N.G.R. TQ 545035, and from the southfacing slope of Mount Caburn, N.G.R. TQ 444087, east of Lewes, Sussex (Fig. 2).

Until recently, the vegetation at both sites was grazed by sheep, but these have, for a short period, been replaced by cattle and horses. Even more important were the large populations of rabbits (*Oryctolagus cuniculus*) which grazed these downs until

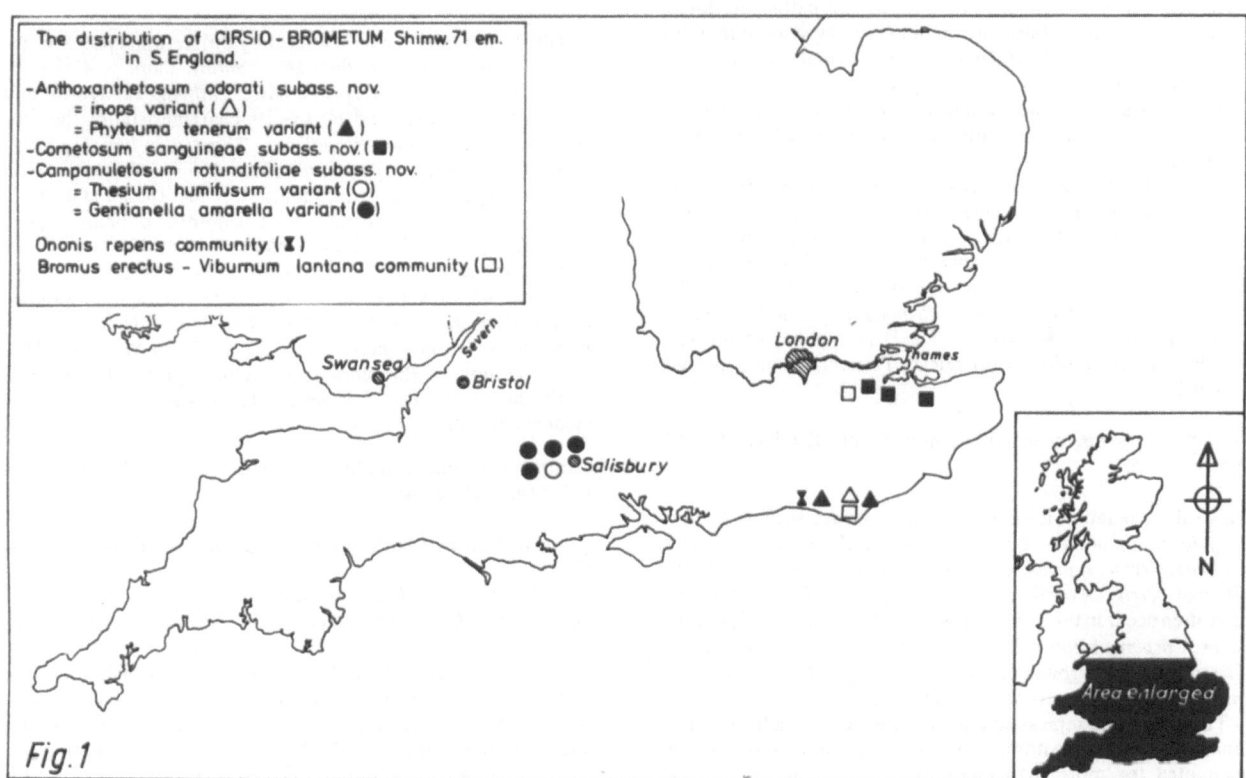

Fig.1

The distribution of CIRSIO - BROMETUM Shimw. 71 em. in S. England.

-Anthoxanthetosum odorati subass. nov.
    = inops variant (△)
    = Phyteuma tenerum variant (▲)
-Cornetosum sanguineae subass. nov. (■)
-Campanuletosum rotundifoliae subass. nov.
    = Thesium humifusum variant (O)
    = Gentianella amarella variant (●)

Ononis repens community (✗)
Bromus erectus - Viburnum lantana community (□)

Fig. 2. View in eastern direction of the slope named 'The Long Man' near Wilmington, Sussex. The chalk grassland occurring on the slope is assigned to the *Cirsio-Brometum anthoxanthetosum odorati* subass. nov. The irregular spots on the slope are patches of *Ulex europaeus. Crataegus monogyna* occurs in the foreground. Photo F. van der Meulen – July 1970.

1954, when myxomatosis arrived in this area. Thomas (1960, 1963) has described the enormous influence these animals had on the chalk vegetation, one of his study areas being adjacent to Wilmington.

In addition to being grazed by sheep and cattle, the lower slopes of Mount Caburn are manured each year with both farmyard manure and artificial fertilizer. *Centaurea scabiosa* is mostly restricted to these lower slopes.

The *Phyteuma tenerum* variant is found on moderately steep to steep (15–45) slopes at altitudes of from 55 to 145 m above sea-level.

East of Lewes a type of vegetation intermediate between the *inops* variant and the *Phyteuma tenerum* variant was investigated on a west-facing slope (relevés 19–21). This community is characterized by both the absence of species such as *Viola hirta, Filipendula vulgaris, Anthoxanthum odoratum, Achillea millefolium, Helictotrichon pratense, Primula veris*, and *Anthyllis vulneraria*, and the occurrence of a large number of mosses, which varies from 6 to 13 per relevé. This vegetation is grazed by cattle.

Cornetosum sanguineae subass. nov. (Table 1, relevés 51–60; type relevé 52)

This sub-association is characterized by shrubs such as *Cornus sanguinea, Viburnum lantana, Acer pseudoplatanus, Corylus avellana, Prunus spinosa, Ligustrum vulgare*, and the introduced *Quercus cerris*. Several species of the genera *Rosa* and *Rubus* also often occur in this community along with saplings of *Quercus robur, Carpinus betulus, Betula pendula*, and *Fraxinus excelsior*. The spontaneous growth of shrubs in this vegetation is due to the disappearance of sheep or cattle grazing.

This vegetation represents a stage in the succession from grassland to woodland. Although the shrubs in our samples never accounted for more than one-third of the total cover, their influence on the community was considerable. Still, I consider this vegetation as a grassland because of the presence of many typical grassland plants, a.o. *Bromus erectus, Cirsium acaulon, Thymus drucei, Brachypodium pinnatum, Carex flacca, Dactylis glomerata*, and *Leontodon hispidus*. The height of the shrubs is mostly less than one meter.

Differential species for this community are *Hypericum perforatum, Origanum vulgare, Teucrium scorodonia*, and *Anomodon viticulosus. Festuca rubra* occurs more frequently and is more abundant in this sub-association than in the other communities studied. *Festuca ovina*, if present at all, is not very abundant in the ungrazed plant community.

The number of phanerogams varies from 22 to 40, with an average of 29, and the number of cryptogams varies from 2 to 9, with an average of 8. The community was met with mostly on south-facing slopes.

This vegetation was found on several localities on the North Downs in Kent: near Wichling, E. of Maidstone, N.G.R. TQ. 916563 (relevés 51, 54, 55, and 60), near Snodland, N.G.R. TQ. 653614 (relevés 49, 50, and 56 58), at Wrotham, N.G.R. TQ. 604598 (relevés 53 and 59), and near Kemsing, N.G.R. TQ. 562594 (relevé 52), W. of Maidstone. The area covered by this sub-association was mostly less than 1 ha, and this may explain to some degree why these areas are no longer grazed.

A number of relevés made in the North Downs (relevés 49 and 50) as well as in the South Downs (relevés 43–48) represent a transitional stage between the sub-associations *anthoxanthetosum odorati* and *cornetosum sanguineae*, both belonging to the *Cirsio-Brometum*. I treat them together as the *Bromus erectus-Viburnum lantana* community.

This community is characterised by differential species of the sub-association *anthoxanthetosum odorati*, namely *Anthoxanthum odoratum, Hippocrepis comosa, Anthyllis vulneraria*, and *Pimpinella saxifraga*, and by species characteristic of the sub-association *cornetosum sanguineae*. These are: *Brachypodium pinnatum, Hypericum perforatum*, and *Viburnum lantana*. The number of species in this community is smaller than that in either of the sub-associations mentioned. The number of phanerogams varies from 11 to 23, with an average of 18 and the number of cryptogams varies from 2 to 10 with an average of 5.

Relevés 43–48 were made on Windover Hill, a slope situated on the same line of hills of which Wilmington Heath forms a part. In 1970, when the relevés were made on Windover Hill, the vegetation on this slope had not been grazed by sheep or cattle during the past five years. There was, however, some evidence of rabbit grazing.

Campanuletosum rotundifoliae subass. nov. (Table 2, relevés 1–55; type relevé 24).

Synonym: *Carex humilis* grassland (*Caricetosum humilis*) Wells 1975.

Within the *Cirsio-Brometum* the sub-association *campanuletosum rotundifoliae* is characterized by the differential species *Campanula rotundifolia, Agrostis tenuis, Euphrasia nemorosa, Cynosurus cristatus*, and *Festuca pratensis*, which are very frequent. Noteworthy is the complete absence of *Bromus erectus* and only the rare occurrence of *Brachypodium pinnatum* in the relevés. Both species are faithful species of the *Cirsio-Brometum*. In the table of this community presented by Wells (1975) both

Additional material from *Plant Species and Plant Communities,*
ISBN 978-90-6193-591-9 (978-90-6193-591-9_OSFO1), is available at http://extras.springer.com

species are present, but only infrequently.

Phytocoenoses belonging to this sub-association were found only on the southern border of Salisbury Plain, West of Salisbury (Fig. 1).

This sub-association can be divided into two variants:

1. *Thesium humifusum* variant (relevés 1–15)

This variant is characterized by the differentials *Thesium humifusum*, *Lolium perenne*, *Trifolium repens*, *Leontodon autumnalis*, *Crepis nicaeensis*, and *Cerastium holosteoides*. The number of phanerogams varies from 39 to 49, with an average of 40 per relevé, the number of cryptogams varies from 0 to 11, with an average of 5. The altitude at which this plant community is found, varies from 90 to 140 m on slopes varying from 15 to 30 .

The *Thesium humifusum* variant occurred in a U-shaped valley, called The Range, about 1 km in length situated S. of the village of Burcombe, N.G.R. SU. 068298, 3 km West of Wilton, Wiltshire (Fig. 3).

The phytocoenoses belonging to the *Thesium humifusum* variant are grazed at intervals throughout the year by hundreds of sheep. Additionally, intensive rabbit grazing occurs. In order to increase productivity the grassland is fertilized, especially the valley bottoms. The high frequency of *Lolium perenne*, *Festuca pratensis*, *Veronica chamaedrys*, and *Phleum pratense* in the relevés is probably a result of this practice.

Anacamptis pyramidalis and *Gymnadenea conopsea* were found during a survey in June 1970 but later, when the chalk grasslands of The Range were studied in detail, neither could be found. Their absence then is perhaps due to sheep grazing or selective rabbit grazing.

Two subvariants of the *Thesium humifusum* variant may be distinguished:

a. *inops* subvariant (relevés 1–5)

In this community the following species are almost or completely absent: *Festuca pratensis*, *Leontodon autumnalis*, *Crepis*

Fig. 3. The Range (Wilton, Wiltshire) looking south. The grassland belongs to the *Cirsio-Brometum campanuletosum rotundifoliae* subass. nov. *Urtica dioica* occurs in patches in the valley bottom, with *Crataegus monogyna* scrub on the slopes. *Carduus nutans* can be seen left foreground. On the horizon a group of trees *(Fagus sylvatica)* surround a prehistoric tumulus. Photo F. van der Meulen – August 1970.

nicaeensis, *Cerastium holosteoides*, *Carlina vulgaris*, and *Fissidens taxifolius*.

This community was found near the entrance of the valley in the extreme northern part of The Range, and on both east- and west-facing slopes. In this part of the valley, especially on the lower slopes and in valley bottoms the influence of sheep grazing is evident.

b. *Leontodon autumnalis* subvariant (relevés 6–15)

This type of vegetation is characterized by the high frequency of *Crepis nicaeensis*, *Leontodon autumnalis*, and *Cerastium holosteoides*. The community is chiefly met with on the east-facing slope of The Range and all relevés containing this variant were made there.

2. *Gentianella amarella* variant (relevés 16–54)

This variant is characterized by the differentials *Gentianella amarella*, *Carex humilis*, and *Koeleria gracilis*. The number of phanerogams varies from 30 to 43, with an average of 36 per relevé, the number of cryptogams from 1 to 10, with an average of 8. It was found at altitudes between 90 -170 m and on slopes of 15–40 .

Phytocoenoses referable to this variant are grazed either by sheep or cattle or by both.

Three subvariants can be distinguished, viz.:

a. *inops* subvariant (relevés 16–24)

This community is characterized by the absence of differential species of high frequency. Only *Neckera crispa* and *Sieglingia decumbens* have a slightly higher degree of presence in this community, than in the other communities within the sub-association *campanuletosum rotundifoliae*. This *inops* subvariant was found on north- and west-facing slopes of The Range at altitudes between 90 m and 140 m, and on slopes of 15–40 .

b. *Succisa pratensis* subvariant (relevés 25–41)

Differential species for this subvariant are *Succisa pratensis*, *Coeloglossum viride*, and *Hippocrepis comosa*. This community is the only one within the sub-association *campanuletosum rotundifoliae* where species such as *Cynosurus cristatus*, *Trisetum flavescens*, *Daucus carota*, and *Veronica chamaedrys* are absent. On the other hand, *Camptothecium lutescens* and *Ctenidium molluscum* have a high frequency in this community.

This subvariant was met with on the highest parts of both the west- and east-facing slopes of The Range (relevés 25–28), and on West Hill, Wylye Down, N.G.R. SU. 915357, West of Wilton (relevés 29–41). On West Hill this community was present on a north-west facing slope at an altitude of about 125–150 m.

c. *Dactylorhiza fuchsii* subvariant (relevés 42–54)

Within the sub-association *campanuletosum rotundifoliae* this subvariant is characterized by the differential species *Dactylorhiza fuchsii*, *Helictotrichon pubescens*, and *Anthoxanthum odoratum*. Furthermore, this community is distinguished from others belonging to the sub-association by the absence of species as *Carlina vulgaris*, *Picris hieracioides*, *Chrysanthemum leucanthemum*, and *Centaurea nigra* (Table 2). *Rhytidiadelphus triquetrus* and *R. squarrosus* have a high frequency in this subvariant, as they also have in the *Succisa pratensis* subvariant.

The *Dactylorhiza fuchsii* subvariant was encountered at two localities, both near Wilton. Firstly at Burcombe Ivers, N.G.R. SU. 045290 (relevés 42 and 43), and secondly at Compton Down, N.G.R. SU. 030280 (relevés 44–54). The altitude of these sites varies from 125 m to 175 m. The downs mainly face north or north-west, and the slope varies from 20 to 35 .

Ecological notes

Notes on succession

When calcareous grasslands are neither mown nor grazed by cattle, sheep, or rabbits, they develop into a woodland vegetation via a shrubby stage. This was already established more than half a century ago as a result of enclosure experiments done by Tansley (1922).

Although most of the calcareous grasslands investigated in this study were still grazed in 1970, some localities had been left untouched for a shorter or longer time. Two communities, both belonging to the *Cirsium-Brometum*, could be distinguished within those grasslands which were no longer affected by sheep grazing: (a) *Bromus erectus-Viburnum lantana* community and (b) the sub-association *cornetosum sanguineae*.

In the *Bromus erectus-Viburnum lantana* community grazing by sheep and cattle had been stopped about five years before the moment of the present investigation. Although only a few shrubs had appeared during these five years, some of the characteristic species of the *Meso-bromion* appeared to be less frequent, for example, *Helicto-trichon pratensis*, *Plantago media*, *Scabiosa columbaria*, *Primula veris*, and *Anthyllis vulneraria*.

While this vegetation remains unmanaged, the establishment and growth of shrubs and trees will continue. After an initial decrease in the number of species during the first years in the *Bromus erectus-Viburnum lantana* community, the number of species in the sub-association *cornetosum sanguineae* will increase again as a result of the establishment of sciophytic species, for example, *Fragaria vesca* and *Anomodon viticulosus*, and by tall herbs such as *Origanum vulgare*, *Teucrium scorodonia*, and *Hypericum perforatum*. The latter species usually occur on neutral or basic soils at the edge of scrub communities.

The increase of *Brachypodium pinnatum* in ungrazed calcareous grasslands is striking. Tansley (1939 : 536) already characterized this species as "an agressive invader". Wells (1975), too, noticed the invading of many chalk grasslands by *Brachypodium pinnatum*. The increase of this grass is supposed to be connected with the decrease of grazing by sheep and also by rabbits (a.o. Lousley 1950).

The above findings agree with the data given in this paper in Table I, where *Brachypodium pinnatum* has the highest frequency in ungrazed calcareous grassland. This phenomenon can be observed also, in most of the communities found in continental Europe, although there are exceptions.

Presence or dominance of *Brachypodium pinnatum* is not always due to undergrazing or the absence of grazing (Thomas et al. 1957). Table I (relevés 1 and 2) in this paper shows the dominance of this species in a vegetation grazed by cattle. The relevés, however, are situated at the foot of the slope, where the soil is deep and more humid: this may account for the dominance of *Brachypodium pinnatum* at these sites.

The sub-association *Cirsio-Brometum cornetosum sanguineae* is closely related to the *Bromus erectus* variant of the *Viburno-Cornetum* Rauschert 1968, described by Knapp & Reichkoff (1975) and occurring in the Leutratal in the vicinity of Jena (G.D.R.). In contrast to the sub-association *cornetosum sanguineae* described in this paper, the shrubs in the *Bromus erectus* variant of the Leutratal have a cover value of between 50 and 90 %, and I regard this as a scrub vegetation not grassland. However, not only the structure but also the floristic composition of these two communities differ so much from each other that the community studied in southern England has to be considered as a sub-association belonging to the *Cirsio-Brometum*.

Ulex europaeus scrub (Table III)

Scrub in which *Ulex europaeus* is the dominant species occurs on the slope named "The Long Man" near Wilmington, Sussex (Fig. 2). This scrub may have originated at a time when there was a temporary decrease in grazing intensity. As the scrub has grown older and higher, the the prickly nature of *Ulex europaes* has made it unattractive to cattle (Tansley 1939). Older *Ulex europaes* scrub, therefore, provides a protection against browsing by cattle to other shrubs, for example *Viburnum lantana*, *Sambucus nigra*, and *Crataegus monogyna*, which grow up among the *Ulex*.

As long as the *Ulex europaeus* scrub remains fairly open, many characteristic chalk grassland plants are able to maintain themselves in this vegetation, for example, *Sanguisorba minor*, *Primula veris*, and *Bromus erectus*. Other grassland species disappear at an early stage for example, *Pimpinella saxifraga* and *Brachypodium pinnatum* (Table III, relevés 4–7).

As the *Ulex europaeus* scrub continues to develop and becomes taller and denser, most of the grassland species vanish, and sciophytic species appear, for example, *Poa nemoralis* and *Brachypodium sylvaticum*.

ULEX EUROPAEUS SCRUB — Table III	1	2	3	4	5	6	7	Presence
Number of relevé	1	2	3	4	5	6	7	
Date July 1970	17	18	17	18	18	18	17	
Quadrat size (m²)	10	40	25	50	15	50	30	
Aspect	N	N	NW	NW	NW	N	N	
Slope°	30	25	30	25	30	30	25	
Altitude in m	115	115	110	150	150	145	115	
pH of A layer	7	6	6	6-7	7	5-6	–	
Grazed (+) or not grazed (–)	+	+	+	+	+	+	+	
Total cover %	100	95	95	100	85	100	100	
Cover % shrubs	90	95	95	85	85	95	80	
" herbs	5	5	5	10	45	10	40	
" mosses	–	1	–	5	–	3	15	
Height of shrubs in m	2	2	2	1,7	1	2	1,5	
Species number of phanerogams	6	7	8	16	17	19	20	
idem of cryptogams	–	1	–	2	–	1	7	
Ulex europaeus	5		2	5	4	4	4	V
Viburnum lantana		5	+	3	1	3	+	V
Rubus div. sp.	+	2	1	2	2	3	+	V
Dactylis glomerata	+			1	+	1	+	IV
Helictotrichon pratense	2		1		1	1	+	IV
Viola hirta			+	+	1	+	+	IV
Deschampsia caespitosa		+		1	1	+		III
Helictotrichon pubescens			1	1		1		II
Vicia cracca	+					+	+	III
Pseudoscleropodium purum		x		x		x	x	III
Solanum dulcamara			+			2		II
Poa nemoralis	+	+						II
Brachypodium sylvaticum			1					I
Sambucus nigra			+					I
Sanguisorba minor				+	1	r	2	III
Primula veris				+	r	r	+	III
Bromus erectus				+	1		1	II
Holcus lanatus				+		+		II
Sonchus arvensis				r			1	II
Cirsium sp.					r	r		II
Phyteuma tenerum				r	r			II
Carex flacca				+			+	II
Anthoxanthum odoratum					+		+	II
Brachythecium rutabulum				+				I
Cerastium holosteoides				r				I
Asperula cynanchica					r			I
Pimpinella saxifraga					r			I
Rosa sp.					+			I
Brachypodium pinnatum						+		I
Achillea millefolium						+		I
Crataegus monogyna						+		I
Cirsium arvense						+		I
Chamaenerion angustifolium						+		I
Rhinanthus minor							+	I
Centaurea nigra							r	I
Lotus corniculatus							+	I
Agrostis stolonifera							+	I
Fissidens adianthoides							x	I
Brachythecium velutinum							x	I
Calliergonella cuspidata							x	I
Ctenidium molluscum							x	I
Rhytidiadelphus squarrosus							x	I
R. triquetrus							x	I

Tansley (1939) postulated a relationship between the presence of *Ulex europaeus* and soil disturbance, a point investigated and further confirmed by Tubbs & Jones (1964) in studies in the New Forest, England. Disturbance of the soil occurred on the slope named "The Long Man" in the form of cattle tracks parallel to the contour lines of the slope. More cattle tracks were found on this slope than elsewhere. Probably this kind of soil disturbance explains why *Ulex europaeus* was more abundant at this locality than at other slopes studied.

Concerning the synsystematic position of the *Ulex europaeus* scrub on the chalk hills in southern England, it is postulated that in all likelihood this vegetation has a separate position within the eu-atlantic order of the *Ulicetalia* Quantin 1935, emend. Schubert 1960. The number of relevés given in Table III, however, is not enough to determine the correct position of the *Ulex europaeus* scrub described in this paper. However, it is certain that this community is not identical with the "*Ulex* scrub association" published in Hopkinson (1927). In the latter community, described from Nottinghamshire, many acidophytic plant species were present.

At Lullington Heath, situated about 2 km from "The Long Man" near Wilmington, Grubb et al. (1969 : 175) studied the ecology of chalk heath: "a plant community growing over chalk and containing an intimate mixture of usual chalk grassland species with heather (*Calluna vulgaris*) or bellheather (*Erica cinerea*)".

Based on both floristic composition and structure of the communities, it can be said that the vegetation studied in detail by Grubb et al. and the one presented here are not identical. Grubb et al. describe a species-rich community, 35–50 species per sample are recorded, in which *Ulex europaeus* is not the dominant species. At the slope "The Long Man" *Ulex europaeus* is dominant and the total cover percentage of the shrub layer in this community varies from 85 to 100%. Besides *Calluna vulgaris* and *Erica cinerea* several other acidophytic species occur in the chalk heath described by Grubb et al., for example, *Sieglingia decumbens*, *Potentilla erecta*, *Hypericum pulchrum*, and *Ulex minor*.

An *Ulex europaeus* scrub with some floristic and structural similarity with the one described in this paper, is reported by Wells et al. (1976) from the eastern limits of Salisbury Plain, Wiltshire.

Discussion

The chalk grasslands studied in 1970 in southern England, clearly belong to the association *Cirsio-Brometum* Shimwell 1971, emend. This is based upon the differential species *Thymus drucei*, *Filipendula vulgaris*, *Viola hirta*, and *Pseudoscleropodium purum*. This association, belonging to the *Mesobromion*, can be divided into three sub-associations:

I. *anthoxanthetosum odorati*, subass. nov. Of frequent occurrence on the South Downs of Sussex.

II. *cornetosum sanguineae* subass. nov. Mainly found on the North Downs of Kent on slopes on which the vegetation has not been grazed by sheep or cattle for some years.

III. *campanuletosum rotundifoliae* subass. nov. This grassland type is described from several localities in Wiltshire.

The sub-association *campanuletosum rotundifoliae* is characterized a.o. by the low frequency of *Bromus erectus* and *Brachypodium pinnatum*. In the phytocoenoses studied in 1970, these species were almost totally absent. This sub-association is possibly the richest in species of the several communities belonging to the *Mesobromion* in the British Isles (Wells 1975). The species composition of this community is probably influenced by the higher average rainfall and a higher relative air humidity (Wells l.c.) than is found elsewhere on the chalk in England and by an accumulation of raw humus on slopes with south-west to west aspects (Shimwell 1971).

The *Dactylorhiza fuchsii* subvariant, belonging to the sub-association *campanuletosum rotundifoliae*, can be considered as a transitional community to other more acidophytic associations. The differential species of the *Cirsio-Brometum*, *Thymus drucei*, *Filipendula vulgaris*, and *Viola hirta*, are still present but not frequent. Only *Pseudoscleropodium purum* remains present in all the relevés made in this community.

The *Dactylorhiza fuchsii* subvariant is similar to the "*Agrostis tenuis/Anthoxanthum odoratum/Viola riviniana* grasslands", described by Wells et al. (1976) from Porton Ranges, an area south-east of the eastern limits of Salisbury Plain, Wiltshire.

Pseudoscleropodium purum is the most common moss of calcareous grasslands in southern England. Watson (1960) attributed this to its so-called broad ecological amplitude. However, if this is the correct explanation, it is remarkable that other moss species, which likewise have a broad ecological amplitude, e.g. *Brachythecium rutabulum*, *Calliergonella cuspidata*, and *Rhytidiadelphus squarrosus*, although sometimes present, do not have such a high frequency in these grasslands as *Pseudoscleropodium purum*. In fact, the concept of "broad ecological amplitude" cannot be used as an explanation of the occurrence of a species anywhere.

The existence of acidophytic mosses in chalk grassland, for example *Dicranum scoparium*, *Pohlia nutans*, and *Dicranella heteromalla*, is probably a result of the accumulation of raw humus on some slopes. These mosses grow in direct rhizoidal connection with the humus.

Comparing the English chalk grasslands belonging to the *Mesobromion* with those on the European continent, the low frequency of lichens in the former is striking. During the investigation in 1970 *Cladonia squamosa* was found twice, in the *inops* variant of the sub-association *anthoxanthetosum odorati* on High-and-Over (Sussex) and in the *Succisa pratensis* variant belonging to the *campanuletosum rotundifoliae* of The Range (Wiltshire). Comparing this with the number of lichen species found in *Mesobromion* communities in the western part of the Continental Europe, it must be regarded as very low. In the French Jura (Willems 1973) and in the southernmost part of the Dutch province of Limburg (Barkman 1953), 8 and 12 lichen taxa were found respectively. Several of these taxa are frequent in the calcareous grasslands, for instance, *Peltigera canina*, *Cladonia furcata* var. *subrangiformis*, and *C. pyxidata* var. *chlorophaea*. Shimwell (1971b) reported three lichen taxa occurring in the *Cirsio-Brometum*, namely *Squamaria crassa*, *Cladonia pyxidata*, and *Cladonia furcata*, the latter being rather frequent in the sub-association *typicum*. Hope-Simpson (1941) reported 7 lichen taxa occurring in the chalk grassland of the South Downs. About the frequency Hope-Simpson noted: "Lichens, although not uncommonly present, are rarely conspicuous in chalk grassland" (page 108). Not one of the taxa mentioned by Hope-Simpson from the South Downs was found in 1970 in this study.

Although lichens in chalk grassland are not very frequent in southern England, some lichen-rich vegetation on basic soils exist. Wells et al. (1976) describe a "lichen-rich grassland – *Festuca ovina/Hieracium pilosella/Cladonia* spp. grassland" from Porton Ranges, Wiltshire, in which some tens of lichen species occur. At this locality a mosaic of lichens is found, the cover of which reaches 80% in some patches, interspersed with tufts of grasses and dicotyledonous plants, mainly typical calcareous grassland species.

Most of the chalk grasslands in southern England are still grazed, either by sheep or by cattle. Cattle grazing has been increasing during the last decade. On the European continent calcareous grasslands are no longer of any importance for agriculture and are rarely grazed. This change started after World War II and has continued since.

Besides grazing, the other major factor influencing the floristic composition of chalk grasslands in southern England has been scrub clearance. Scrub clearance, in which all shrubs and seedling trees are cut down and removed, is mainly carried out when shrubs threaten to gain the upperhand, mostly during periodes when grazing intensity is low.

Concerning the age of existing chalk grassland, Wells & Morris (1970), Shimwell (1971), and Wells et al. (1976) have drawn attention to the correlation between archaeological remains and calcareous grassland on many sites in southern England. Similar observations were made during this investigation. On the slope near Wilmington, named "The Long

Man", and on the one called High-and-Over in the vicinity of Alfriston, the vegetation as well as the upper soil layer down to the underlying white chalk had been removed, shaping two gigantic figures, a human shape and horse respectively.

These early-historic figures (Dyer 1973) are visible from a great distance, due to the absence of woodland or high scrub on the slopes on which the figures have been carved.

These figures are shaped in such a way that, seen from the valley, they are proportionally correct. Since they date from the early-historic past, the conclusion can be drawn that the grass and other vegetation was already present when the figures were made. Otherwise the figures would not have been functional.

A similar argument can be applied to the Wiltshire situation. Prehistoric tumuli are often situated on the top of a slope on which chalk grassland occurs (Fig. 3). These tumuli are only visible from the valley as silhouettes against the sky when woodland is absent on the slope.

Summary

In 1970 chalk grasslands were studied in southern England, according to the principles of the French-Swiss School. The communities described belong to the association *Cirsio-Brometum* Shimwell 1971, emend., alliance *Mesobromion erecti* (Br. Bl. & Moor 1938) Oberd. 1957, order *Brometalia* W. Koch (1926 n.n.) Br.-Bl. 1936, of the class *Festuco-Brometea* Br.-Bl. 1936, emend. Tüx. 1961. Within the *Cirsio-Brometum* three new sub-associations are distinguished: 1) subass. nov. *anthoxanthetosum odorati*; 2) subass. nov. *cornetosum sanguineae*; and 3) subass. nov. *campanuletosum rotundifoliae*.

Special attention is paid to the relationship that exists between the management of these communities and the seral stages that lead to the development of scrub following the decreasing influence of grazing.

Ulex europaeus scrub occurring on calcareous soil was also studied. The cryptogams of the chalk grasslands, mainly Musci, were investigated in detail.

References

Barkman, J.J. 1953. De kalkgraslanden van Zuid-Limburg. B. Cryptogamen. Publ. Nat. Hist. Gen. in Limburg VI: 21 30.

Barkman, J.J. 1966. Systematiek en gegevens van de kenmerken en de standplaats. In: J. Landwehr, Atlas van de nederlandse bladmossen. K.N.N.V., Amsterdam.

Barkman, J.J., J. Moravec & S. Rauschert. 1976. Code of Phytosociological Nomenclature. Vegetatio 32: 131 185.

Bornkamm, R. 1960. Die Trespen-Halbtrockenrasen im oberen Leinegebiet. Mitt. flor.-soz. Arbeitsgem. N.F. 8: 181 208.

Braun-Blanquet, J. & M. Moor. 1938. Verband des Bromion erecti. In: Prodromus der Pflanzengesellschaften 5: 1 64. Com. Int. du Prodrome Phytosoc. S.I.G.M.A., Montpellier.

Braun-Blanquet, J. 1964. Pflanzensoziologie. 3. Aufl. Springer. Wien New York.

Clapham, A.R., T.G. Tutin & E.F. Warburg. 1968. Excursion Flora of the British Isles. Univ. Press, Cambridge.

Dyer, J. 1973. Southern England: An archeological Guide. Faber & Faber, London.

Grubb, P.J., H.E. Green & R.C.J. Merrifield. 1969. The ecology of chalk heath: its relevance to the calcicole-calcifuge and soil acidification problems. J. Ecol. 57: 175 212.

Heukels, H. & S.J. van Oostostroom. 1970. Flora van Nederland. Noordhoff, Groningen.

Hopkinson, J.W. 1927. Studies on the vegetation of Nottinghamshire. J. Ecol. 15: 130 171.

Knapp, H.D. & L. Reichhoff. 1975. Die Vegetation des Naturschutzgebietes 'Leutratal' bei Jena. Arch. Naturschutz und Landschaftsforsch. 15: 91 124.

Korneck, D. 1974. Xerothermvegetation in Rheinland-Pfalz und Nachbargebieten. Schriftenreihe für Vegetationskunde 7: 1 196.

Lousley, J.E. 1950. Wild flowers of chalk & limestone. Collins, London.

Margadant, W.D. 1959. Mossentabel. N.J.N., Amsterdam.

Moss, C.E. 1911. The sub-formation of the Older Limestones. In: A.G. Tansley (ed.), Types of British Vegetation pp. 146 160. Univ. Press, Cambridge.

Oberdorfer, E. 1957. Süddeutsche Pflanzengesellschaften. Pflanzen-soziologie Band 10. Fischer, Jena.

Scherrer, M. 1925. Vegetationsstudien im Limmattal. Veröff. Geob. Inst. Rübel 2: 1 116.

Shimwell, D.W. 1971a. Festuco-Brometea Br.-Bl. & R. Tüx. 1943 in the British Isles: The phytogeography and phytosociology of limestone grasslands. Part 1 (a) General Introduction; (b) Xerobromion in England. Vegetatio 23: 1 28.

Shimwell, D.W. 1971b. Festuco-Brometea Br.-Bl. & R. Tüx. 1943 in the British Isles: The phytogeography and phytosociology of limestone grasslands. Part II. Eu-Mesobromion in the British Isles. Vegetatio 23: 29 60.

Tansley, A.G. (ed.) 1911. Types of British Vegetation. Univ. Press, Cambridge.

Tansley, A.G. & W.M. Rankin. 1911. The sub-formation of the Chalk. In: A.G. Tansley (ed.), Types of British Vegetation pp. 161 180. Univ. Press, Cambridge.

Tansley, A.G. 1922. Studies of the vegetation of the English chalk. II Early stages of redevelopment of woody vegetation on chalk grassland. J. Ecol. 10: 168 177.

Tansley, A.G. & R.S. Adamson. 1925. Studies of the vegetation of the English chalk. III The chalk grasslands of the Hampshire-Sussex border. J. Ecol. 13: 177 223.

Tansley, A.G. 1939. The British Isles and their Vegetation. Univ. Press, Cambridge.

Thomas, A.S., M. Rawes & W.J.L. Banner. 1957. The vegetation of the Pewsey vale escarpment, Wiltshire. J. Brit. Grassland Soc. 12: 39 48.

Thomas, A.S. 1960. Changes in vegetation since the advent of myxomatosis. J. Ecol. 48: 287–306.

Thomas, A.S. 1963. Further changes in vegetation since the advent of myxomatosis. J. Ecol. 51: 151–186.

Tubbs, C.R. & E.L. Jones. 1964. The distribution of gorse (Ulex europaeus L.) in the New Forest in relation to former land use. Proc. Hampshire Fld. Club 23: 1–10.

Tüxen, R. 1928. Bericht über die pflanzensoziologische Exkursion der floristisch-soziologischen Arbeitsgemeinschaft nach dem Pleszwalde bei Göttingen. Mitt. flor.-soz. Arbeitsgem. Niedersachsen 1: 25–51.

Tüxen, R. 1937. Die Pflanzengesellschaften Nordwestdeutschlands. Mitt. flor.-soz. Arbeitsgem. Niedersachsen 3: 1–170.

Watson, E.V. 1960. A quantitative study of the Bryophytes of chalk grassland. J. Ecol. 48: 397–414.

Wells, T.C.E. & M.G. Morris. 1970. Conservation research and management of calcareous grassland. In: Guide of post-symposium tour no. 5, of Int. Symp. on the scient. management of animal and plant commun. for conservation. 50 pp. Univ. Norwich, Norwich.

Wells, T.C.E. 1975. The floristic composition of chalk grassland in Wiltshire. In: L.F. Stearn (ed.), Supplement to the Flora of Wiltshire, pp. 99–125. Wiltsh. Archaeol. Nat. Hist. Soc., Wiltshire.

Wells, T.C.E., J. Sheail, D.F. Ball & L.K. Ward. 1976. Ecological studies on the Porton Ranges: Relationships between vegetation, soils and land-use history. J. Ecol. 64: 589–626.

Willems, J.H. 1973. Limestone grassland vegetation in the central part of the French Jura, South of Champagnôle (dept. Jura). Proc. Kon. Ned. Ak. Wetensch. Series C, 76 (3): 231–244.

Willems, J.H. & F.G. Blanckenborg. 1975. Kalkgraslandvegetaties van de St. Pietersberg ten Zuiden van Maastricht. Publ. Nat. Hist. Gen. in Limburg, XXV-1: 1–24. (with English summary).

Wijk, R. van der, W.D. Margadant & P.A. Florschütz. 1959–1969. Index Muscorum. Int. Ass. Plant Tax. Utrecht. Vol. I-V.

VERGLEICH VERSCHIEDENER CHRYSOPOGON GRYLLUS-REICHER TROCKENWIESEN DES INSUBRISCHEN KLIMABEREICHES UND ANGRENZENDER GEBIETE[*, **]

Martin MEYER[***]

Aeschstrasse 35, CH-8127 Forch, Schweiz

Stichwörter (Keywords):
Affinität (Affinity), *Chrysopogon gryllus, Diplachnion, Festuco-Brometea,* Insubrisch, Minimum spanning tree, Norditalien (N. Italy) *Orno-Ostryon,* Stetigkeitstabelle (Constancy table), Südschweiz (S. Switzerland)

Einleitung

In einer vorangegangenen Arbeit (Meyer 1976), in welcher Trockenwiesen karbonathaltiger Standorte des insubrischen Klimabereiches (das Gebiet zwischen dem Garda See und dem Lago Maggiore) beschrieben wurden, stellte sich die Frage der Verbandeszugehörigkeit der *Chrysopogon gryllus*-reichen Gesellschaften.

Aus den inneralpinen Tälern der Alpensüdseite beschrieb Braun-Blanquet (1961) ebenfalls Trockenrasen mit *Chrysopogon gryllus.* Auch aus Nordostitalien (Lorenzoni 1967, Ferlan & Giacomini 1955, Pedrotti 1963) sind solche Rasen bekannt. Aus dem angrenzenden Jugoslawien werden Saumgesellschaften beschrieben (Van Gils et al. 1975), in denen diese *Poaceae* dominant vorkommt. Schliesslich sei noch auf die Arbeiten von Antonietti (1968, 1970) und Hofer (1967), in welchen lichte Buschwälder sowie Felsheiden erläutert werden, hingewiesen.

Ein Vergleich der Gesellschaften aller oben genannten Autoren untereinander sowie mit einigen *Bromus erectus*-wiesen der Alpennordseite (Zoller 1954) sollte uns helfen, die Zugehörigkeit des *Carici humilis-Chrysopogonetum grylli* sowie des *Holco-Chrysopogonetum grylli* prov. (Meyer 1976) abzuklären.

* Contribution to the Symposium on Plant species and plant communities, held at Nijmegen, 11–12 November 1976, on the occasion of the 60th birthday of Professor Victor Westhoff.
** Die Artennamen wurden alle, soweit dort vorhanden, der Namengebung von Hess, Landolt & Hirzel (1967–1972) angepasst.
*** Herrn Prof. Dr. E. Landolt, Zürich, danken wir an dieser Stelle für die Durchsicht des Manuskriptes sowie für seine Anregungen bestens.

Methodik

Die zum Vergleich herangezogenen Assoziationen und Subassoziationen der verschiedenen Autoren wurden nach der Braun-Blanquet (1964)-Methode aufgenommen. Wo nicht schon Stetigkeitstabellen vorlagen, wurden solche angefertigt. In der Übersichtstabelle (*Tabelle 1*) wurden alle Arten der Stetigkeitsklasse I, welche nur einmal vorkamen und nicht zur charakteristischen Artenkombination gehörten, weggelassen. Dabei wurde eine Ausnahme für diejenigen Arten gemacht, welche nur im *Carici humilis-Chrysopogonetum galietosum* vorkommen. Die in der Stetigkeitstabelle angegebenen pflanzensoziologischen Zugehörigkeiten der Arten wurden Oberdorfer (1970) entnommen.

Zur Ermittlung der Affinität der einzelnen Gesellschaften zueinander wurde die Formel von Kulcynski, wie sie in Knapp (1971) angegeben wird, verwendet. Als Ausgangspunkt für die Berechnungen diente die Stetigkeitstabelle mit den Klassen I–V.

Affinität (%)
$$= \frac{100}{2}\left[\frac{\sum S_1 + \sum S_2 - \sum D}{2\sum S_1} + \frac{\sum S_1 + \sum S_2 - \sum D}{2\sum S_2}\right]$$

wobei $\sum S_1$ = Summe der Stetigkeitswerte in der einen,
$\sum S_2$ = Summe der Stetigkeitswerte in der anderen Gesellschaft,
$\sum D$ = Summe aller Differenzen (Absolutwert) der Stetigkeitswerte aller Arten.

Pignatti (1962), der auch mit dieser Formel gearbeitet hat, bemerkte, dass es bei diesem Vergleich wesentlich besser sei, sich auf die floristisch charakteristische Arten-kombination zu beschränken und nicht sämtliche Arten zu berücksichtigen. Zu einem ähnlichen Schluss gelangte auch Raabe (1952). Daher werden hier Begleitarten mit der Stetigkeit I und zum Teil auch II bei der Berechnung nicht berücksichtigt. In unserem Vergleich wurden die-jenigen Arten, welche mit Stetigkeit I unter den Begleit-pflanzen auftraten und sonst in keiner anderen Vergleichs-gesellschaft mit einer höheren Stetigkeit vorkamen, nicht berücksichtigt. Hingegen sind Arten mit Stetigkeit I, welche in einer Artengruppe enthalten sind, berücksichtigt worden.

Die Affinitäten zwischen den einzelnen Gesellschaften wurden mittels eines Computers berechnet und in einer Dreiecksmatrix dargestellt. In einem zweiten Schritt wurde mittels eines leicht modifizierten Rechenprogrammes von Gower & Ross (1964) der sogenannte 'minimum spanning tree' berechnet. Dieser ermöglicht es, die Verknüpfungen der einzelnen Gesellschaften untereinander graphisch dar-zustellen und zwar so, dass jeder Gesellschaft diejenigen noch freien Partner zugeordnet werden, welche zu dieser die grösstmögliche Affinität haben.

Vergleich der verschiedenen Gesellschaften

Die Vergleichsgesellschaften

Wie in der Einleitung erwähnt, soll das *Carici humilis-Chrysopogonetum grylli* (Meyer 1976), in der Tabelle 1 Nummern 5 und 6, und das *Holco-Chrysopogonetum grylli* prov. (Meyer 1976), Nr. 7, mit nahestehenden, meist *Chrysopogon gryllus*-reichen Gesellschaften verglichen wer-den.

Im Folgenden soll kurz auf die Vergleichsgesellschaften eingegangen werden.

Aus den inneralpinen Tälern wurden drei von Braun-Blanquet (1961) beschriebene Gesellschaften herangezogen:
– das *Fumano-Andropogonetum contorti*, Tab. 1, Nr. 1, auf Porphyrhängen zwischen Bozen und Meran, welches an Rän-dern und in Lichtungen des *Orneto-Ostryetum* zu finden ist; – das *Diplachno-Festucetum vallesiacae*, Nr. 2, von den Tal-hängen des   welches über silikatischer Unterlage wächst und ebenfalls von *Orneto-Ostryetum*-Buschwäldern um-geben ist; – das *Contorteto-Diplachnetum*, Nr. 3, im Val di Susa.

Zu diesen drei, dem *Diplachnion* angehörenden Gesellschaften (Braun-Blanquet 1961) ist auch das *Ischaemo-Diplachnetum*, Nr. 4, von Pedrotti (1962) zu zählen. Letzterer Autor hat seine Aufnahmen an Hängen in der kalkreichen Umgebung von Trento gemacht. Aus dem *Orno-Ostryon* wurden ebenfalls drei Gesellschaften herangezogen. Es handelt sich um das auf Kalkhängen vorkommende *Helleboro-Ornetum asteretosum*

prov. von Antonietti (1968), Nr. 8, sowie um zwei **Buschwald**-typen, Nr. 9 und 10, des *Orno-Ostryetum* von Hofer (1967). Der erste Typ, Nr. 9, stammt aus dem Comer- und Luganersee-becken, der zweite, Nr. 10, aus dem Gardaseegebiet.

Gesellschaften mit *Chrysopogon gryllus*, welche zum *Saro-thamnion*-Verband gehören, wurden im insubrischen Bereich ebenfalls beschrieben. Für unseren Vergleich wurden drei Varianten des *Gryllo-Callunetum* prov., *Allium senescens*-Sub-assoziation (Antonietti 1970) herangezogen: die *Aster linosyris*-Variante, Nr. 11., die *Andropogon contortus*-Variante, Nr. 12, und die Typische Variante, Nr. 13. Auch von Hofer (1967) wurden zwei Felsheidegesellschaften übernommen: die Suk-kulenten-Felsheide, Nr. 14, aus dem Langenseegebiet und die Pfeifengras-Felsheide, Nr. 15, aus dem gleichen Gebiet.

Vom weiter entfernten jugoslawischen Karst, aus der Gegend zwischen Triest und Rijeka, stammen die Saumgesellschaften von Van Gils et al. (1975). Es wurden folgende, dem *Dictamno-Ferulagion* (Van Gils et al. 1975) zugeordnete Aufnahmen berücksichtigt:
– Gesellschaft mit *Potentilla alba* und *Laserpitium siler*, Nr. 16, aus der Umgebung von Rupa,
– das *Cirsio pannonicae-Clematidetum rectae*, Nr. 17, aus der Umgebung von Sezana,
– die Subassoziation mit *Genista sagittalis*, Nr. 18, und die-jenige mit *Senecio lanatus*, Nr. 19 des *Veronicetum-spicato-jaquinii*, welche beide auf flachen Landschaftsteilen über sauer reagierender Terra rossa vorkommen,
– das *Cirsio pannonicae-Peucedanetum cervariae*, Nr. 20, über Terra fusca und flachgründigen Kalkböden,
– das *Libanoto daucifoliae-Laserpitietum siler*, Nr. 21, über flachgründigen Kalkböden der submontanen Stufe und
– das *Origano vulgaris-Cnidietum silaefoliae*, Nr. 22, aus dem Vipavatal über neutral reagierender Mullrendzina.

Als nächste Vergleichsgesellschaft ist das *Chrysopogoneto-Centauretum cristatae*, Nr. 23, von Ferlan & Giacomini (1955) zu nennen. Diese, zum *Saturejon subspicatae*-Verband zählende Gesellschaft wurde an den Hängen des Monfalcone-Gebietes aufgenommen.

Als letzte *Chrysopogon gryllus*-reiche Gesellschaft wurde das *Chrysopogonetum grylli* (Lorenzoni 1965) mit seiner Typischen Subassoziation, Nr. 24, und der *Molinia coerulea*-Subassozia-tion, Nr. 25 in der Stetigkeitstabelle, zugezogen. Die Aufnahmen stammen meist von Standorten in Flussnähe aus der Ebene des Friaul, wo Gletscherablagerungen und Flussschotter den Un-tergrund bilden.

Um den Vergleich zu vervollständigen, wurden noch einige Grünlandgesellschaften ohne 'Chrysopogon gryllus mitberück-sichtigt. Dabei handelt es sich um Gesellschaften, die nörd-lich der Alpen vorkommen, nämlich um Vertreter des *Xero-bromion* und des *Mesobromion* aus dem Schweizer Jura. Aus der Arbeit von Zoller (1954) wurden folgende vier Gesellschaften beigezogen:
– des *Teucrio-Xerobrometum* mit seiner Subjurassisch-süd-westschweizerischen Fazies; Nr. 26 in der Stetigkeitstabelle,
– das *Cerastio-Xerobrometum* mit der *Trifolium dubium-Trifoli-um striatum*-Subassoziation, Nr. 27, aus dem südwestschweiz-erischen Jura;
– das *Teucrio-Mesobrometum* mit der *Globularia cordifolia-Coronilla vaginalis*-Subassoziation, Nr. 28, hauptsächlich aus dem Berner- und Solothurner Jura sowie

22

Additional material from *Plant Species and Plant Communities,*
ISBN 978-90-6193-591-9 (978-90-6193-591-9_OSFO2), is available at http://extras.springer.com

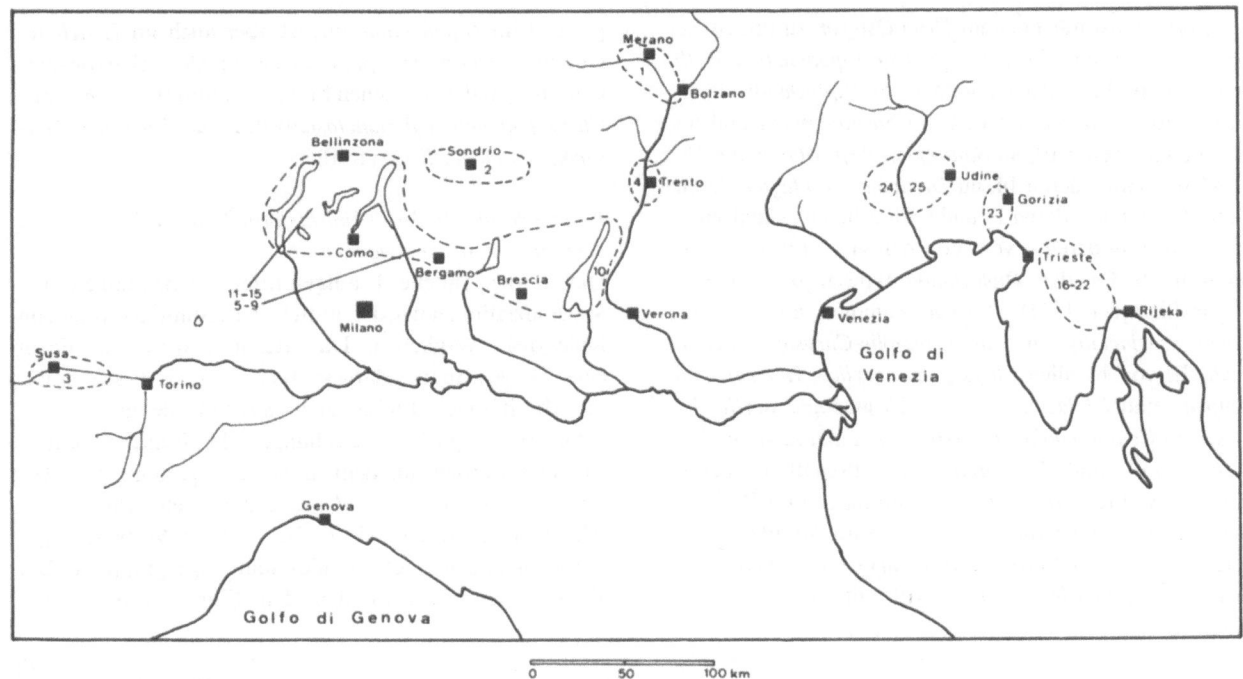

Abb. 1. Geographische Lage der verschiedenen Untersuchungsgebiete. Die Nummern bezeichnen die Einheiten der einzelnen Autoren und sind im Text erläutert.

– das wechselfeuchte *Colchico-Mesobrometum*, Nr. 29, mit Aufnahmen aus dem gesamten Schweizer Jura.

Als letzte Vergleichsgesellschaft wurde das *Centaureo dubiae-Arrhenatheretum*, Nr. 30, von Oberdorfer (1964) aus dem insubrischen Bereich in die Tabelle aufgenommen.

Abbildung 1 gibt die grobe geographische Lage der Einheiten 1–25 an. Es lässt sich daraus auch erkennen, dass nur die Einheiten 5–9 und 11–15 im eigentlichen insubrischen Bereich liegen, während die restlichen Aufnahmen aus inneralpinen Tälern (Nr. 1–3) oder aus dem submediterranen (Nr. 4, 10, 16–25) Klimabereich stammen.

Vergleich mittels einer Stetigkeitstabelle

Die im vorangehenden Abschnitt genannten Assoziationen und Subassoziationen wurden alle in Tabelle 1 aufgenommen. Die Aufnahmen wurden nach Verbänden zusammengefasst und in der Tabelle so dargestellt, dass Artengruppen entstanden, die für die einzelnen Subassoziationen, Assoziationen und Verbände charakteristisch oder dann gemeinsam sind. Die beiden *Xerobromion* (Nr. 26 und 27) und *Mesobromion*-Gesellschaften (Nr. 28 und 29) sowie das *Centaureo dubiae Arrhenatheretum* (Nr. 30) wurden als Gegenüberstellung der Tabelle angehängt.

Fast allen Subassoziationen, Assoziationen und Verbänden gemeinsam ist nur die Artengruppe mit *Brachy-*

podium pinnatum. Überall, ausser im *Colchico-Mesobrometum* (Nr. 29) und im *Centaureo dubiae-Arrhenatheretum* (Nr. 30) kommt die Artengruppe mit *Teucrium chamaedrys* und auch *Chrysopogon gryllus* vor. Ausser im *Arrhenatherion* und etwas schwächer im *Diplachnion* ist die Artengruppe mit *Bromus erectus*, überall gut vertreten. Als letzte, ausser im *Diplachnion* und *Arrhenatherion*, gut vertretene Artengruppe sei die Gruppe mit *Dactylis glomerata* genannt. Die übrigen Artengruppen können der Tabelle 1 entnommen werden.

Der Vergleich soll schliesslich auch Aufschluss über die Stellung des *Carici humilis-Chrysopogonetum grylli* geben. Zu diesem Zwecke wurden in der Stetigkeitstabelle jene Arten unterstrichen, welche zur charakteristischen Artenkombination (Meyer 1976) dieser Einheit gehören. Es zeigte sich, dass ausser *Polygala pedemontana* und *Bromus condensatus* alle Arten auch in anderen Einheiten vorkommen. Es muss allerdings erwähnt werden, dass die Autoren der in unserem Vergleich aufgeführten Einheiten die beiden genannten Arten gar nicht als solche unterschieden haben.

Interessant ist es in diesem Zusammenhang zu sehen, wo diese charakteristischen Arten sonst hauptsächlich auftreten. Sonst nur noch im *Diplachnion* sind *Ononis pusilla* und *Centaurea bracteata* anzutreffen, während *Leontodon*

23

tenuiflorus sonst nur noch im *Orno-Ostryon* zu finden ist. Von den 33 für das *Carici humilis-Chrysopogonetum grylli* charakteristischen Arten kommen im *Diplachnion* deren 29, im *Orno-Ostryon* deren 24, im *Sarothamnion* und im *Xerobromion* deren 19, im *Saturejon subspicatae* deren 17, im *Mesobromion* deren 14, im *Dictamno-Ferulagion* deren 13, im *Stipo-Poion* deren 11 und schliesslich im südalpinen *Arrhenatherion* deren 6 vor. Von den vier in ihrer Kombination als für das *Diplachnion* typisch bezeichneten (Braun-Blanquet 1961) *Poaceae* kommen ausser *Danthonia provincialis* im *Carici humilis-Chrysopogonetum grylli* alle vor, nämlich *Chrysopogon gryllus, Heteropogon contortus* und *Diplachne serotina. Chrysopogon gryllus* ist ausser im *Centaureo dubiae-Arrhenatheretum* sowie in den *Xerobromion-* und *Mesobromion-*Gesellschaften überall vertreten, während *Heteropogon contortus* nur im *Diplachnion-*, schwach im *Orno-Ostryon-* und im *Sarothamnion-*Verband sowie im *Carici humilis-Chrysopogonetum grylli* vorkommt. *Diplachne serotina* tritt mit ihrem Schwer-

gewicht im *Diplachnion* auf, ist aber auch im *Helleboro-Ornetum asteretosum* prov. aus dem *Orno-Ostryon-*Verband und in der trockenen Subassoziation (*Carici humilis-Chrysopogonetum fumanetosum*) des *Carici humilis-Chrysopogonetum grylli* anzutreffen.

Vergleich mittels der Affinitätsformel von Kulczynski und dem 'minimum spanning tree'
Die in der Tabelle 1 aufgeführten Assoziationen und Subassoziationen wurden mittels der Affinitätsformel von Kulczynski verglichen. Das Resultat wurde in Form einer Dreiecksmatrix dargestellt (*Tabelle 2*); diese erlaubt, die Affinität jeder Einheit zu anderen abzulesen.

Die bestmöglichen Beziehungen der Einheiten untereinander wurden mit dem 'minimum spanning tree' berechnet und sind in *Abbildung 2* dargestellt. Zur Veranschaulichung wurden die Einheiten nach Verbandszugehörigkeit und je nach Standortsunterlage gruppiert. Die Verbindungslinien zwischen den Einheiten zeigen die

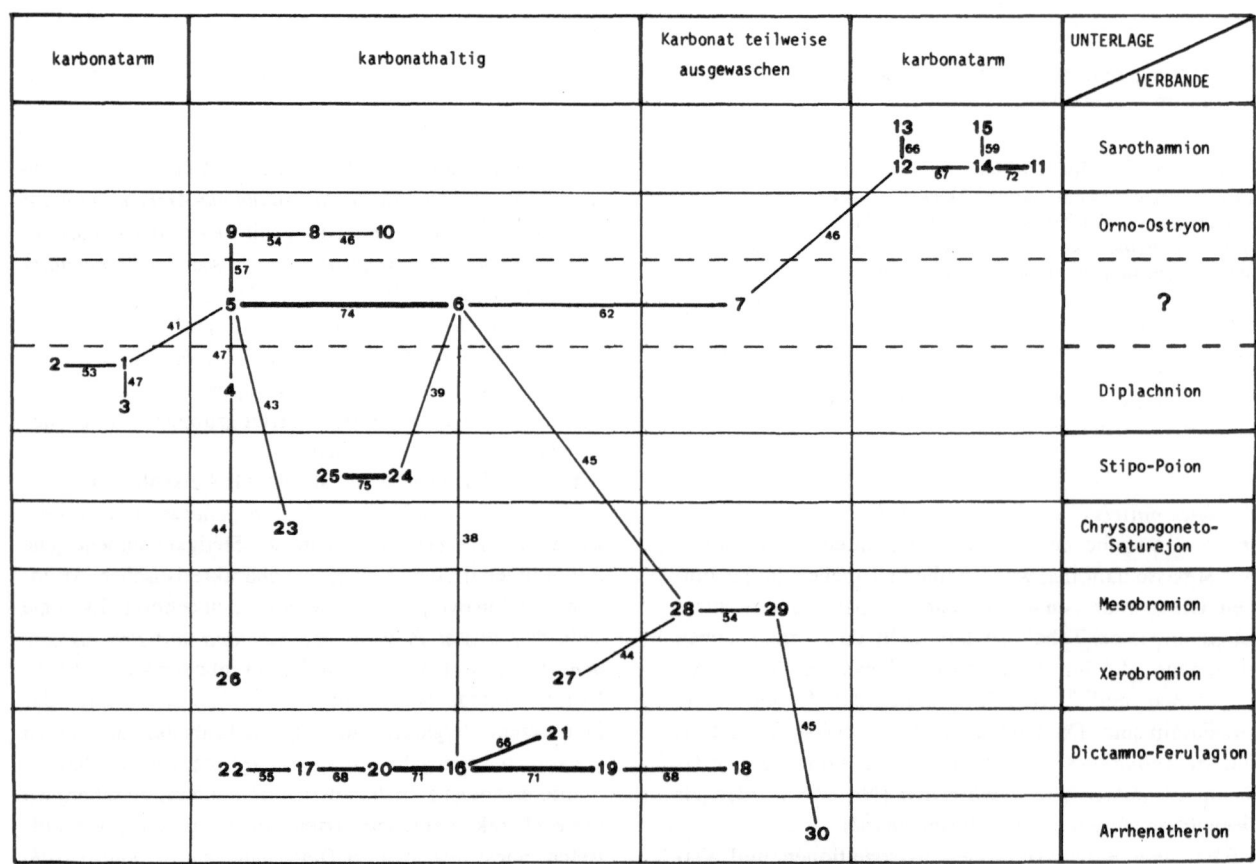

Abb. 2. Minimum spanning tree mit den dazugehörigen Affinitätswerten der Vergleichseinheiten.

Tabelle 2. Affinitäten (nach Kulcynski) zwischen den Vergleichseinheiten

	Einheiten Nr.																													
	1	2	3	4	5	6	7	8	9	10	11	12	13	14	15	16	17	18	19	20	21	22	23	24	25	26	27	28	29	30
1	100																													
2	54	100																												
3	47	40	100																											
4	42	33	34	100																										
5	41	35	36	47	100																									
6	29	28	27	36	73	100																								
7	13	16	11	18	46	62	100																							
8	23	17	18	30	50	43	32	100																						
9	39	28	28	37	57	45	28	55	100																					
10	19	14	17	28	36	30	16	46	45	100																				
11	28	23	20	23	35	34	36	23	25	15	100																			
12	26	24	21	24	45	43	46	37	37	26	62	100																		
13	16	11	10	19	36	32	36	30	27	18	53	66	100																	
14	24	21	19	20	37	36	39	27	31	19	72	67	58	100																
15	9	15	9	12	34	32	42	31	20	12	46	58	57	59	100															
16	15	8	10	24	36	38	30	33	28	19	18	26	18	18	17	100														
17	8	7	9	19	32	33	26	32	22	18	16	25	17	16	21	66	100													
18	11	9	10	17	27	30	24	29	21	19	19	23	14	14	17	65	63	100												
19	13	10	10	19	31	34	27	27	27	20	20	25	16	18	16	71	60	68	100											
20	13	10	10	22	35	35	27	33	28	20	19	29	19	21	22	71	68	60	65	100										
21	10	8	8	20	33	34	25	30	28	21	15	23	17	17	18	66	56	50	59	66	100									
22	7	9	5	12	27	30	27	34	21	16	17	24	17	17	20	50	55	53	48	53	47	100								
23	32	23	20	41	43	36	24	28	33	22	19	26	17	20	16	27	23	24	26	22	24	17	100							
24	22	15	16	27	34	39	35	23	18	13	23	24	17	22	19	23	21	24	20	18	18	19	27	100						
25	21	15	16	25	32	38	34	23	17	14	19	22	16	18	23	21	20	20	17	18	16	17	21	75	100					
26	35	26	27	44	44	40	27	32	37	24	23	27	16	23	14	23	18	22	23	21	18	16	30	26	25	100				
27	17	17	13	23	30	35	30	16	15	12	17	20	12	17	14	16	16	12	13	9			25	34	32	41	100			
28	16	11	11	22	37	45	38	26	18	17	21	27	17	20	19	31	26	28	30	25	23	25	24	38	35	38	44	100		
29	6	4	4	11	24	35	37	15	9	9	14	17	14	14	14	21	20	21	21	16	16	19	16	26	26	17	30	54	100	
30	4	3	2	4	15	30	36	9	7	7	8	13	10	8	9	12	11	9	13	7	10	13	9	17	15	7	18	23	45	100

bestmöglichen Verknüpfungen an, die Zahlen bei diesen Verbindungslinien bedeuten die Affinitäten (in Prozenten) zwischen den Einheiten. Ohne an dieser Stelle in die Einzelheiten dieses Vergleichs eingehen zu wollen – in der Diskussion soll dies nachgeholt werden – möchten wir hier einige Bemerkungen anbringen. Die beiden von Zoller (1954) beschriebenen *Xerobromion*-Gesellschaften, das *Teucrio-Xerobrometum* (Nr. 26) und das *Cerastio-Xerobrometum* (Nr. 27) weisen bei dieser Vergleichsmethode nicht etwa die engste Verknüpfung untereinander, also im selben Verband auf, sondern sind enger mit der *Diplachnion*-Gesellschaft Nr. 4 (*Ischaemo-Diplachnetum*) bzw. mit der *Mesobromion*-Gesellschaft Nr. 28 (*Teucrio-Mesobrometum*) verknüpft.

Die drei Varianten des *Gryllo-Callunetum* prov., *Allium senescens*-Subassoziation (Nr. 11, 12 und 13) von Antonietti (1970) sind ebenfalls nicht untereinander am engsten verknüpft. Die *Aster linosyris*-Variante wird durch die Sukkulenten-Felsheide vom Langensee (Nr. 14) von den beiden übrigen Varianten getrennt. Die Affinitätswerte zwischen den fünf *Sarothamnion*-Einheiten bestätigen jedoch die von Antonietti (1970) aufgestellte Felsheidegesellschaft, das *Gryllo-Callunetum* prov., zu welchem auch die beiden Felsheidentypen von Hofer (1967) gehören.

Es zeigt sich zudem, dass innerhalb des *Dictamno-Ferulagion* suball. nov. (Van Gils et al. 1975) teilweise sehr hohe Affinitäten erreicht werden. Dies bedeutet ja nichts anderes als dass die Artengarnitur solcher Einheiten sehr ähnlich ist. Wir neigen daher zur Annahme, dass es sich bei den von Van Gils et al. (1975) beschriebenen Assoziationen wohl eher um Subassoziationen oder sogar Varianten handelt als um eigenständige Gesellschaften. Als relativ eigenständig betrachten wir das *Origano vulgaris-Cnidietum silaefoliae* (Nr. 22) sowie das *Libanoto daucifoliae-Laserpitium siler*, Nr. 21. (vergleiche dazu auch die Stetigkeitstabelle). Die Einheiten Nr. 16–20 sollten unseres Erachtens jedoch zu einer Assoziation zusammengefasst werden.

Werden schliesslich noch die Affinitäten zwischen den verschiedenen Verbänden betrachtet, so zeigt sich, dass der *Orno-Ostryon*-Verband die grösste Affinität zum *Carici-humilis-Chrysopogonetum grylli* aufweist. Die nächstbeste Beziehung zu dieser Gesellschaft geht über das *Ischaemo-Diplachnetum* (Nr. 4) zum *Diplachnion*-Verband. Das *Centaureo dubiae-Arrhenatheretum* (Nr. 30) sowie die beiden *Xerobromion*-Gesellschaften (Nr. 26 und 27) weisen als einzige keine direkte Verknüpfung im Sinne einer grösstmöglichen Affinität zum *Carici humilis-Chrysopo-*

Tabelle 3: Prozentuales Vorkommen der Vertreter von pflanzensoziologischen Einheiten (nach Oberdorfer 1970) innerhalb der Vergleichseinheiten

Klassen / Ordnungen / Verbände / Unterverbände	Abkürzung	Diplachnion	Carici hum.-Chrysopog. gr.	Holco-Chrysopogenetum gr.	Orno-Ostryon	Sarothamnion	Dictamno-Ferulagion	Saturejon subspicatae	Stipo-Poion	Xerobromion	Mesobromion	Arrhenatherion
Festuco-Brometea	FB	36	18,9	15	18	14,8	12	27	34	28	15	11,1
Festucetalia vallesiacae	V	16	6,4	2,3	4,7	1,7	2,4	5,4	8,5	3,5		
Stipeto-Poion carniolicae	s	5,2	0,9							0,9		
Diplachnion												
Brometalia	B	12	4,3	3,8	4,7	5,2	3,2	12,5	6,8	5,3	4,4	
Chrysopogoneto-Saturejon subspicatae	j						3,2	10,7				
Bromion erecti	b		1,7	1,5	3,3	1,7	0,8	1,8	1,7	1,8	0,7	
Xerobromion	x	3	1,7	1,5	1,3	1,7		1,8	1,7	4,4		
Mesobromion	m	0,8	2,6	3,0	0,7	0,9	2,4		5,1	5,3	6,7	6,6
Sedo-Scleranthethea	SS	3	1,3		1,3	1,7				6,2	0,7	
Festuco-Sedetalia	S	6	2,2	2,3	2,7	4,4	0,8	1,8	3,4	3,5	1,5	
Sedo-Scleranthetalia	T					0,9						
Alysso-Sedion	e									4,4		
Asplenieta rupestris	AR		0,9		0,7	1,7						
Potentilletalia caulescentis	P	0,8	0,9		0,7	1,7	0,8	1,8				
Androsacetalia Vandellii	D					0,9						
Nardo-Callunetea	NC		2,2	2,3	1,3	5,2	2,4	1,8	1,7	3,5	5,2	3,7
Nardetalia	N		0,4		0,7	1,7					1,5	
Nardion	n		0,4			0,9						
Calluno-Ulicetalia												
Sarothamnion												
Calluno-Genistion	c					0,9						
Molinio-Arrhenatheretea	MA	1,5	3	5,3	2,7	4,4	0,8	3,6	6,8	2,6	11	22,2
Arrhenatheretalia	A	1,5	3,9	3,8	0,7	3,5	1,6	1,8	3,4	3,5	7,4	14,8
Arrhenatherion	a		0,4	0,8			0,8			1,8	3,0	1,8
Cynosurion	y			1,5						0,9	1,5	9,2
Molinietalia	M										3,7	1,8
Molinion	m		2,2	3,8	2,7	4,4	4	3,6	3,4		2,2	1,8
Trifolio-Geranietea	TG				0,7	0,9	0,8					1,8
Origanetalia	O	1,5	1,7	3	2,7	3,5	2,4			0,9	3	1,8
Trifolion medii	r										2,2	1,8
Geranion sanguinei	g	5,2	5,6	3,8	5,3	3,5	11,2	5,4	5,1	8	6	3,7
Dictamno-Ferulagion												
Querco-Fagetea	*GF	0,8	2,2	0,8	4,7	0,9	2,4	1,8			2,2	
Prunetalia	*P		1,3		2,7	0,9	3,2					
Berberidion	*b	0,8	1,7		3,3	3,5	0,8					
Quercetalia pubescenti	*Q	0,8	3,0	2,3	8	1,7	5,6			2,6	3	
Quercion pubescenti	*q		0,4									
Orno-Ostryon	*o	3	1,3		2		1,6	1,8	1,7			
Fagetalia silvaticae	*F		0,4		2		0,8				1,5	3,7
Fagion	*f				0,7		0,8					
Carpinion betuli	*c	1,5	1,7	2,3	1,3	1,7	0,8	1,8		0,9	0,7	
Erico-Pinetea												
Erico-Pinetalia	*E				1,3	0,9						
Erico-Pinion	*e				2,7	0,9	2,4					
Vaccinio-Piceetea												
Vaccinio-Piceetalia	*V				0,7	0,9						
Total Anzahl der Arten		135	233	133	150	115	125	56	59	133	135	54

26

gonetum grylli (Nr. 5 und 6) oder zum *Holco-Chrysopogonetum* prov. (Nr. 7) auf. Ihre besten Affinitäten laufen zum *Mesobromion* bzw. zum *Diplachnion*.

Vergleich mittels des pflanzensoziologischen Artenspektrums

In der Tabelle 1 wurden in der ersten Kolonne links die pflanzensoziologischen Zugehörigkeiten der Arten nach Oberdorfer (1970), soweit dort angegeben, eingetragen. Aufgrund dieser Angaben wurde, ausgehend von der totalen Anzahl Arten pro Verband, das prozentuale Vorkommen der Klassen-, Ordnungs- und Verbandscharakterarten berechnet. Die sich ergebenden Daten können der *Tabelle 3* entnommen werden. Die Reihenfolge der Verbände von links nach rechts entspricht der Anordnung in der Tabelle 1.

Das *Carici humilis-Chrysopogonetum grylli* sowie das *Holco-Chrysopogonetum grylli* prov. sollen anschliessend genauer betrachtet werden. Die meisten Arten des *Carici humilis-Chrysopogonetum grylli* gehören der Klasse *Festuco-Brometea* an. Innerhalb dieser Klasse ist der Anteil der *Festucetalia vallesiacae*-Arten etwas grösser als derjenige der *Brometalia*-Ordnungsarten. *Arrhenatheretalia*-Arten sind nur leicht weniger vorhanden als *Brometalia*-Arten. Interessant ist auch die relativ gute Präsenz der *Querco-Fagetea*-, insbesondere der *Quercetalia pubescentis*-Arten. Ein erster, grober Vergleich dieser Gesellschaft mit dem soziologischen Artenspektrum der anderen Verbände zeigt einerseits eine gewisse Übereinstimmung mit dem *Orno-Ostryon*, andrerseits aber auch eine mit dem *Diplachnion* und dem *Xerobromion*.

Auch beim *Holco-Chrysopogonetum* gehören die meisten Arten der Klasse *Festuco-Brometea* an. Aber innerhalb dieser Klasse ist die *Brometalia*-Ordnung (mit dem *Mesobromion*) etwas besser vertreten als die *Festucetalia vallesiacae*-Ordnung. Die Arten, welche bezüglich ihrer Häufigkeit an zweiter Stelle liegen, gehören zur Klasse *Molinio-Arrhenatheretea*. Hier sind die beiden Ordnungen *Arrhenatheretalia* und *Molinietalia* zu gleichen Anteilen vertreten. Der grobe Vergleich dieser provisorischen Gesellschaft mit den in der Tabelle aufgeführten Verbänden lässt gewisse Parallelen mit dem *Mesobromion* erkennen.

Diskussion und Schlussfolgerungen

Die Tabelle 1 zeigt, dass die Aufnahmen aus dem insubrischen Klimabereich, also *Carici humilis-Chrysopogonetum grylli*, *Holco-Chrysopogonetum grylli*, *Orno-Ostryon* und *Sarothamnion*, sich doch recht deutlich von den weiter östlich vorkommenden *Chrysopogon gryllus*-reichen Wiesen unterscheiden. Das eher inneralpine *Diplachnion*, welches nach Oberdorfer (1964) als Ersatzeinheit des *Orno-Ostryon* zu betrachten ist, steht insofern den aus dem insubrischen Bereich stammenden Aufnahmen nahe, als diese nicht dem insubrischen Vegetationskomplex, sondern dem submediterranen, angehören. Da das *Carici humilis-Chrysopogonetum grylli* ebenfalls als Ersatzgesellschaft des *Orno-Ostryon* angesehen werden muss (Meyer 1976), ist dessen Beziehung zum *Diplachnion* offensichtlich. Das *Carici humilis-Chrysopogonetum fumanetosum* könnte sicherlich dem *Diplachnion*-Verband zugeordnet werden.

Der Vergleich mittels der Affinitätsformel und dem anschliessenden 'minimum spanning tree' bestätigt die engeren Beziehungen zwischen dem *Carici humilis-Chrysopogonetum fumanetosum* und den Aufnahmen des *Orno-Ostryon* einerseits und jenen des *Diplachnion* andrerseits (Abbildung 2). Sehr deutlich kommt aber auch zum Ausdruck, dass das *Carici humilis-Chrysopogonetum grylli* bei dieser Vergleichsart eine zentrale Stellung einnimmt und zu nahezu allen Vergleichsverbänden eine bestmögliche Beziehung aufweist. Daraus wird auch ein Nachteil dieser Methode ersichtlich: Vergleichseinheiten, welche eine hohe Anzahl Arten aufweisen, werden mit einer grösseren Wahrscheinlichkeit die korrespondierenden Arten in Vergleichseinheiten mit niedriger Artenzahl vorfinden als umgekehrt. Da das *Carici humilis-Chrysopogonetum grylli* mit gesamthaft 233 Arten in der Tabelle 1 vertreten ist (siehe auch Tabelle 3), d.h. rund doppelt soviele Arten aufweist wie die übrigen Einheiten, ist dessen zentrale Stellung verständlich. Die ausserordentliche Artenvielfalt dürfte auch einer der Gründe sein, dass diese Gesellschaft nicht ohne weiteres eine systematische Zuordnung findet. Ein weiterer Nachteil dieser Methode besteht darin, dass die Häufigkeit des Vorkommens der Arten nicht berücksichtigt wird. Arten welche zum Beispiel auf 10 Aufnahmen immer mit + vorkommen, erhalten genau den gleichen Stetigkeitswert wie jene, die mit Dominanz-, Abundanz-Werten von beispielsweise 3 oder 5 (Braun-Blanquet 1964) aspektprägend sind.

Auch aus dem Vergleich mittels des pflanzensoziologischen Artenspektrums ist hervorgegangen, dass das *Orno-Ostryon* sowie das *Diplachnion* mit dem *Carici humilis-Chrysopogonetum grylli* eine gute Übereinstimmung zeigen·(Tabelle 3). Der Klasse *Querco-Fagetea* und somit dem *Orno-Ostryon*-Verband dürfen jedoch unseres Erachtens die beschriebenen *Chrysopogon gryllus*-Wiesen nicht zugeordnet werden, da es sich ja nicht um Wald- oder Buschwaldgesellschaften handelt, sondern diese sich höchstens gegen eine solche hin entwickeln. Die Fragestellung lautet somit, zu welcher Ordnung innerhalb der Klasse *Festuco-Brometea* die insubrischen *Chrysopogon gryllus*-Wiesen gestellt werden sollen.

Wird von den Vergleichsverbänden ausgegangen, so scheint das *Diplachnion* dem *Carici humilis-Chrysopogonetum grylli* doch näher zu stehen als etwa das *Xerobromion*. Obschon zahlreiche Vertreter der Klassen *Querco-Fagetea* und *Molinio-Arrheatheretea* auf die *Brometalia*-Ordnung hinweisen, neigen wir heute dazu, die Vermutung von Braun-Blanquet (1961) und Pedrotti (1963) zu bestätigen und das insubrische *Carici humilis-Chrysopogonetum grylli* zum *Diplachnion* und somit in die *Festucetalia vallesiacae*-Ordnung zu stellen. Bei dieser Zuordnung müsste der inneralpine *Diplachnion*-Verband allerdings etwas weiter gefasst werden. Zusätzlich zu den von Braun-Blanquet (1961) vorgeschlagenen charakteristischen Poaceae des *Diplachnion* (*Chrysopogon gryllus*, *Heteropogon contortus*, *Danthonia provincialis* und *Diplachne serotina*) sollten in diesem Falle noch die Arten *Koeleria gracilis* und *Carex humilis* gezählt werden.

Summary

A comparison of different *Chrysopogon gryllus*-rich grasslands of the region between Lake Garda and Lake Locarno (Lago Maggiore) and of neighbouring regions was carried out to show the relationship between these grasslands and shrub-associations and to ascertain the alliance to which the *Carici-humilis Chrysopogonetum grylli* (Meyer 1976) could belong. *Mesobromion* and *Xerobromion* of the northern part of the Alps were also involved in this comparison.

The three methods of comparison applied, viz. comparison with a constancy table, comparison of affinities with a minimum spanning tree, and inspection of the phytosociological spectrum of the species, led to the following conclusions:
- The phytosociological units of the region between Lake Garda and Lake Locarno differ significantly in their spectrum of species from the other units involved in this comparison.
- The *Chrysopogon gryllus*-rich grasslands of the region between Lake Garda and Lake Locarno show partly a large spectrum of development stages which goes from the *Diplachnion* to the *Orno-Ostryon* alliance.
- The *Carici humilis Chrysopogonetum grylli* belongs to the Diplachnion-alliance.
- To the species which characterize the *Diplachnion*-alliance according to Braun-Blanquet (1961), *Koeleria gracilis* and *Carex humilis* should be added.

Literatur

Antonietti, A. 1968. Le associazioni forestali dell' orizonte submontano del Cantone Ticino su substrati pedogenetici ricchi di carbonati. Mitt. Schweiz. Anst. Forstl. Versuchswes. 44: 81–226.

Antonietti, A. 1970. Su un'associazione di brughiera del piede meridionale delle alpi. Iser. Geobot. Inst. ETH, Stiftung Rübel, Zürich, 40: 9–27.

Braun-Blanquet, J. 1961. Die inneralpine Trockenvegetation. Geobot. selecta 1, 273 S.

Braun-Blanquet, J. 1964. Pflanzensoziologie. 3. Aufl., Wien, 865 S.

Ferlan, L. & V. Giacomini. 1955. Appunti fitosociologici su esempi di 'Pascolo carsico': Chrysopogoneto-Centaureetum cristatae. Atti I Conv. Friulano Sc. Nat. Udine, S. 159–183.

Gils, H. van, E. Keysers & W. Launspach. 1975. Saumgesellschaften im klimazonalen Bereich des Ostryo-Carpinion orientalis. Vegetatio 31: 47–64.

Gower, J. C. & G. J. S. Ross. 1969. Minimum spanning trees and single linkage cluster analysis. Appl. Statist. 18: 54–64.

Hess, H. E., E. Landolt & R. Hirzel. 1967–1972. Flora der Schweiz. 3 Bde., Basel.

Hofer, H. R. 1967. Die wärmeliebenden Felsheiden Insubriens. Bot. Jb. 87: 176–251.

Knapp, R. 1971. Einführung in die Pflanzensoziologie. Stuttgart, 388 S.

Lorenzoni, G. G. 1967. Ricerche sui prati a 'Chrysopogon gryllus' della pianura Friulana. Boll. Bibl. Musei Udine 4: 5–21.

Meyer, M. 1976. Pflanzensoziologische und ökologische Untersuchungen an insubrischen Trockenwiesen karbonathaltiger Standorte. Veröff. Geobot. Inst. ETH, Stiftung Rübel, Zürich, 57: 145 S.

Oberdorfer, E. 1964. Der insubrische Vegetationskomplex, seine Struktur und Abgrenzung gegen die submediterrane Vegetation in Oberitalien und in der Südschweiz. Beitr. Naturk. Forsch. SW-Deutschl. 23: 141–187.

Oberdorfer, E. 1970. Pflanzensoziologische Exkursionsflora für Süddeutschland und die angrenzenden Gebiete. 3. Aufl. Stuttgart, 987 S.

Pedrotti, F. 1963. Nota sulla vegetazione steppica (Stipeto-Poion xerophilae e Diplachnion) dei dintorni di Trento. St. Trent. Sc. Nat. 60: 288–301.

Pignatti, S. 1962. Associazioni di alghe marine. Mem. Ist. Veneto Sc. M. Nat. 32: 3.

Raabe, E. W., 1952. Über den Affinitätswert in der Pflanzensoziologie. Vegetatio 4: 53–68.

Zoller, H. 1954. Die Typen der Bromus erectus-Wiesen des Schweizer Juras. Beitr. Geobot. Landesaufn. Schweiz 33, 309 S.

PLANNING AN ADAPTIVE NUMERICAL CLASSIFICATION*

Michael B. DALE**

Division of Geobotany, University of Nijmegen, Toernooiveld, Nijmegen, The Netherlands

Keywords:
Classification, Heuristic search, Semantics, Two-parameter, Taxon rank

Introduction

Although the application of numerical classification methods is becoming common, there remains considerable confusion over the value of the results obtained. Thus in comparing numerical methods with the tabular sorting of the Braun-Blanquet system, Stanek (1973), Adam et al. (1975) and Kortekaas, van der Maarel & Beeftink (1976) stress the similarities of the results obtained with preference for the numerical approach, whereas Moore & O'Sullivan (1970), Moore et al. (1970), and Coetzee & Werger (1973, 1975) stress discrepancies and prefer the traditional approach. Such differences might simply reflect differences in the data used, as noted by Hogeweg (1976) but, in my view, there is a more fundamental reason for the difference. Present numerical methods rigidly formalise one part of the analytic process; specifically that concerned with the organisation of floristic data using intrinsic criteria. In contrast, the human classifier is considerably more flexible, primarily because he works in a wider context and can, therefore, choose between alternative procedures during the course of his analysis. As Mackay (1969) phrases it, patterns are for agents and the human agent can select, reject and reconsider patterns as he chooses. In this paper I want to examine some few means of similarly increasing the flexibility of numerical methods, although I shall also indicate in passing a larger number of alternatives which might be fruitful.

* Contribution to the Symposium on Plant species and plant communities, held at Nijmegen, 11–12 November 1976, on the occasion of the 60th birthday of Professor Victor Westhoff.
** Present address: C.S.I.R.O. Division of Tropical Crops and Pastures, Cunningham Laboratory, Mill Road, St Lucia, Queensland 4067, Australia.

Classification

Classification is, loosely, concerned with finding and characterising sets, and I shall first concentrate on the characterisation. There are two ways of describing a set, which I shall call extensive and intensive. An extensive definition is just an enumeration of the members of the set, for example, the set of written vowels is {a, e, i, o, u}. In an intensive definition we provide, instead, an explicit procedure for determining if any item is, or is not, a member of the set.

Such a procedure represents an embedding of knowledge and can be used even with infinite sets. It will obviously require knowledge of certain property values for the unknown item in order to determine membership, the values providing necessary and/or sufficient conditions for assignment to the set.

In classifying vegetation, we start by obtaining an extensive definition of a type based on some finite selection of stands. Often, as Lambert & Dale (1964) noted, this alone may be sufficient for our purposes. However, we may further seek, by abduction, to identify an intensive definition which can of course be applied more widely. We might seek, for example, faithful species to provide sufficient, or constant species to provide necessary, conditions. In practice our sets are fuzzy (that is, their members show degrees of belonging (Zadeh 1965)) so that simple monothetic rules may be inappropriate and polythetic assignment necessary, but this does not alter the basic procedure. It may also be desirable that the rules reflect the processes generating the patterns, the syntax of the production, since such rules are often of very general applicability, but again this does not change the basic procedure and other rules may be useful for specific purposes. The intrinsic definition of a type is thus intimately related with the properties used to describe the members, and these properties must be comparable between members. However, a stand of vegetation consists of a collection of individual plants which, being unique, is obviously not immediately comparable with any other such collection. We must, then, associate similar plants into homologous groups and traditionally the taxonomic species has fulfilled this associative role, though there is in fact no explicit rule for deciding when a collection of individuals should be considered a species. Where the species are unknown, or do not provide comparability, phytosociologists have adopted other means of obtaining homologies. Webb et al. (1970) used structural description a surrogate for species, van der Maarel (1972) used taxonomic orders, to relate idiotaxonomic and syntaxonomic characters, while Dale & Clifford (1976) showed that species, subgenera and genera could all be used to examine spatial patterns with little change in results, although subfamily and family levels were less effective. In some cases the purposes of the analysis demand yet other homologies, as in the agronomical use of parts of plants rather than species, which Lambert (1972) has widened with her eco-organ concept to permit more general ecological studies.

I have recently extended the previous study of Dale & Clifford (1976) up to Class level, using quantitative data and a more complex analytic method, the two parameter method of Dale & Anderson (1973). The taxonomic ranks form a nested hierarchy so that it is possible to start at a high level, say the Class, and move down towards the species without ambiguity. The lower ranks can be expected to provide more detailed information, so that moving down the hierarchy is rather like increasing the magnification of a microscope. Obviously, the major problem is to decide when to change from one level to another, and here the two parameter method has some considerable advantages, for the decision to change can be made in a data dependant manner. It is for this reason that the method will be explicated.

The two parameter method

Earlier I noted that classification concerns finding groups, i.e. it involves a search for subsets which conform to some model being sought. To classify I must first define the model being sought and second define procedural rules for searching. The distinguishing feature of numerical classification is that the degree to which some subset conforms to the model in use is expressed as a number.

Most numerical methods presently in use search for the so-called minimal variance model and differ in details of the methods used for searching. It is in fact possible, for a fixed number of groups and the minimal variance model, to obtain an optimal subdivision but most methods employ heuristic criteria for directing the search, the reasons for which need not detain us here. The two parameter method still uses heuristic search but it employs a more general model. If the probability of finding species j in stand i is x_{ij}, when the model fits then

$$x_{ij} = a_i b_j / (1 + a_i b_j)$$

where a_i is a parameter measuring stand richness and b_j is a parameter measuring species abundance. In practice we use estimates of a_i and b_j and we do not expect perfect fit, only adequate fit. Now the minimal variance model either implies

$$\forall a_i \quad a_i = a \text{ (normal analysis)}$$

or

$$\forall b_j \quad b_j = b \text{ (inverse analysis)}$$

and this clearly shows why normal and inverse analyses are incompatible. In the two parameter model there is no such incompatability, since neither the a's nor the b's are constrained in value.

For any subset of the given data, the fit to the model can be measured by the discrepancy between the observed x_{ij} and the estimates of them based on the a_i and b_j values. This measure of fit tells us when to stop searching but it still does not tell us how to search. At present we use a simple monothetic search in the following way. Given some subset we obtain a value for the fit of the model, say I_0. We then divide the subset on each species in turn and for the subgroups formed by dividing on the jth species we obtain values of fit of I_{1j} and I_{2j}. The effectiveness of species j can then be measured by

$$\Delta j = I_0 - I_{1j} - I_{2j}$$

and this can be used to rank the species in order of effectiveness. We simply select the best species and impose the division, recursively dividing the subgroups thus formed until all fit adequately. Note that we could equally well divide on the ith stand to obtain an inverse analysis, and we can actually choose between the best species and best stand division as in the inosculate analysis of Dale & Anderson (1973), a choice which these authors argued was equivalent to a choice between classification and ordination.

I do not claim that monothetic search is optimal, simply that it is practicable. There are a number of modifications which can be used to improve the effectiveness, for example the use of error correction techniques (Pohl 1969), reallocation, greater depth of search, and polychotomous division but I do not want to pursue these here. Instead I shall follow Dale & Webb (1975) in considering that a change from normal to inverse mode of division, a modulation, is important, and describe an alternative search procedure I call laminate search.

Let us start with data such that the stands are described initially using some high taxonomic level. We can divide the stands using normal analysis into classes until either the model fits adequately or the division modulates to inverse. The stand groups at this level form a layer, and we can at this point change our data to use a lower taxonomic level, increasing the resolution. In effect we peel off this layer having determined its existence in a data-dependant manner and proceed to further examine the subsequent structure using increasingly precise data. Thus we proceed in effect through a series of layers or lamina.

Using higher taxa

Before employing the laminate search procedure it seems sensible to examine if each taxonomic level gives information in its own right. To do this I have employed data on the presence of 321 species in 55 stands from the C.S.I.R.O. Narayen Research Station at Mundubbera, Queensland. The area has been mapped by Coaldrake et al. (1972) and I am grateful to Dr. Tothill for permission to use these data. For the higher taxa, the stands were described by a richness measure, the number of species per taxon, and analyses were carried out for Subgenus, Genus, Subfamily, Family, Order, Superorder and Class levels. I shall briefly outline the results obtained for each level.

For Class data, the two parameter analysis indicated that no division was required, but if one was forced it was an inverse division separating areas with high monocotyledonous richness, associated with granitic rocks from the remainder. The Superorder level also divided first in an inverse mode, separating a group of taxa which seem to lack resistance to even light grazing. The remaining taxa were then used to identify 3 stand groups reflecting the 3 major rock types: granitic, sandstone and basaltic. Neither Class nor Superorder levels would give any

divisions in a laminate analysis since they start with inverse divisions.

Order, Family and Subfamily data all agree in forming 6 or 7 stand groups, again distinguishing the 3 rock types but with additional groups reflecting soil depth and drainage. The groups correspond to the major subdivisions of the vegetation noted on the map, though with some discrepancies. Genus and Subgenus data again form the 3 major groups but then modulate to inverse divisions.

I have not given much detail of these results for three reasons. First, the primary concern is simply to establish that the results do have some ecological interpretation. Second, at the highest levels taxonomic confusion is rife. For example, at the Class level the Dicotyledons were divided into 2 types for this analysis, but various alternatives exist, and it is certainly n̲ ̲ ̲ ̲ ̲ for all species to which type they should be allocated. Third, as a consequence of this taxonomic uncertainty, detailed interpretation with high order taxa is difficult, so that with our present inexperience, detailed argument would seem of doubtful value.

For the species data two analyses were made. The first permitted only normal divisions and this resulted in 28 stand groups which correspond quite well with the mapped groupings; at least the workers who drew the map are prepared to agree that the discrepancies reflect alternative organisations which seem acceptable. The second analysis was made with inosculate search, and gave only inverse divisions for the first 13 subdivisions. Even the normal division which then occurred was marginal and could well reflect inefficiency in the monothetic search as noted by Coetzee & Werger (1975) for Association analysis. In all, some 23 species groups were formed, the analysis being presently incomplete at the final levels. These groups can be arranged to form a network of gradients (*Figure 1*) and seem to support Gleason's (1926) individualistic model of vegetation, though this is not the only possible model consistent with the evidence.

Considering the laminate analysis, we start at the Order level with seven groups. Family and Subfamily data further divide one group each, suggesting that they have little extra information to give. Generic data causes three groups to divide, subgeneric none. The species analysis is still incomplete but few divisions have occurred yet. Thus Order and Genus seem, with these data, to be effective levels, the others doing little beyond cleaning up the results a little. On the evidence of one analysis I do not wish to make exaggerated claims, though, especially as the higher levels are not well defined taxonomically.

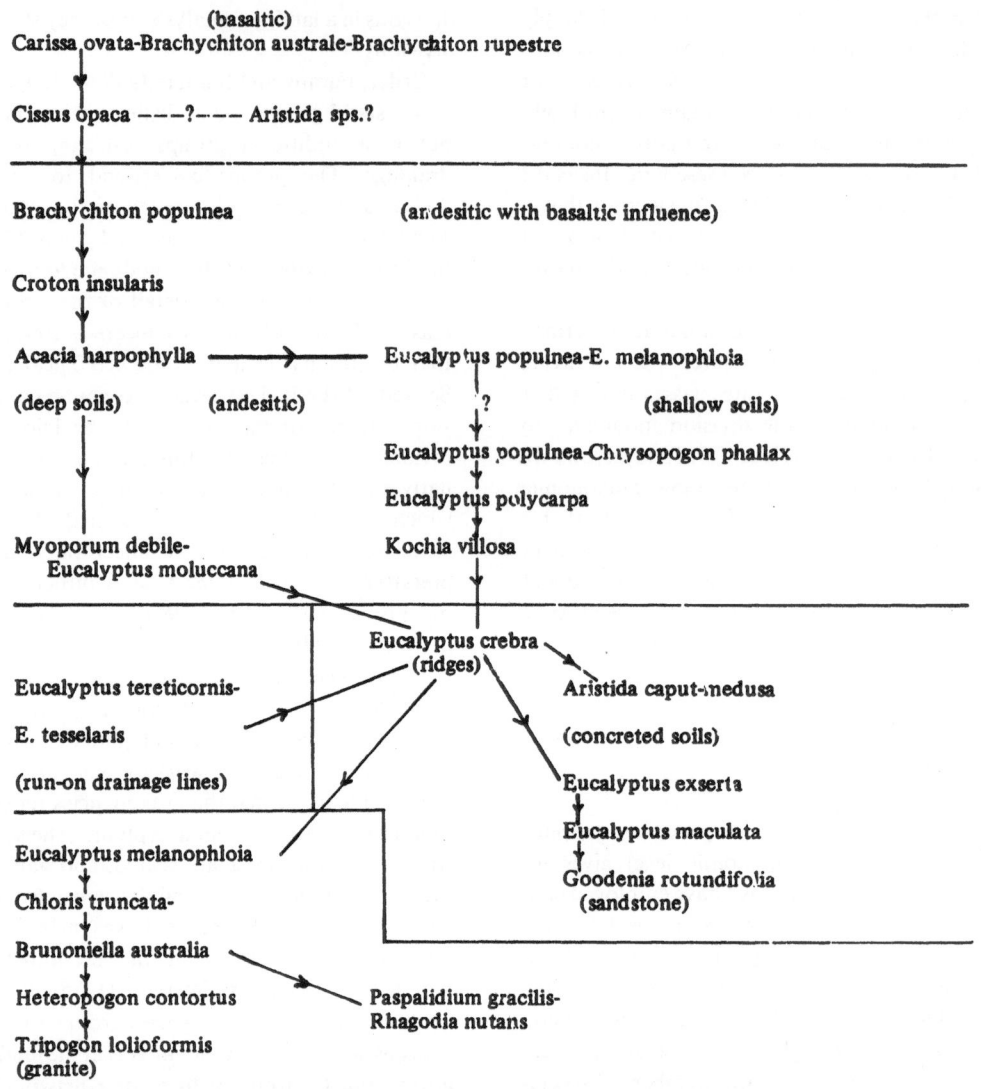

(basaltic)
Carissa ovata-Brachychiton australe-Brachychiton rupestre

Cissus opaca ———?——— Aristida sps.?

Brachychiton populnea   (andesitic with basaltic influence)

Croton insularis

Acacia harpophylla ————————→ Eucalyptus populnea-E. melanophloia

(deep soils)   (andesitic)   ?   (shallow soils)

                                        Eucalyptus populnea-Chrysopogon phallax

                                        Eucalyptus polycarpa

Myoporum debile-   Kochia villosa
    Eucalyptus moluccana

                          Eucalyptus crebra
                              (ridges)   Aristida caput-medusa

Eucalyptus tereticornis-   (concreted soils)

E. tesselaris

(run-on drainage lines)   Eucalyptus exserta

                                                          Eucalyptus maculata

Eucalyptus melanophloia   Goodenia rotundifolia
    ↓   (sandstone)
Chloris truncata-

Brunoniella australia

Heteropogon contortus   Paspalidium gracilis-
                               Rhagodia nutans
Tripogon lolioformis
(granite)

Fig. 1. Diagrammatic relationships between floristic groups from the twoparameter analysis. Species names are those presently in use in the Botany Branch Queensland Department of Primary Industries. In parentheses are indications of rock type and soil conditions for the groups. The letter E is used as an abbreviation for *Eucalyptus*, and the ? indicates uncertain affinities.

Discussion

The laminate search procedure is one example of changing the context of an analysis during its execution, using the hierarchical structure of the taxonomic system. Kelly (1971) has suggested that, for samples from a systematic grid, a similar nested approach would be possible. This would allow examination of the effects of sample size on species association, an effect noted by Pielou (1969), and Noy-Meir & Anderson (1971). Similarly the effects of precision of measurement can be investigated, for presence, rooted frequency and shoot frequency (or cover) form another nested series suitable for adaptive analysis.

The use of these hierarchies still relies, however, on intrinsic properties of the floristic data, but, as Coetzee & Werger (1975) state, vegetation types must be ecologically interpretable to be useful. This involves an evaluation of the meaningfulness of the types, the semantics as Dale (1971) calls it, and this is a task of daunting magnitude. The kind of problem can be illustrated as follows using

Icelandic data. Dale & Anderson (1972) identified a group of seven species: *Dryas octopetala* L., *Juncus trifidus* L., *Thalictrum alpinum* L., *Festuca vivipara* (L)Sm., *Silene acaulis* (L), Jacq., *Luzula spicata* (Huds) Gaud and *Galium pumilum* Mull. Of these, five are listed by Conolly & Dahl (1970) as limited by high summer radiation load, and the species seem to be reacting similarly even when they are not near their southern limits. Inspection of the British distributions shows that six of the seven species are characteristically associated with the montane-oceanic areas of the north and west. Only *Galium pumilum* does not fit and on checking the specimens it was found to be misnamed in the flora used. In fact the species was *Galium normanii* O. C. Dahl, an Icelandic endemic apart from two Norwegian records. Thus the analysis has permitted the correction of an error, allowed a physiological similarity to be exposed, and even suggests where to look for *Galium normanii* in Britain. The problem of course is how to program a machine to associate these disparate sources of information into a coherent interpretation, and still more how to make it learn the appropriate information. In theory it seems possible but the practical difficulties are immense, and the only available solution is to include a human supervisor as in Dale & Quadraccia (1973). Vickers (1964) has suggested that we all 'cling to a self spun web obscurely moored in vacancy'. Perhaps as we tackle the problems of assigning meanings, we shall learn how to avoid the spider.

Summary

While numerical methods of classification seek to strictly formalise one part of the classificatory process, a human classifier uses a variety of different approaches to obtain a satisfactory organisation of data. In this paper some means of increasing the flexibility of numerical methods are discussed, in particular the possible uses of the taxonomic hierarchy. It is first necessary to show that higher taxonomic units contain ecologically interesting information. Then, using a two parameter classification method a means of progressing from higher to lower taxonomic units, during analysis and in a data dependent manner, is outlined. A similar procedure is possible with nested samples. However, since the value of a classification resides in its ecological interpretation, an automatic classification requires some means of ascribing meanings to classes. While such a semantic analysis seems theoretically possible, its practical attainment presents considerable difficulties.

References

Adam, P., H. J. B. Birks, B. Huntley & I. C. Prentice. 1975. Phytosociological studies at Malham Tarn moss and fen, Yorkshire, England. Vegetatio 30: 117–132.

Coaldrake, J. E., J. C. Tothill, G. W. McHarg & J. N. G. Hargreaves. 1972. Vegetation map of the Narayen Research Station, South-East Queensland. Techn. Paper 12, Div. Tropical Pastures, CSIRO, Brisbane.

Coetzee, B. J. & M. H. A. Werger. 1973. On hierarchical syndrome analysis and the Zürich-Montpellier table method. Bothalia 11: 159–164.

Coetzee, B. J. & M. J. A. Werger. 1975. On association analysis and the classification of plant communities. Vegetatio 30: 201–206.

Connolly, A. P. & E. Dahl. 1970. Maximum summer temperature in relation to the modern and quaternary distribution of certain arctic-montane species in the British Isles. In: D. Walker & R. G. West (eds), Studies in the vegetation history of the British Isles, pp. 159–223. Cambridge University Press, Cambridge.

Dale, M. B. 1971. Validity and utility of information theory in ecology. Proc. Ecol. Soc. Australia 6: 639–653.

Dale, M. B. & D. J. Anderson. 1972. Qualitative and quantitative information analysis. J. Ecol. 60: 639–653.

Dale, M. B. & D. J. Anderson. 1973. Inosculate analysis of vegetation data. Austr. J. Bot. 21: 253–276.

Dale, M. B. & H. T. Clifford. 1976. On the effectiveness of higher taxonomic ranks for vegetation analysis. Austral. J. Ecol. 1: 37–62.

Dale, M. B. & L. Quadraccia. 1973. Computer-assisted tabular sorting of phytosociological data. Vegetatio 28: 57–73.

Dale, M. B. & L. J. Webb. 1975. Numerical methods for the establishment of associations. Vegetatio 30: 77–87.

Gleason, H. A. 1926. The individualistic concept of the plant association. Bull. Torrey Bot. Club 53: 7–26.

Hogeweg, P. 1976. Topics in biological pattern analysis. Thesis Utrecht. 230 pp.

Kelly, M. D. 1971. Edge detection in pictures by computer using planning. In: B. Meltzer & D. Michie (eds), Machine Intelligence 6, pp. 379–404. Edinburgh University Press, Edinburgh.

Kortekaas, W. M., E. van der Maarel & W. G. Beeftink. 1976. A numerical classification of European Spartina communities. Vegetatio 33: 51–60.

Lambert, J. M. & M. B. Dale. 1964. The use of statistics in phytosociology. Adv. Ecol. Res. 2: 59–99.

Lambert, J. M. 1972. Theoretical models for large-scale vegetation Survey. In: J. R. Jeffers (ed), Mathematical Models in Ecology. pp. 87–110. Blackwell, Oxford.

Maarel, E. van der. 1972. Ordination of plant communities on the basis of their plant genus, family and order relationships. In: E. van der Maarel & R. Tüxen (eds), Grundfragen und Methoden in der Pflanzensoziologie, pp. 183–190. Junk, The Hague.

Mackay, D. M. 1969. Recognition and action. In: S. Watanabe (ed.), Methodologies of pattern recognition. pp. 409–416. Academic Press, London.

Moore, J. J., P. Fitzsimmons, E. Lambe & J. White. 1970. A comparison and evaluation of some phytosociological techniques. Vegetatio 20: 1–20.

Moore, J. J. & A. O'Sullivan. 1970. A comparison between the results of the Braun-Blanquet method and cluster analysis. In: R. Ruxen (ed.), Gesellschaftsmorphologie, pp. 26–30. Junk, The Hague.

Noy-Meir, I. & D. J. Anderson. 1971. Multiple pattern analysis or multiscale ordination: pathway to a vegetation hologram. In: G. P. Patil, E. C. Pielou & H. E. Water (eds), Statistical Ecology Vol. 3, pp. 207–234. Pennsylvania State University Press, London.

Pielou, E. C. 1969. An introduction to mathematical ecology. Wiley, New York. 286 pp.

Pohl, I. 1969. First results on the effect of error in heuristic search: In: B. Meltzer & D. Michie (eds.), Machine Intelligence 5, pp. 219–236. Edinburgh University Press, Edinburgh.

Stanek, W. 1973. A comparison of Braun-Blanquet's method with sum of squares agglomeration for vegetation classification. Vegetatio 27: 323–345.

Vickers. G. 1964. The psychology of policy making and social change. Brit. J. Psychol. 110: 143–167.

Webb, L. J., J. G. Tracey, W. T. Williams & G. N. Lance. 1970. Studies in the numerical analysis of complex rain forest communities. V. A comparison of the properties of floristic and physiognomicstructural data. J. Ecol. 58: 203–232.

Zadeh, L. 1965. Fuzzy sets. Information and Control 8: 338–353.

CALLUNA AND ITS ASSOCIATED SPECIES: SOME ASPECTS OF CO-EXISTENCE IN COMMUNITIES*,**

C.H. GIMINGHAM***

Department of Botany, University of Aberdeen, St. Machar Drive, Aberdeen, Great Britain

Keywords:
Co-existence, Complementary strategies, Heath communities

Introduction

In recent years, research concerned with plant communities has been focussed very largely upon the study of variation in floristic composition, and its causes. In this paper, however, a complementary approach is adopted – that which considers the mechanisms of community organization and the ability of certain species to co-exist within a community. An attempt is made to identify some of the characteristics or properties of these species which either make them intolerant of the company of certain other species as neighbours, or on the contrary confer an ability to co-exist in this way. Conversely, some characteristics or properties of stands and communities which tend either to permit or prevent invasion by species not originally present, are also touched upon.

Clearly, competition and other related interactions between species have much to do with their ability to co-exist. Complementary 'strategies' (i.e. life-forms and life-histories) may enable species to avoid competitive interactions to a greater or lesser degree, and hence to occur in close association in communities. Emphasis will be placed on this aspect of the subject.

Calluna vulgaris as a dominant species

Communities in which *Calluna vulgaris* is dominant afford valuable opportunities for investigating this subject. There is an extensive literature on *Calluna*-dominated heathlands in western Europe (Gimingham 1972), from which information on the species physiologically capable of occurring in any particular heathland habitat is available. From amongst these, the selection of species which comprises any example of a *Calluna*-dominated stand must have characteristics which permit their co-existence and, more importantly, their close association with the dominant. *Calluna* exerts powerful influences on its immediate environment and hence on the conditions under which any associated species must survive or establish. However, these influences are not constant throughout the life of an individual *Calluna* plant. Its growth and development are such that marked changes in its properties occur. This makes the task of identifying some of the characteristics which enable other species to become its associates rather easier.

Furthermore, because (especially in Britain) *Calluna* heathlands are frequently managed by burning, comparisons are possible between even-aged stands of varying age, resulting from regeneration after fire, and uneven-aged stands which have not been managed in this way.

* Contribution to the Symposium on plant species and plant communities, held at Nijmegen, 11–12 November 1976, on the occasion of the 60th birthday of Professor Victor Westhoff – to whom the author conveys the good wishes of Scottish botanists and conservationists, and particularly those of the University of Aberdeen.
**Nomenclature follows Clapham, Tutin & Warburg (1962) for vascular plants, Watson (1968) for bryophytes, and Duncan (1970) for lichens. Following common practice *Calluna vulgaris* is referred to simply as *Calluna*.
***Grateful acknowledgement is made to several present and former research students who have kindly permitted the use of their data: particularly P. Barclay-Estrup, Edith M. French, Gong Wooi Khoon, C.J. Legg, Evelyn W. Paterson, S.D. Ward. I am especially grateful to C.J. Legg for valuable discussions on the subject of this paper.

Growth-phases of Calluna

The changes mentioned above in the properties of *Calluna* as a dominant, which take place during its life-span, were first described by A.S. Watt in 1955, and have been further examined by Barclay-Estrup & Gimingham (1969) and Barclay-Estrup (1970, 1971). Four 'phases' have been recognised, though the progression from one to another is gradual and not abrupt:

Pioneer phase, in which young individuals are more or less pyramidal in shape, usually scattered, and neither height nor cover are at their maximum.

Building phase, in which the plant is at its most vigorous, becoming hemispherical in shape (unless growing in dense stands), and reaching maximum height and cover.

Mature phase, in which the plant is becoming somewhat less active in terms of extension growth and its canopy becomes less dense in the centre, due in part to a tendency of the older branches to collapse sideways, forming a gap.

Degenerate phase, in which all the central frame-branches tend to collapse and die, extending the gap. Eventually, the only parts which remain alive are the outermost branches which have rooted in litter, and sustain a ring of foliage-bearing shoots surrounding the gap.

The age at which individuals begin to pass from one phase to the next varies considerably with habitat, but to give a very general indication based on lowland habitats in the centre of the heathland region, the pioneer phase may pass to building at about 6 years, building to mature about 15 years, and mature to degenerate at over 20 years, while plants may survive in the degenerate phase until they are more than 30 years old.

Uneven-aged stands

Watt (1955) first demonstrated a relationship between the phasic behaviour of *Calluna* and the ability of other species to co-exist with it. He showed that where *Pteridium aquilinum* was present in an uneven-aged stand of *Calluna*, in which individuals of all phases occurred side by side in a mosaic, the fronds were dispersed with significantly greater numbers amongst *Calluna* in the pioneer and degenerate phases, than in the building and mature phases (Table 1).

This type of observation can be extended by consider-

Table 1. Relationship between density of fronds (numbers m⁻²) of *Pteridium aquilinum* and growth-phase of *Calluna vulgaris* in mixed stands (data derived from Watt 1955)

Growth-phase of *Calluna vulgaris*	Pioneer	Building	Mature	Degenerate
A. In *Calluna* stand being invaded by *Pteridium*	1.68	1.53	1.24	3.04
B. In centre of mixed stand	4.65	1.42	0.92	2.11

ing not merely one species associated with *Calluna*, but also the contribution of all other species in the community. Fig. 1 shows the cover-contribution of associated species of vascular plants in a *Calluna-Vaccinium* heath community on peat, in north-eastern Scotland (Aberdeenshire), which is highest in areas occupied by *Calluna* in the pioneer phase, less among *Calluna* in the building and mature phases, and greater again amongst degenerate *Calluna*.

It may be concluded that, as regards the ability of other species to co-exist, the adverse effects of *Calluna* are greatest in the building and mature phases, and that these effects are relatively slight while it is in the pioneer phase and also are relaxed in the degenerate phase. However, some niches are available for associated species even during the stages in which *Calluna* is at its most vigorous.

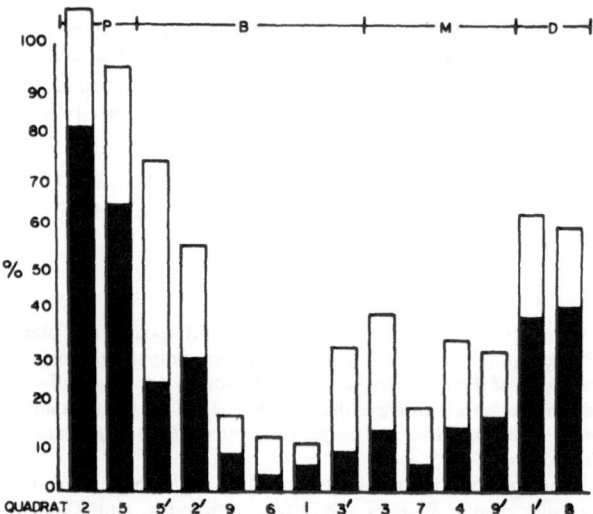

Fig. 1. Percentage cover of all vascular plants (excluding *Calluna vulgaris*) in areas occupied by *Calluna* in each of its growth-phases, in an uneven-aged stand on peat in Aberdeenshire, north-eastern Scotland. *Calluna* growth-phases: P, pioneer; B, building; M, mature; D, degenerate. Whole column (solid + open portions) – total cover; solid portion – upper stratum only. Data from Barclay-Estrup 1966)

36

Detailed examination of the species which co-exist with *Calluna,* and the growth-phases with which they are particularly associated, reveals some of the properties and characteristics which enable them to occupy the various niches available. A preliminary list follows:

A) Competitors – species having a general similarity in size, form, structure and development to *Calluna* itself, existing alongside it throughout the sequence of growth-phases: e.g. *Vaccinium myrtillus.*

B) Species with 'complementary strategies'
(i) Subordinates – shade-tolerant species forming lower strata beneath a *Calluna* canopy, present throughout the sequence of growth-phases: e.g. *Erica cinerea, Hypnum cupressiforme.*
(ii) Species requiring open phases (e.g. pioneer or degenerate *Calluna*) for establishment, capable of surviving throughout the sequence but with reduced vigour in the building and mature phases: e.g. *Festuca ovina.*
(iii) Species surviving throughout the whole sequence in one spot by dying back each winter to an underground perennating organ, from which climbing or straggling shoots are produced every year, bringing foliage to the periphery of the *Calluna* canopy: e.g. *Potentilla erecta.*
(iv) Species requiring open phases (e.g. pioneer or degenerate) for establishment, but spreading as the *Calluna* canopy develops by means of rhizomes or creeping stems which extend laterally into new gaps: e.g. *Vaccinium vitis-idaea, Arctostaphylos uva-ursi, Pteridium aquilinum.*
(v) Small, low-growing species which appear only in the open phases (pioneer or degenerate, or both): e.g. *Polygala serpyllifolia,* certain lichens and bryophytes.

(iv) and (v) are the 'strategies' particularly associated with the gap phases; the remainder to a greater or lesser degree confer ability for co-existence with *Calluna* throughout the whole sequence of phases.

Even-aged stands

The interpretation suggested above of the relationships between *Calluna* and associated species may be tested by examining even-aged stands of *Calluna* produced by burning. In this case, whole stands proceed uniformly through the sequence of phases. However, there is a major difference in the sequence as displayed by even-

aged stands from that observed in individual bushes growing in a stand of mixed ages. During the period immediately following a fire a relatively large area is for a time devoid of vegetation, and subsequently passes through a stage in which various species, including *Calluna,* are either colonizing or regenerating vegetatively. The habitat in this post-burn 'phase' may differ considerably from that of the small gaps in an uneven-aged stand which may be colonized by pioneer *Calluna.*

Despite this difference, the numbers of species associated with *Calluna* in even-aged stands of increasing age varies as expected, with the larger numbers occurring in the post-burn and degenerate stands. An example from heathland on podzol in central Scotland (Perthshire) is given in Table 2.

The various ways in which the contribution of individual species to the community relates to the sequence of growth-phases in *Calluna* may be exemplified from a study of *Calluna-Arctostaphylos uva-ursi* heathland at Dinnet (Aberdeenshire, north-eastern Scotland). Here, on a podzolized brown soil with very thin A_0 horizon, the sequence of phases proceeds rather faster than indicated earlier in this paper. Following a fire the *Calluna* stand regenerates relatively rapidly, passes through the building stage and within about 16 years the canopy is already thinning out and giving the appearance of the mature phase. With this in mind, the contribution of various associated species shown in Fig. 2, a–g and Fig. 3, a–d may be interpreted in relation to the 'strategies' listed above.

In view of the rapid re-establishment of dominance by *Calluna,* following vigorous and uniform vegetative regeneration after fire, species in the category 'Competitors' are poorly represented. However, niches are present for species which may be described as having 'Complementary strategies'. Some are present in quantity throughout at least the greater part of the sequence. *Erica cinerea* (Fig. 2, a), for example, reappears rapidly after a fire and remains as a subordinate to *Calluna* throughout its development. *Hypnum cupressiforme*

Table 2. Numbers of species of vascular plants occurring in areas of 4 m² in even-aged *Calluna* stands of increasing age after burning: Moulin Moor, Perthshire, Central Scotland

Age of *Calluna*-stand (years since burning)	Growth phase of *Calluna*	No. of species
2-3 } 5-6	Post-burn	15 22
ca.8	Building	16
13-14	Mature	11
> 22	Degenerate	14

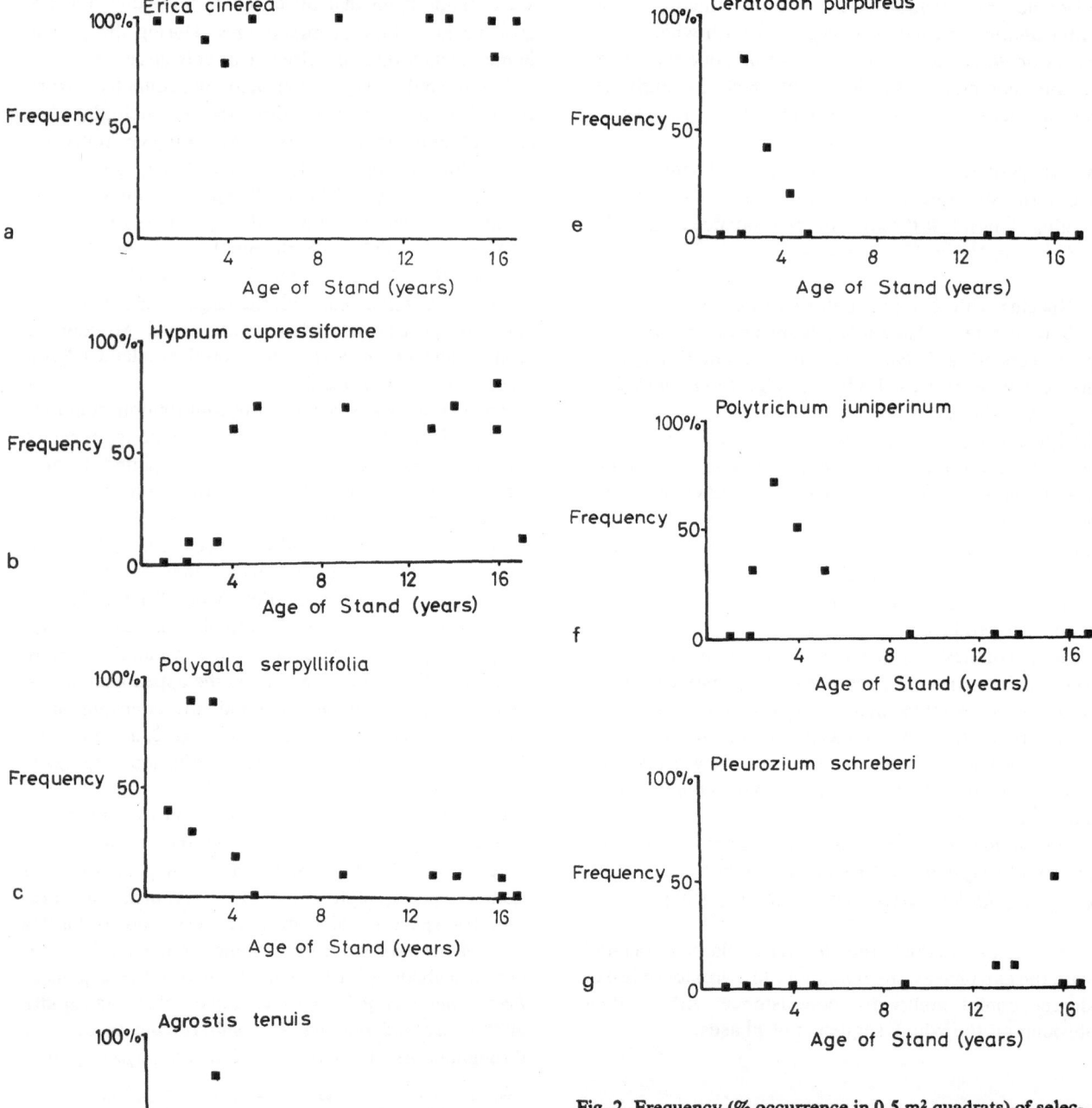

Fig. 2. Frequency (% occurrence in 0.5 m² quadrats) of selected species of vascular plants and mosses in relation to age of stand in even-aged stands of *Calluna* resulting from management by burning on Dinnet Moor, Aberdeenshire, north-eastern Scotland. (a) *Erica cinerea*, (b) *Hypnum cupressiforme*, (c) *Polygala serpyllifolia*, (d) *Agrostis tenuis*, (e) *Ceratodon purpureus*, (f) *Polytrichum juniperinum*, (g) *Pleurozium schreberi*. (Data supplied by Edith M. French).

38

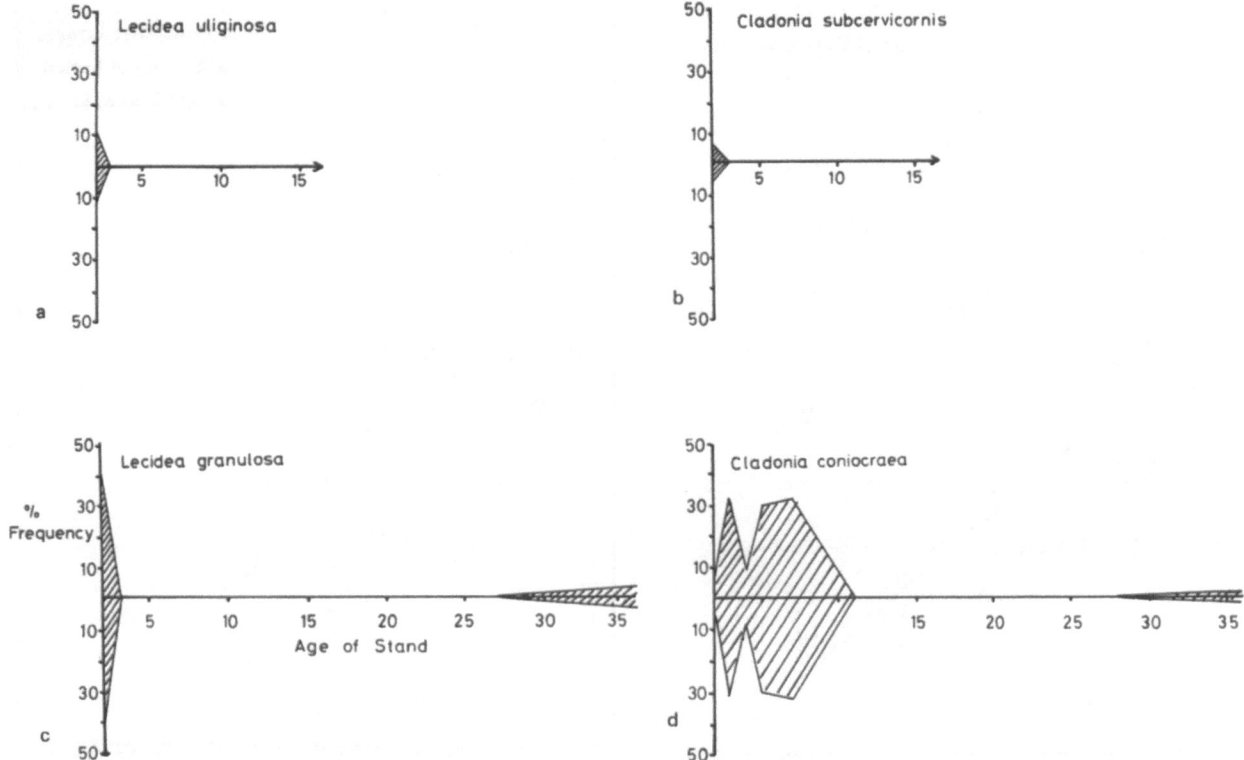

Fig. 3. Frequency (% occurrence in 44.5 cm² quadrats) of selected species of lichens in even-aged stands of *Calluna* resulting from management by burning on Dinnet Moor, Aberdeenshire, north-eastern Scotland. (a) *Lecidea uliginosa*, (b) *Cladonia subcervicornis*, (c) *Lecidea granulosa*, (d) *Cladonia coniocraea*. (Note: frequencies are shown in the form of 'kite' diagrams; hence for example a frequency of 20% is indicated by extension of the hatched area to 10 units above and 10 units below the horizontal line). (Data supplied by Evelyn W. Paterson).

(Fig. 2, b) is slower to re-establish, but is well represented during the period in which *Calluna* is most vigorous. However, this species appears to lose its niche, presumably owing to competition from other species in the same stratum, as the *Calluna* canopy thins out and the sequence passes into the mature and degenerate phases. These two species illustrate strategy B(i).

Certain species of low growth-form are largely confined to the early stages of the sequence:
(a) herbs such as *Polygala serpyllifolia* (Fig. 2, c), *Veronica officinalis,* and some grasses, e.g. *Agrostis tenuis* (Fig. 2, d):
(b) mosses such as *Ceratodon purpureus* (Fig. 2, e) and *Polytrichum juniperium* (Fig. 2, f);
(c) lichens requiring open habitats, e.g. *Lecidea uliginosa* (Fig. 3, a), *Cladonia subcervicornis* (Fig. 3, b).

These species illustrate strategy B(v), as do those which are not only present in the open conditions of the

early stages following burning, but also reappear as the sequence proceeds and the canopy thins or gaps develop: e.g. *Lecidea granulosa* (Fig. 3, c), *Cladonia coniocreaea* Fig. 3, d). One species, *Pleurozium schreberi* (Fig. 2, g), appears only in the later stages.

Hence, in general, as even-aged stands get progressively older, the capacity of various species to accompany *Calluna* in its several growth phases is similar to that observed in respect of individual bushes in uneven-aged stands. However, there are some differences, particularly the fact that there are certain species which belong only to the post-burn stage, and some which belong only to the degenerate stage (whereas in uneven-aged stands there is little difference between the floras associated with pioneer and degenerate *Calluna*).

Notable among the considerable number of species belonging only to the post-burn stage are the low-growing herbs and a variety of lichens, while among

39

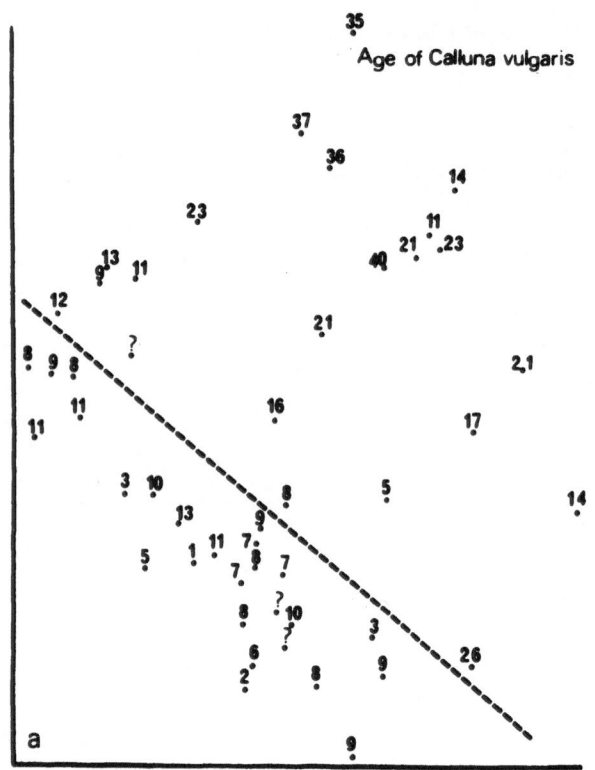

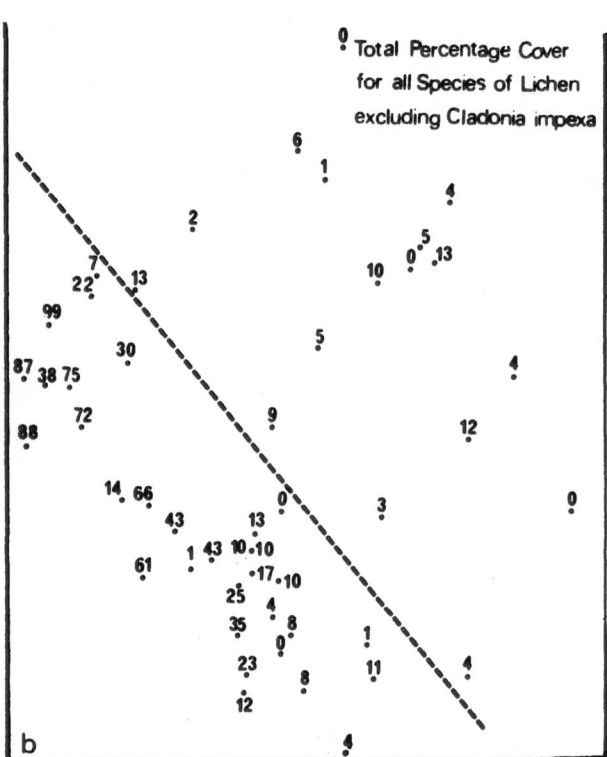

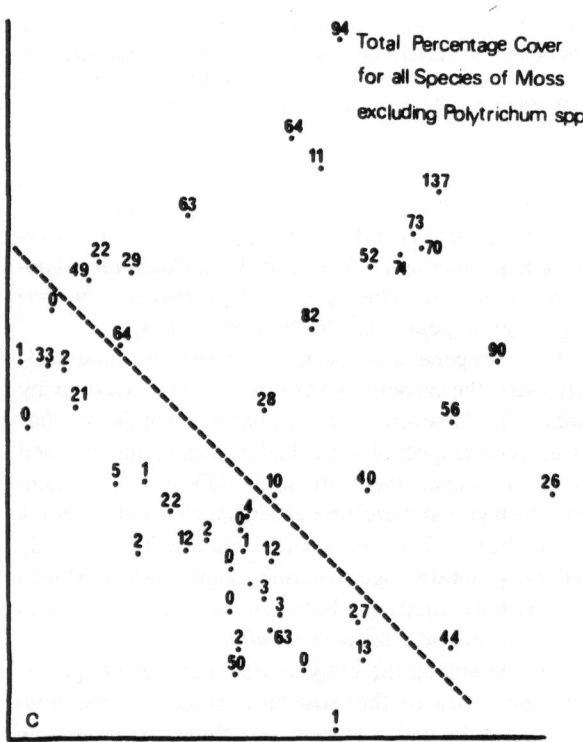

Fig. 4. Ordination of *Calluna* stands at Dinnet Moor, Aberdeenshire, north-eastern Scotland, on first two axes of a Principal Components Analysis, using cover data for plant species. The following data are plotted onto the ordination: (a) age of *Calluna* stand since burning; (b) total percent cover for all species of lichen (excluding *Cladonia impexa*); (c) total percent cover for all species of moss (excluding *Polytrichum* spp.). (After Ward 1970).

species related only to the degenerate phase are a number of bryophytes. These relationships are well illustrated by the results of a Principal Components Analysis of a number of stands at Dinnet (Fig. 4). The stands are plotted in relation to the first 2 principal component axes, and the age of each stand is shown (Fig. 4, a). The cover of lichens (4, b) and mosses (4, c) is shown against the position of each stand. (*Cladonia arbuscula* and *Polytrichum* spp. are excluded as species which contribute considerable cover in their own right, the former mainly in older stands, the latter in young stands.)

It may be concluded that where there is a marked distinction between species associated with pioneer *Calluna* and those associated with degenerate stands, a management factor (burning) is operating.

Invasion of Calluna communities

Finally, brief reference may be made to the entry to heathland communities of species which have the capacity eventually to eliminate *Calluna* – mainly tree species. This is governed to a large degree by the same properties of *Calluna* as those which affect the capacity of various species to co-exist in the community. In Scotland, the trees *Betula* spp. and *Pinus sylvestris* serve as examples. There is evidence that in an uneven-aged stand of *Calluna* the entry of *Betula* spp. is confined to the gap phases. In the managed heathlands, however, the post-burn stage permits the establishment of large numbers of seedlings (Table 3). These are progressively reduced in numbers as the stands increase in age. Only those which grow rapidly enough to overtop the *Calluna* canopy survive, and little or no further recruitment takes place. It has also been shown experimentally (C.J. Legg), by introducing seed of *Pinus sylvestris* to *Calluna* stands of varying age, that germination in this

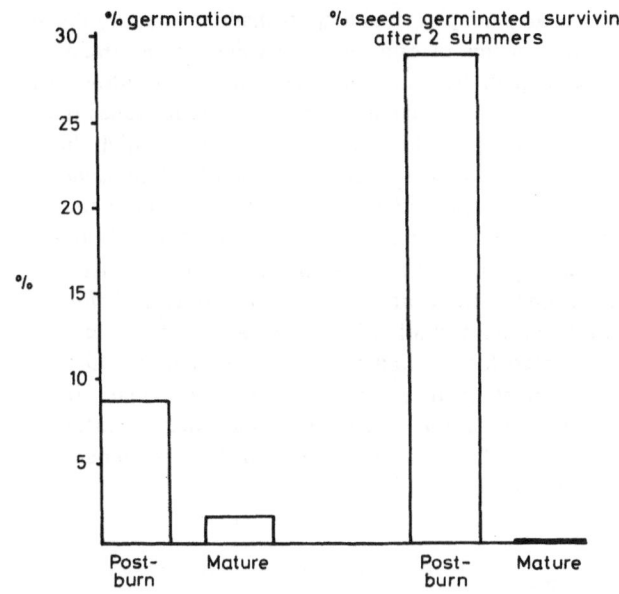

Fig. 5. Percentage germination of seeds of *Pinus sylvestris* introduced experimentally at the post-burn stage and in mature *Calluna* stands, and seedling survival, on Dinnet Moor, Aberdeenshire, north-eastern Scotland. (Data from C.J. Legg).

species is more successful in the post-burn stage than under older *Calluna* and that establishment is largely confined to the open phases (Fig. 5).

Conclusion

Numerous species are capable of occupying the habitats available in acid heathland areas. Where *Calluna* is dominant, however, the community composition is considerably restricted and the mechanics of community formation are based, at least in part, on the complementary strategies outlined above. The properties of *Calluna* are such that the niches available for other species change during its life-span, so determining the ability of other species to co-exist, according to their own characteristics and properties. Hence, in a managed area composed of even-aged stands of *Calluna* the community composition varies markedly according to the age of the stand.

Summary

Using heath communities (dominated by *Calluna vulga-*

Table 3. Numbers of individuals (seedlings or small trees) of *Betula* spp. (*pubescens + pendula*) in relation to growth-phase of *Calluna* in 20 m² samples of even-aged *Calluna* stands resulting from management by burning on Dinnet Moor, Aberdeenshire, north-eastern Scotland (data from Gong 1976)

Growth-phase of *Calluna*	Number of individuals of *Betula* spp.	
	Site A	Site B
Post-burn (prior to *Calluna* regeneration)	55.0 ± 21.0	268.0 ± 162.0
Post-burn, passing to young building-phase	6.5 ± 0.5	6.5 ± 0.5
Building	3.0 ± 0.0	2.5 ± 0.5
Mature	0.5 ± 0.5	1.0 ± 0.0
Degenerate	1.0 ± 1.0	0.5 ± 0.5

ris) as an example, an attempt is made to identify those characteristics of other species wiich enable them to co-exist with the dominant. In the course of its life-span, *Calluna* passes through a series of growth phases (pioneer, building, mature, degenerate) during which there are marked changes in the condition affecting associated species. These display a variety of 'complementary strategies' which enable them to co-exist with *Calluna* during a part, or the whole, of the sequence. Changes with time in the flora of even-aged stands of *Calluna*, produced as a result of management by burning, are accounted for in similar terms. It is also shown that invasion of heath communities by tree species (e.g. *Betula* spp., *Pinus sylvestris*), which may eventually eliminate *Calluna*, is related to the sequence of growth-phases.

References

Barclay-Estrup, P. 1966. Interpretation of cyclical processes in a Scottish heath community. Ph.D. Thesis, University of Aberdeen. 378 pp.

Barclay-Estrup, P. 1970. The description and interpretation of cyclical processes in a heath community. II. Changes in biomass and shoot production during the Calluna cycle. J. Ecol. 58: 243–249.

Barclay-Estrup, P. 1971. The description and interpretation of cyclical processes in a heath community. III. Micro-climate in relation to the Calluna cycle. J. Ecol. 59: 143–166.

Barclay-Estrup, P. & Gimingham, C.H. 1969. The description and interpretation of cyclical processes in a heath community. I. Vegetational change in relation to the Calluna cycle. J. Ecol. 57: 737–758.

Clapham, A.R., Tutin, T.G. & Warburg, E.F. 1962. Flora of the British Isles (2nd Ed.) Cambridge University Press. 1269 pp.

Duncan, U.K. 1970. Introduction to British lichens. Buncle, Arbroath. 292 pp.

Gimingham, C.H. 1972. Ecology of heathlands. Chapman and Hall, London. 266 pp.

Gong Wooi Khoon, 1976. Birch regeneration in heathland vegetation. Ph.D. Thesis, University of Aberdeen. 191 pp.

Ward, S.D. 1970. The phytosociology of Calluna-Arctostaphylos heaths in Scotland and Scandinavia. I. Dinnett Moor, Aberdeenshire. J. Ecol. 58: 847–863.

Watson, E.V. 1968. British mosses and liverworts. (2nd Ed.) Cambridge University Press. 495 pp.

Watts, A.S. 1955. Bracken versus heather, a study in plant sociology. J. Ecol. 43: 490–506.

ASPECTS OF POPULATION DYNAMICS IN HALIMIONE PORTULACOIDES COMMUNITIES*·**

W.G. BEEFTINK, M.C. DAANE, W. DE MUNCK & J. NIEUWENHUIZE***

Delta Institute for Hydrobiological Research, Yerseke, The Netherlands****

Keywords: *Aster tripolium*, Environmental disturbances, *Halimione portulacoides*, Population dynamics, *Puccinellia maritima*, Salt-marsh vegetation, Species strategies, *Suaeda maritima*, Succession

Introduction

Plant succession after interference owing to environmental disturbance has often been interpreted as a process whereby a plant community regenerates as to its inherent properties towards a semblance of its original composition and structure, analogous to the recovery of an organism from injury (Clements 1916). Recent workers, however, view succession in the context of adaptations of individual species forming spatial and temporal patterns in population development independent of any transcendent properties of the whole community (Horn 1976). Teleologically, and moreover system-theoretically, both changes in single populations and in integrated ecosystems may be considered relevant.

However that may be, succession continues to raise a number of basic questions, such as:

a) Do orderly and predictable succession patterns exist?
b) Are there both convergent and divergent series in succession, and under what conditions do they occur?
c) Does succession reach a stable end point, either via a series of environmentally equilibrated intermediate stages, or is a mature final stage reached in the long run via a more gradual development?
d) To what extent can succession trends be interpreted as a process of internally integrated species fluctuations whithin ecosystems that are 'externally stable'. In this respect one may raise the question whether some stages of succession are more fragile than others, i.e. more susceptible to ex-

ternal environmental changes. According to a recent elaboration of the Relation theory of van Leeuwen (1966, 1973, also van der Maarel 1976) a dynamic equilibrium develops in an ecosystem when it is in relative constancy with the environmental conditions of its surroundings. Changes in these conditions cause changes in the dynamic equilibrium within the ecosystem; the latter are progressively more drastic the less dynamic their starting point is. In terms of Holling (1973) resilience is gradually replaced by resistance.

Answers to these questions are of basic importance for understanding dynamics of vegetation and for managing nature and natural resources. In this paper some aspects of these questions are dealt with based on studies of *Halimione portulacoides* communities occurring in estuarine salt marshes of the S.W. Netherlands.

Material and methods

The vegetational data have been obtained from studies on permanent sample plots established in selected salt-marsh communities occurring in the Rhine-Meuse estuary. Some of these permanent plots had undergone a sort of natural environmental disturbance shortly before or after their establishment. Others were established to trace the influence of disturbances induced by man, either patchy or large-scale.

The first group of data to be dealt with in this paper concerns patchy disturbances in *Halimione portulacoides* communities occurring in the polyhaline to euhaline zones of the estuary. The salt marshes concerned are situated in the Grevelingen, i.e. the Springersgors salt marsh, and in the Eastern Scheldt, i.e. the Stroodorpepolder salt marsh (Fig. 1). In these marshes the following disturbances occurred:

* Contribution to the Symposium on Plant species and plant communities held at Nijmegen, 11–12 November 1976, on the occasion of the 60th birthday of Professor Victor Westhoff.
** Nomenclature follows Heukels-van Ooststroom. Flora van Nederland, 18e druk, 1975. Wolters-Noordhoff, Groningen.
*** The authors are greatly indebted to Dr K.F. Vaas (Yerseke) for reviewing the English text.
**** Communication Nr. 160.

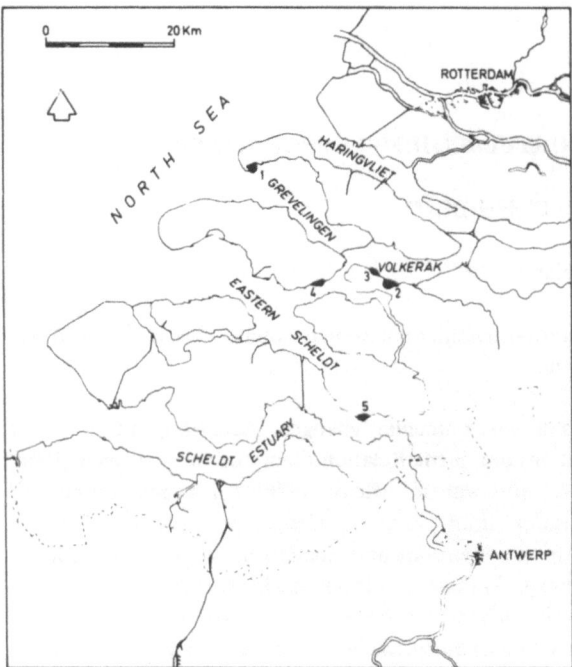

Fig. 1. Situation of the salt marshes investigated in the estuaries of the rivers Scheldt, Meuse and Rhine. 1. Springersgors, 2. Slikken van de Heen, 3. Kramerspolder, 4. Oosterlandpolder, and 5. Stroodorpepolder salt marsh.

a) Extremely waterlogged conditions occurring in some depressions during 1961 (Springersgors salt marsh).
b) Very low temperatures (to −16 C) to which snow-free patches had been exposed during the severe winter frost and snow period of 22 December 1962 to 2 March 1963 (Springersgors salt marsh).
d) Destruction of the vegetation by spraying with a mixture of the herbicides paraquat and diquat in May 1964 (Stroodorpepolder salt marsh).

The second data set is concerned with the influence of a sudden increase in flood level in the Volkerak-Eastern-Scheldt estuarine branch upon the population dynamics in communities evolving into *Halimione portulacoides* stands. This sudden increase in flood level was a consequence of the building of the Volkerak Dam, April 1969, closing the estuarine arm at its riverside and inducing a depth gradient from about 50 cm near the new dam to a few cm in the Eastern Scheldt. The effects of this rise in flood level has been studied in a range of salt marshes of which here only the Slikken van de Heen, the Kramerspolder, and the Oosterlandpolder salt marshes are relevant (Fig. 1).

The permanent plots established in all these marshes vary from 10 to ca. 40 m² in surface. Species performance expressed as coverage and number of individuals was analysed each year according to the estimation method of Doing Kraft (1954).

The third data set is obtained from an experiment with different types of management or environmental impacts induced by man, most of them borrowed from traditional cultural practices. In an extensive *Halimione portulacoides* community growing on a highly accreted and well-developed creek bank levee a series of plots of 3 × 4 m, enclosed by a buffer zone of ½–1 m, was established in autumn 1971. The types of environmental impact employed were:
– raking – once a year, in autumn
– treading – once a month, from May to October
– 'hay making' (mowing) – once a year, in autumn
– spraying with a mixture of paraquat and diquat (0.1 g/m² active ingredients of both) according to directions for use
– once, at the beginning of the experiment
– cutting sods – as the former treatment
– tilling – as the former treatment
– covering the soil surface with clay from the same creek bank levee to a thickness of about 15 cm after having mowed the vegetation – as the former treatment
– covering the soil surface in the same way but now after having tilled the original top-soil – as the former treatment
– excavating to about 20 cm, and preventing the basin to catch stagnant water by cutting a drainage channel – as the former treatment.
One plot was left undisturbed as a control. The species composition was analysed in the same manner as in the other sample plots.

Results

Figure 2 shows the population dynamics of the relevant species under conditions of apparent undisturbedness in these two sample plots, which can serve as a basis for comparison with the verified disturbed sample plots. Coverage of *Halimione portulacoides* fluctuates very little (from 90 to 100% in nine years). Other species, however, such as the annual *Suaeda maritima* and the mainly biannual *Aster tripolium*, seem to fluctuate more, at least concerning numbers of individuals as well as frequencies, than those of *Halimione*. It should be noted that the perennial *Puccinellia maritima* is absent in sample plot Nr. 6 (Fig. 2A), but present in Nr. B 1 (Fig. 2B). A more important difference between the two sample plots seems that in Nr. 6 the population density of both *Aster tripolium* and *Suaeda*

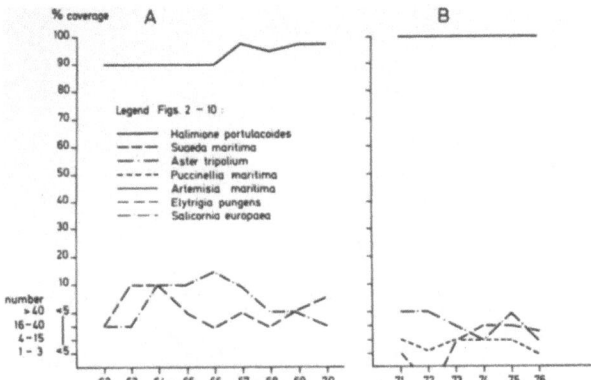

Fig. 2. Population dynamics in *Halimione portulacoides* communities without apparent environmental disturbance. A. Springersgors salt marsh, sample plot Nr. 6, surface 6 × 7 m. B. Stroodorpepolder salt marsh, sample plot Nr. B 1, surface 3 × 4 m.

maritima is evidently higher than that in Nr. B 1. It is suggested that this difference in population densities between the sample plots is related to a difference in environmental fluctuations interpreted as a difference in disturbance. From this viewpoint sample plot Nr. 6 should be subjected to a higher level of disturbance than sample plot Nr. B 1. This supposition is sustained by a higher fluctuation in coverage of *Halimione* in sample plot Nr. 6.

Under temporarily waterlogged conditions occurring during the growth season in 1961 (see Fig. 3) the *Halimione* communities showed a more dynamic development. During the years in which *Halimione portulacoides* reached minimal values (1962–63 and 1963–64, resp.) or was

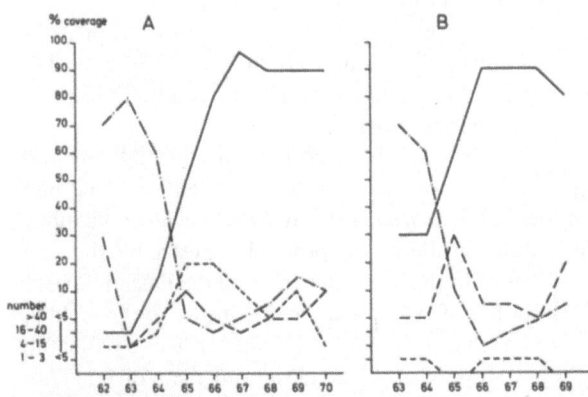

Fig. 3. Population dynamics in *Halimione portulacoides* communities growing in the Springersgors salt marsh after extremely waterlogged conditions, 1961. A. Sample plot Nr. 9, surface 4.5 × 5 m. B. Sample plot Nr. 14, surface 3 × 4 m. For explanation of the lines indicating species behaviour see Fig. 2.

recovering, *Suaeda maritima* and *Aster tripolium* developed strongly, growing with maximal coverage percentages. In sample plot Nr. 9 (Fig. 3A) *Puccinellia maritima* also reached a maximum during the recovery period of *Halimione* (1965–66). In this sample plot some weak secondary density maxima in *Suaeda* (1965), *Aster* (1969) and perhaps in *Puccinellia* (1969) can be observed. The peak in the *Suaeda* population in sample plot Nr. 14 (1965) will also be a secondary one.

After exposure to severe winter frost, generating a patchy structure in the field owing to blow-outs in the snow cover, the decline of *Halimione portulacoides* in salt-marsh depressions was followed by an increase in *Suaeda* and *Aster* populations successively (Fig. 4A).

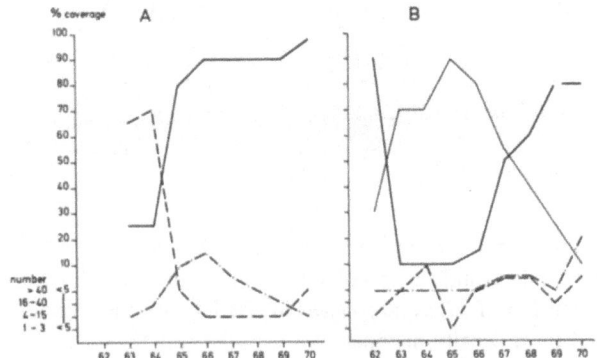

Fig. 4. Population dynamics in *Halimione portulacoides* communities growing in the Springersgors salt marsh after severe winter frost, 1962–1963. A. Sample plot Nr. 17, surface 3 × 5 m. B. Sample plot Nr. 7, surface 6 × 7 m. For explanation of the lines indicating species behaviour see Fig. 2.

On creek bank levees these successive density maxima can be impeded by a peak in the coverage of *Artemisia maritima* when that species was already present before the impact of low temperatures (Fig. 4B). In the latter case, however, both *Suaeda* and *Aster* populations show tendencies to the development of secondary density maxima separated by the primary ones by years in which the number of plants is minimal.

After chemical destruction recovery of the *Halimione* vegetation takes much more time than in the case of the natural patchy disturbances concerned (Fig. 5). In both sample plots the die-back in *Halimione* subsequent to spraying is succeeded by a sequence of density maxima in the *Suaeda*, *Aster* and *Puccinellia* populations successively (1967, 1969, resp. 1969–70). Moreover, each population of these species developed a second weaker maximum before *Halimione* recovered (1971, 1972, resp. 1974–75).

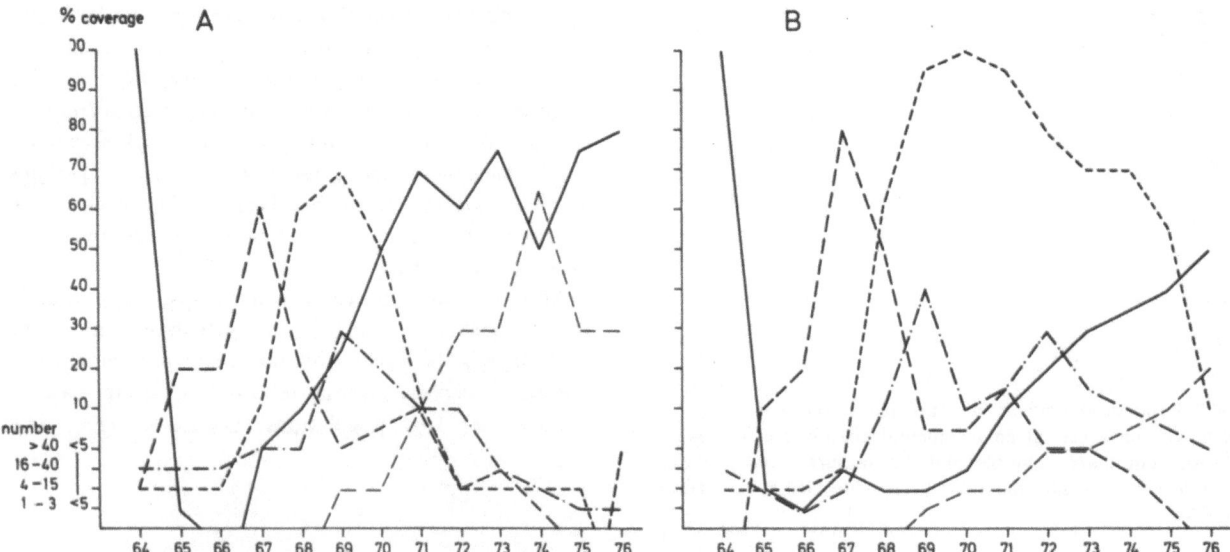

Fig. 5. Population dynamics in *Halimione portulacoides* communities growing in the Stroodorpepolder salt marsh after chemical destruction of the vegetation, 1964. A. Sample plot Nr. 8, surface 2 × 8 m. B. Sample plot Nr. 15, surface 2 × 6 m. For explanation of the lines indicating species behaviour see Fig. 2.

In one case (*Suadea maritima* in Fig. 5A) even a very small third one could be traced (1973). In both examples an expansion of *Elytrigia pungens* interfered with the second and third density maxima.

The data set concerned with the influence of a sudden increase in flood level does not yield essential other results. In the sample plots established in the Slikken van de Heen (Fig. 6), the Kramerspolder (Fig. 7) and the Oosterlandpolder salt marsh (Fig. 8), *Suaeda*, *Aster* and *Puccinellia*

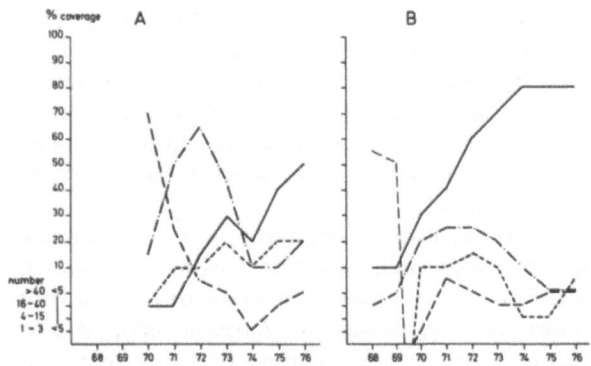

Fig. 6. Population dynamics in communities evolving to *Halimione portulacoides* stands in the Slikken van de Heen salt marsh. A. Sample plot Nr. 23, surface 3 × 7 m. B. Sample plot Nr. 21, surface 2.5 × 9 m. For explanation of the lines indicating species behaviour see Fig. 2.

populations are developing in the same way as in the case of the patchy disturbances, exhibiting a series of subsequent maximal population densities. Traces of secondary peak densities are only found in the Slikken van de Heen salt marsh, where the highest increase in flood level occurred. Where this increase is less than about 10 cm, such as in the Oosterlandpolder salt marsh (Fig. 8), the maximum population densities become less evident. In case of an even smaller increase in flood level, such as in the Stroodorperpolder salt marsh, peak densities remain absent or at least unobserved (Fig. 9A), or have been kept down by other environmental impacts such as washing of plant debris onto the vegetation (Fig. 9B), resulting in an invasion of *Elytrigia pungens*.

In the latter two sample plots environmental dynamics appear to be more complicated. In Figure 9A the high densities in the *Suaeda*, *Aster* and *Puccinellia* populations, culminating in 1969, are probably caused by the high quantities of fresh river water entering the Eastern Scheldt in the springs of 1965–68 as a consequence of the building of the Grevelingendam in 1964 (Table 1). This desalinating influence stopped after the closure of the Volkerakdam, April 1969. After that date the salinity fluctuations became controlled mainly by local climatic conditions instead of by quantities of glacier- and rainwater flowing down from Central Europe. In Figure 9B the peak density of *Suaeda maritima* is perhaps suppressed, or at least influenced by

46

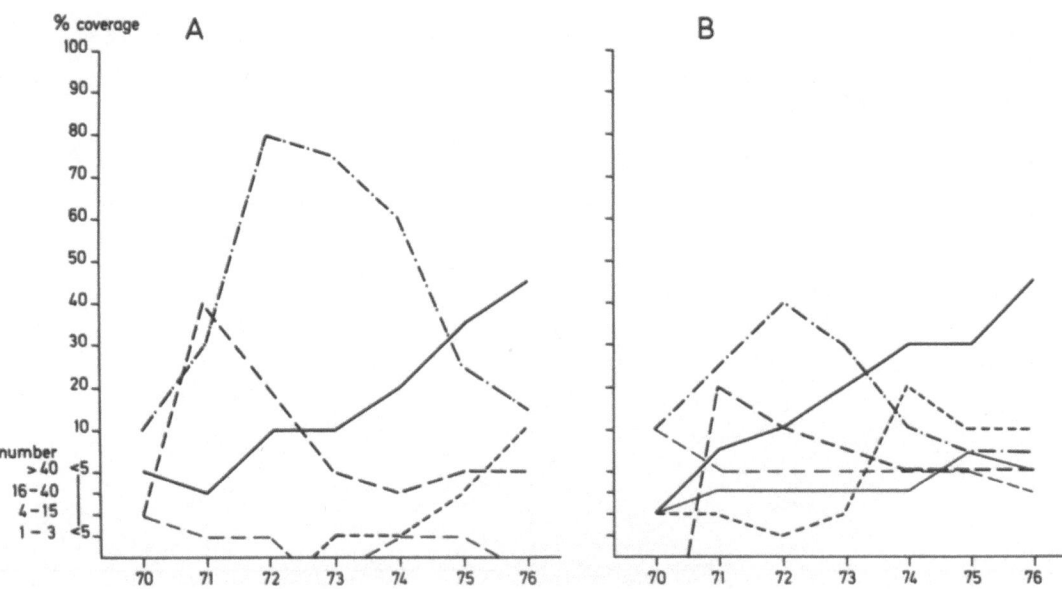

Fig. 7. Population dynamics in communities evolving to *Halimione portulacoides* stands in the Kramerspolder salt marsh. A. Sample plot Nr. 8, surface 3 × 6 m. B. Sample plot Nr. 11, surface 1.5 × 6 m. For explanation of the lines indicating species behaviour see Fig. 2.

Elytrigia pungens favoured by allochthonous plant debris washed ashore onto the sample plot.

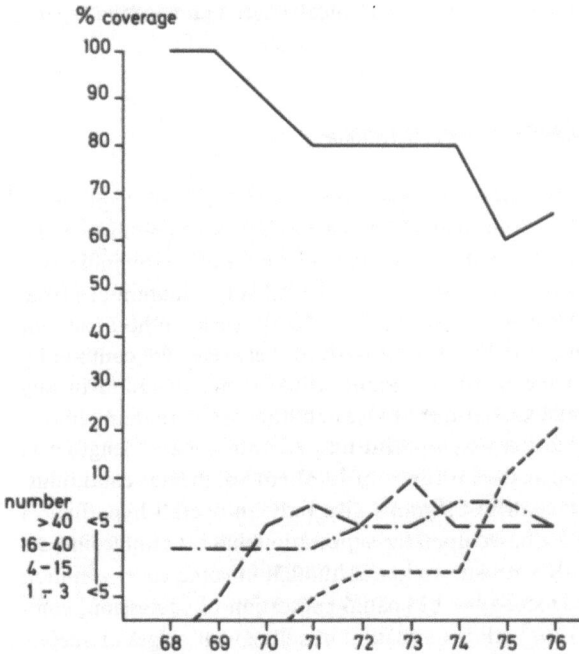

Fig. 8. Population dynamics in a *Halimione portulacoides* community growing in the Oosterlandpolder salt marsh. Sample plot Nr. 1, surface 2 × 5 m. For explanation of the lines indicating species behaviour see Fig. 2.

The results of the experiment with different kinds of environmental impacts imitating traditional cultural practices, are illustrated in Figure 10. The decline of *Halimione portulacoides* reflects the intensity of the environmental disturbance induced. Under the types of disturbances limited to the vegetation itself, raking represents the slightest and spraying the worst impact. The influence of raking appears to be balanced already after 3 or 4 years. The influence of treading is slow but continuous for the time being. Mowing keeps *Halimione* within bounds, reducing its growth to small scattered individuals only, whereas

Table 1. Salinity minima (°/oo chloride) in the Eastern Scheldt at high water during the springs of 1964-1976. Figures from the water off Yerseke (Data collected by P.J. van Boven).

Year	Salinity minimum	Period
1964	16-17	February - May
1965	13-14	April - June
1966	9-12	half December - February
1967	9-12	half December - half February
1968	11-12	half January - March
1969	14	March - half June
1970	14	April - May
1971-74	16-17	all year
1975	15-16	all year
1976	16-17	all year

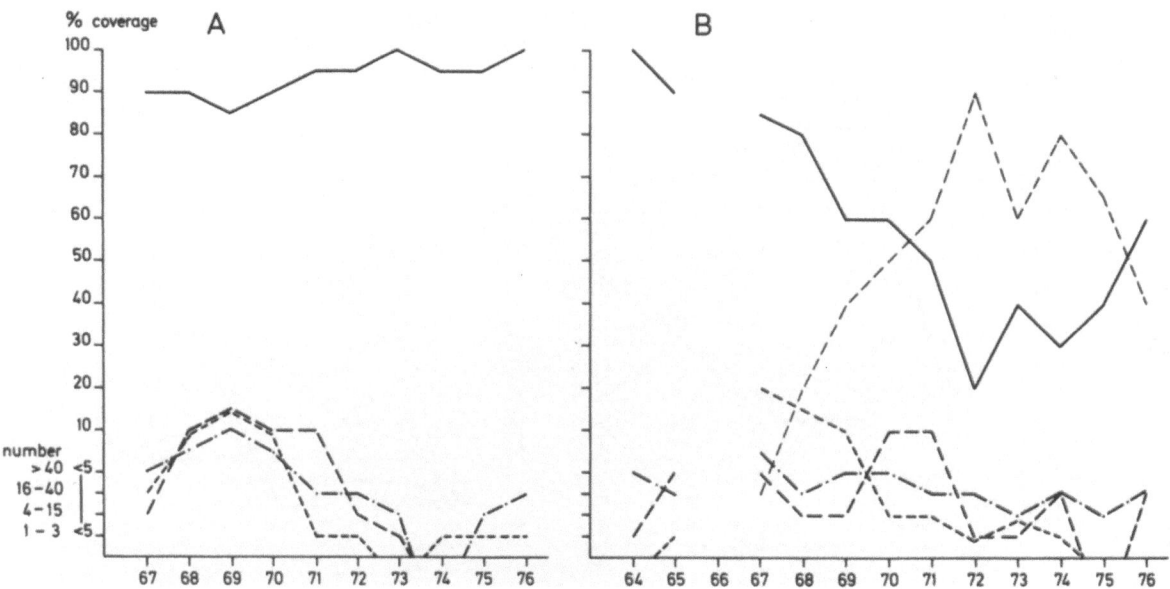

Fig. 9. Population dynamics in *Halimione portulacoides* communities growing in the Stroodorpepolder salt marsh. A sample plot Nr. 10, surface 4 × 6 m. B. Sample plot Nr. 7, surface 2 × 6 m. For explanation of the lines indicating species behaviour see Fig. 2.

spraying destroys this species completely for two years.

Types of minor impacts result in slight and temporary reactions of *Suaeda*, *Aster* and *Puccinellia*. Intense impacts lead to a sharp reaction, a complete dominance of *Suaeda* after 3 years, followed by a comparable density peak in *Puccinellia* after 5 or 6 years. In the latter case the *Aster* populations were backward in comparison with both other species, and developed synchronously with the *Puccinellia* populations.

The types of disturbances in which the top-soil is involved are all more or less intense, although *Halimione* is nowhere totally eradicated except in the excavated sample plot. Cutting sods and, much more, excavation suppressed the development of the *Suaeda* populations, and favoured that of *Puccinellia*. Tilling the top-soil caused *Suaeda* to grow with a maximal density during two years instead of during one year in all other cases, also after covering the surface with about 15 cm soil. Covering with soil only favoured *Suaeda* populations in the same way, but seems to suppress the subsequent *Puccinellia* density. Raising the soil surface allows *Elytrigia pungens* to establish itself with varied density values. Here, too, *Aster tripolium* played a minor part, parallel and concurrently to that of *Puccinellia*.

Tillage, including cutting sods and supply of soil from elsewhere, generally interferes more in the vegetation than disturbances in which only the aerial parts of the vegetation

are involved: all populations reach their maximal densities one year earlier in cases of disturbance of the soil as compared to those of disturbance of aerial parts of the vegetation only.

Discussions and conclusions

Various authors (e.g. Chapman 1960) present diagrams of the zonation pattern of vegetation from bare mud to the upper limit of the salt marsh without evidently indicating whether this pattern is really related to zonation or rather to succession. In these diagrams such authors do not suggest that all relationships between the community depicted in the diagrams will be actually found in any particular salt-marsh area, but they stress that any line of vegetation development may be shortened or lengthened through the influence of local environmental conditions. In fact these diagrams give only an overall hypothetical picture based upon the supposition that zonation coincides highly with succession. Although in some cases zonation may indeed be the spatial expression of succession, comparison with the situation in subsequent stages of succession without conclusive research on the spot over an appropriate period of time neglects the different pathways that succession actually follows in the ecosystems. To give an example (Beeftink 1965): the *Halimionetum portula-*

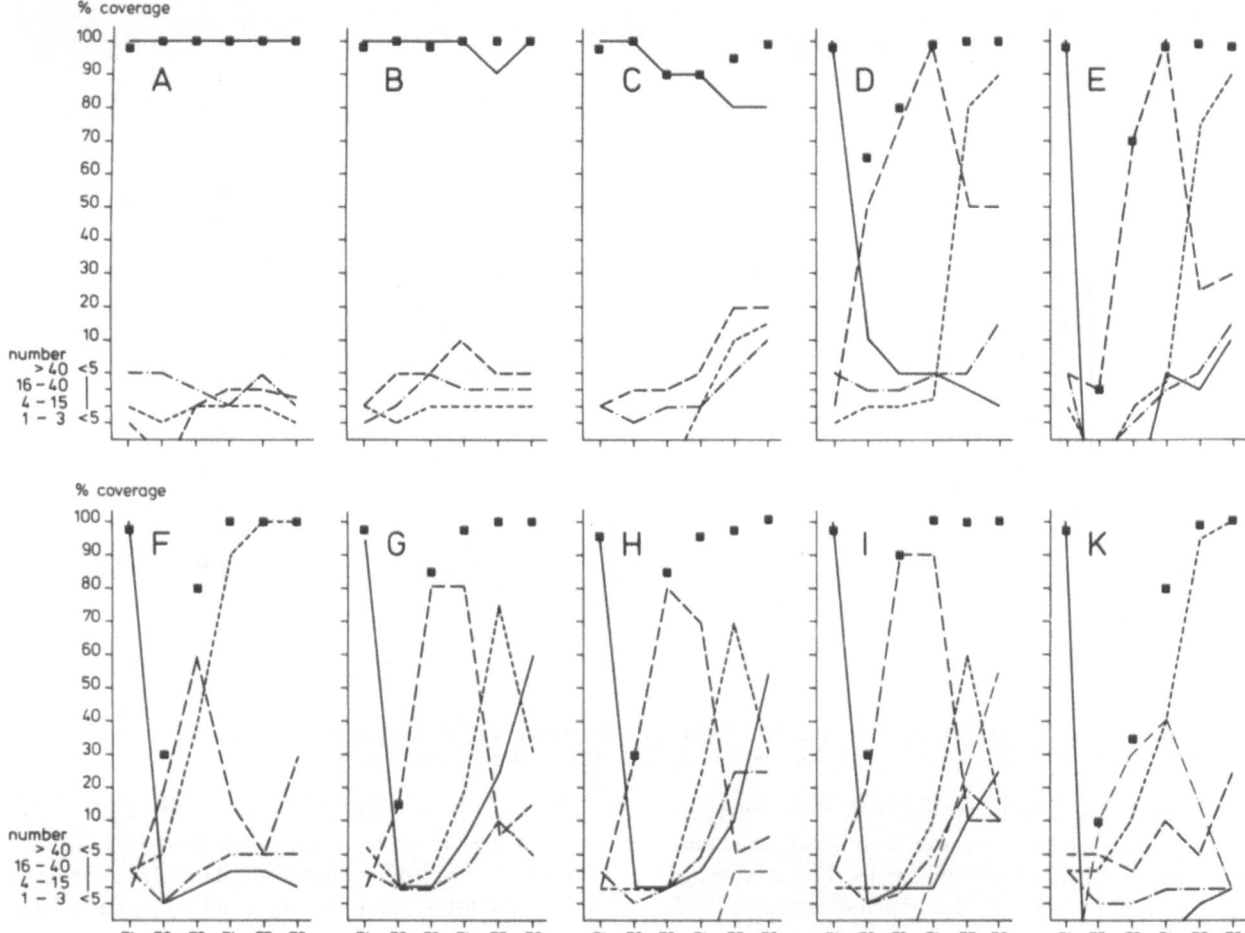

Fig. 10. Field experiment in a *Halimione portulacoides* community with different types of disturbance. Locality: Stroodorpepolder salt marsh. A. No experimental disturbance. B. Raking. C. Treading. D. Mowing. E. Chemical destruction. F. Cutting sods. G. Tilling. H. Mowing and covering the soil surface with 15 cm clay. I. Tilling and covering the soil surface with 15 cm clay. K. Excavating to about 20 cm. For explanation of the lines indicating species behaviour see Fig. 2. Squares indicate total coverage of vegetation.

coidis appeared to develop in geomorphologically different situations (creek-bank levees, basins) and in various developmental series and stages of succession (first on the creek banks, and later in the basins).

The results of studies on the dynamics of *Halimione portulacoides* communities in the SW Netherlands suggest the following conclusions, illustrated by some succession models:

1. There is a fixed pattern of population succession after environmental disturbances, patchy as well as of wider impact, in *Halimione* communities. This pattern consists of a sequence of partly overlapping density fluctuations in *Suaeda maritima*, *Aster tripolium* and *Puccinellia maritima* successively (Fig. 11). In highly accreted stands of

Halimione, such as that where our field experiment was carried out, the *Aster* population drops behind in density maximum and in speed of development, running concurrently with *Puccinellia*.

2. In case of minor environmental impacts this sequence can be limited to the first two species populations mentioned, or even to the first species only (Fig. 11). If full re-establishment of the original dominant *Halimione portulacoides* takes a relatively long time (here as a result of chemical destruction) the high density series of the three subsequent plant populations is repeated weakly, suggesting a redundant capacity of these species for filling the gap between the original and the future equilibrated community (Fig. 5 and 11).

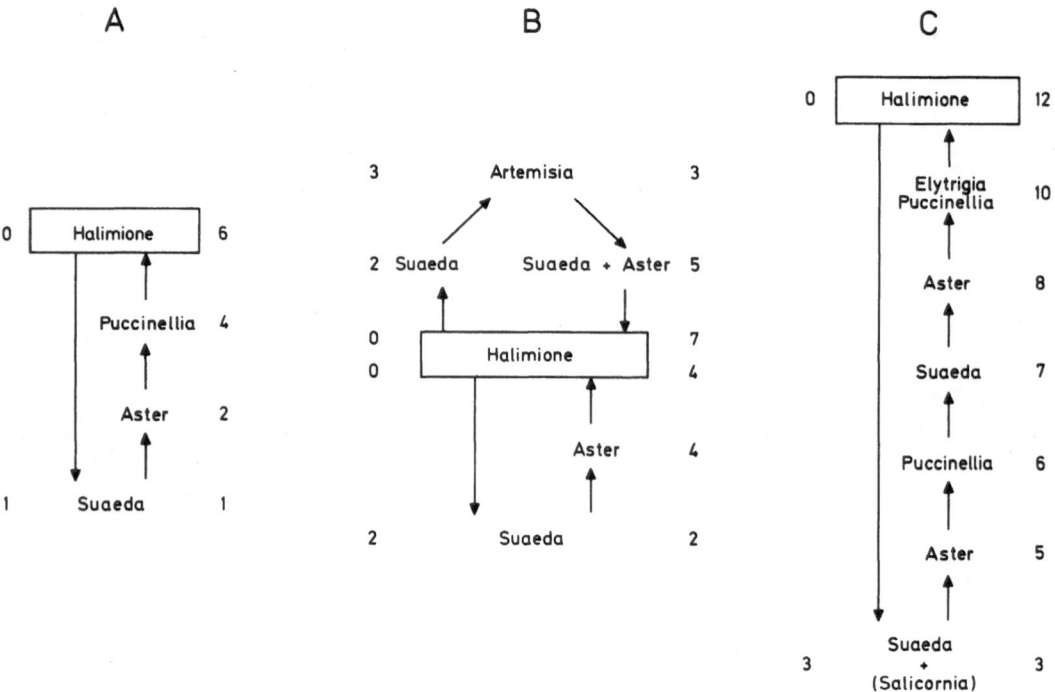

A
0 Halimione 6
Puccinellia 4
Aster 2
1 Suaeda 1
B
3 Artemisia 3
2 Suaeda
Suaeda + Aster 5
0 Halimione 7
0 4
Aster 4
2 Suaeda 2
C
0 Halimione 12
Elytrigia
Puccinellia 10
Aster 8
Suaeda 7
Puccinellia 6
Aster 5
Suaeda
3 + 3
(Salicornia)

Fig. 11. Models of succession following three types of environmental disturbance in *Halimione portulacoides* communities. The numbers indicate the years of maximal population densities counting from the year of disturbance (= year zero).

3. The pattern of population succession can be disrupted when a species already present in the original community was suppressed by the dominant species. This is suggested by the behaviour of *Artemisia maritima* situated on a creek bank levee in the *Halimione* community and having endured the impact of severe winter frost (Fig. 4B and 11). Another form of interference can be recognized when a newly established species emerges suddenly and in large

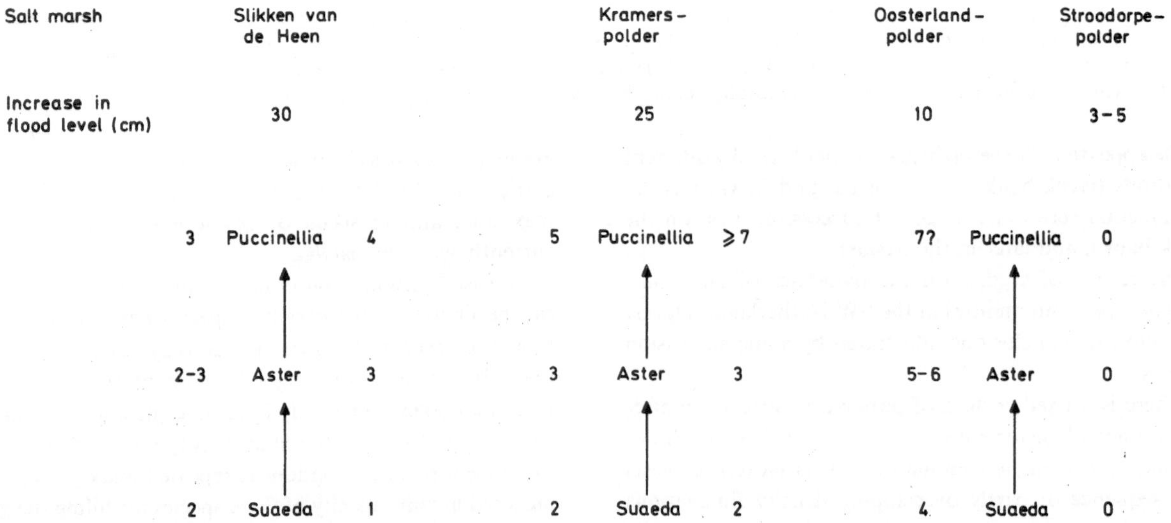

Salt marsh
Slikken van
de Heen
Kramers-
polder
Oosterland-
polder
Stroodorpe-
polder
Increase in
flood level (cm)
30
25
10
3-5
3 Puccinellia 4
5 Puccinellia ≥7
7? Puccinellia 0
2-3 Aster 3
3 Aster 3
5-6 Aster 0
2 Suaeda 1
2 Suaeda 2
4 Suaeda 0

Fig. 12. Model of succession on a gradient caused by an increase of flood level. The numbers indicate the years of maximal population densities counting from the year of disturbance (= year zero).

numbers superseding the less competitive interim species. This is shown in the case of chemical destruction (Fig. 5) and, when the soil surface is covered with clay (Fig. 10). Here *Elytrigia pungens* follows *Halimione* in its development, competing spatially with this species and with *Puccinellia maritima*.

4. As the succession models in Figure 11 illustrate, there is a time-lag between the year in which the environmental impact takes place (year zero), and the year in which the plant populations gain their highest densities. Studies on the influence of the gradient in increased flood level demonstrate that this time-lag is longer the smaller the magnitude of the environmental impact is (Fig. 12). Very slight environmental impacts, such as occurred in the Stroodorpepolder salt marsh (Fig. 9), may induce no observable reaction at all. This phenomenon suggests that under such conditions the impulse is too small to induce a sequence of successive density maxima. Only very locally in the community the opportunity remains to establish individuals preventing within-community extinction of the species. In this connection it is interesting that the within-species time-lag between two subsequent density maxima is 4 years for *Suaeda* and *Puccinellia*, and 3 years for *Aster* (Figs. 3, 4 and 5). If the time-lag between the year in which the environmental impact takes place, and the year of the first density maximum in *Suaeda* would be more than 4 years, density maxima would not occur, as is apparent from a comparison of the data from the Oosterlandpolder and Stroodorpepolder salt marshes (Figs. 8 and 9). This would mean that there is a threshold value in environmental dynamics above which the species populations react by peak densities (dominance in the community), and below which very local (micro)dynamic conditions within the community have to maintain minimal survival. Apart from immigration of diaspores, within-community extinction is imagined to appear below a second, lower threshold value where the dynamics required for the species are too slight or totally absent for any establishment of individuals.

5. More intense impacts lead to sharper reactions of the interim species, as Figure 10 shows. Impacts in which the top-soil is involved are more intense than those that only affect the above-ground parts of the vegetation. The reaction of the interim species does concern both the intensity of dominance and the time-lag between the year of disturbance and the year in which *Suaeda maritima* reaches its maximum population density.

6. It is suggested that in the salt-marsh ecosystem the pattern of spatial variation in densities and that of temporal variation in fluctuations of *Suaeda*, *Aster* and *Puccinellia* populations under natural conditions reflects corresponding patterns of environmental disturbances in the vegetation cover, taking account of a time-lag associated with the magnitude of the impact concerned.

7. To elucidate questions on the functioning of species in the salt-marsh ecosystem, studies on species strategies in life-cycle and in morphological, ecological and physiological plasticity are of primary importance.

At first, life-time strategies should be mentioned. These vary from obligatory annual in *Suaeda* to potentially full-perennial in *Puccinellia maritima*. *Aster* possesses an intermediate strategy. According to Gray (1971) flowering behaviour is associated with a tendency to perennial life-time on the lower marsh, while on the higher marsh annual plants are more frequent. Wijnands (1969), however, found up to 5 year old specimens in a high-lying marsh. So, during stabilization of the vegetation there is a sequence in increasing life-time plasticity from obligate annual via short-lived perennial to full-perennial, corresponding to the sequence in which the species expand successively as a reaction on the environmental disturbance.

As to dissemination capacity, all three species show a great plasticity. According to Westhoff (1947) their seeds can be spread by water, wind as well as by animals. *Aster* also reproduces vegetatively by detachable auxilliary buds (Wijnands 1969, Ranwell 1972). *Puccinellia* is distinct from both other species by a heavy reliance on vegetative reproduction. In this latter species vegetative propagation is favoured and is also related by tendencies to apomictic seed development (Hubberd 1968) and by sheep grazing. The activity of sheep leaves large quantities of discarded fragments that root readily when heeled into damp marsh surfaces by their treading (Ranwell 1972). Thus, the sequence of species populations seems to correspond to a decrease in generative and an increase in vegetative reproduction.

Another striking feature is the remarkable morphological plasticity. *Suaeda* occurs in erect as well as in prostrate forms and can vary considerably in size from about one centimeter in dry, sandy soils poor in nutrients, to more than half a meter on open grounds in freshly embanked salt marshes. *Aster* can range from tiny impoverished non-flowering plants, one or two centimeters high in grazed salt marshes, to vigorous and freely flowering plants up to 180 cm high in brackish marsh soils rich in silt. *Puccinellia* grows in upright as well as in depressed forms, and is characterized by narrow-leaved, widely spaced or broad-leaved, elongated shoots, and stolons. Under intense grazing pressure growth is reduced to a tight mat-like sward of little more than 1 cm high.

The three species appear to indicate also obvious gradient situations in their response to ecological conditions. Field experiences preliminary suggest that the nutrient level required is decreasing in the sequence *Suaeda*, *Aster* and *Puccinellia*. Experiments with clearings in salt-marsh vegetation (not yet fully worked out) show a rise in the mineral nitrogen level of the soil after clearing, which fades away with the revegetation process.

In the experiment in question, where different types of management are compared, mineral nitrogen compounds and soil-moisture contents have been analysed in the fields which were untreated, sprayed, cut for sods, and tilled (Table 2). Soil-moisture contents (SM) were not significantly different in the investigated fields. The reduced mineral nitrogen compounds (RMN), however, show in nearly all cases significantly higher values in the treated fields than in the untreated one. Moreover, cutting sods leads to a significant rise in RMN (NH_4) compared with spraying, while tilling seems to result in higher values compared with cutting sods. The high peak densities in Suaeda and Puccinellia after these treatments may be promoted by these compounds, probably via a preceding nitrification process.

The total and oxidized mineral nitrogen compounds (TMN resp. OMN) show more complex, but parallel patterns: In both compounds tilling had significantly higher amounts compared with spraying. The other correlations were not significant. Tilling may promote nitrification more than the other treatments, just as

covering with clay, and this may be the cause of the two years lasting maximal densities in the Suaeda populations after these treatments (compare Fig. 10).

As to *Suaeda*, Figure 10 illustrates that the killing of *Halimione* plants by spraying and mowing favours this species greatly in comparison to removal of the plants by raking and cutting sods. The same is true for tillage and supply of soil from elsewhere in comparison to removal of the top-soil layer including excavating. Contrary to the behaviour of *Sudeda*, *Puccinellia maritima* reaches complete dominance in cases where top-soils including their nutrients have been removed. *Aster tripolium* is thought to take an intermediate position in this respect.

Physiological and genetic plasticity is also emphasized by Ranwell (1972) for *Suaeda* and *Aster*. Using an enzyme bio-assay technique Stewart et al. (1972) showed an intriguing adaptive behaviour in *Suaeda* as to its nitrate reductase enzyme corresponding with the rate of emergence of the salt-marsh part occupied. Ranwell (1972) holds the view that *Aster tripolium* illustrates well Ehrlich & Raven's (1969) conception of selection, favouring close 'tracking' of the environment and co-adapted genetic combinations related to environmental factors changing with time. Apart from survival selection, its generation time must be sufficiently long for environmental modification to interact with the gene complex, but at the same time sufficiently short to enable it to adapt more closely to specific environmental conditions. In this way *Aster tripolium* is better suited to cope with environmental changes in space and time than longer-lived perennials such as *Puccinellia maritima*.

Table 2. Correlation matrices of soil parameters in experimental fields which were untreated (A), sprayed with herbicides (E), cut for sods (F), and tilled (G). For the treatments compare Fig. 10. Locality: Stroodorpepolder salt marsh. Date of starting the experiment: 11-12 October, 1971. Sampling monthly from May 1972 to April 1973. Soil depth 5-20 cm. Parameters: soil moisture content (SM) in % of fresh soil, total mineral nitrogen content (TMN) in mg N-min./kg dry soil, oxidized mineral nitrogen content (OMN) in mg N-NO_3 + NO_2/kg dry soil, and reduced mineral nitrogen content (RMN) in mg N-NH_4/kg dry soil.

SM ↓	A	E	F	G
A ↓		−0.2319 ns	+0.5670 ns	−0.4196 ns
E	+0.1043 ns		+0.0078 ns	+0.8108 **
F	+0.0272 ns	+0.5745 ns		−0.3584 ns
G	+5.0462 ns	+0.4411 ns	−0.3843 ns	

←— TMN

OMN ↓	A	E	F	G
A ↓		+0.5913 *	+0.7283 **	+0.5020 ns
E	+0.3039 ns		+0.7896 **	+0.6931 *
F	+0.0715 ns	−0.0011 ns		+0.7683 *
G	+0.2425 ns	+0.8833 ***	−0.2193 ns	

←— RMN

ns = non significant, * = P < 0.05, ** = P < 0.01, *** = P < 0.001

Thus, a gradient character can be recognized in various biological attributes of the three species according to the sequence in which they invade successively as a consequence of environmental disturbance. It can therefore be concluded that for understanding this mechanism in the salt-marsh ecosystem a thorough study on the functioning of these and other relevant biological attributes in the local populations is essential.

Returning to the basic questions put in the introduction it is concluded that:

a) There exist orderly and predictable succession patterns in the investigated salt-marsh vegetation.

b) Following occasional disturbances, whatever these may be, the vegetation recovers via a sequence of overlapping interim species populations showing a unilinear or obligatory (Horn 1976) succession pattern.

c) During succession the vegetation reaches its original stable end point with respect to the normal ambient environmental conditions operative. A lower impact of a disturbance results in longer time-lag periods and, possibly, lower density maxima of the interim species, down to smoothing away all fluctuations of that kind. Under the latter circumstances these plant populations can only survive if micro-disturbances within the community allow establishment. If not, new individuals fail to establish, and these species must become locally extinct, sooner or later, depending on their life-time strategies.

d) *Suaeda maritima*, *Aster tripolium* and *Puccinellia maritima* can only survive in *Halimione* communities under the influence of environmental disturbances coming from outside. Their functioning seems to consist mainly in consuming (neutralizing) outbursts of nutrients connected with die-back phenomena, and, indirectly, in covering the soil, protecting it against erosion. When these disturbances persist, for instance through repeated covering by plant debris washed ashore, or by grazing, succession is blocked at the stage which is attuned to that persisting impact. In the first case it may especially be the *Suaeda* stage, in the second one the *Puccinellia* stage. The mechanism described can therefore be interpreted as a complex of mostly well-integrated inherent species strategies capable of absorbing environmental shocks. Measured by the magnitude of the environmental impact (the magnitude of spatial exposure) early succession stages are less fragile than later ones, but measured by the time environmental conditions can be endured, the reverse is true.

Summary

Studies on sample plots in *Halimione portulacoides* communities show that environmental disturbances, either natural or induced by man, start a sequence of partly overlapping density maxima in *Suaeda maritima*, *Aster tripolium* and *Puccinellia maritima* successively, before the original *Halimione* community totally recovers. When succession time before recovering is long enough, there are tendencies in redundancy of this sequence stressing the unilinear character of the succession. Minor environmental impacts induce a longer time-lag period of the *Suaeda* density maximum, suggesting threshold values of these impacts for the species to maintain minimal population densities or to become locally extinct. This sequence of interim species starting after an environmental disturbance, suggests also a gradient character in various biological attributes, for instance in life-time, propagation, nutrient and genetic plasticity strategies. The mechanism described can therefore be interpreted as a complex of mostly well-adapted and well-integrated inherent species strategies capable of absorbing environmental shocks. It is suggested that in the salt-marsh ecosystem the pattern of spatial variation in densities and that of temporal variation in fluctuations of the three species populations under natural conditions reflect corresponding patterns of environmental disturbances in the vegetation taking into account a time-lag associated with the magnitude of the impact concerned.

References

Beeftink, W.G. 1965. De zoutvegetatie van ZW-Nederland beschouwd in Europees verband. Meded. Landbouwhogesch. Wageningen 65(1): 1–167.

Chapman, V.J. 1960. Salt marshes and salt deserts of the world. Hill, London, 392 pp.

Clements, F.E. 1916. Plant succession. An analysis of the development of vegetation. Carnegie Inst. Wash. Publ. 242: 1–512.

Doing Kraft, H. 1954. l'Analyse des carrés permanents. Acta Bot. Neerl. 3: 421–424.

Ehrlich, P.R. & Raven, P.H. 1969. Differentiation of populations. Science 165: 1128–1232.

Gray, A.J. 1971. Variation in Aster tripolium L., with particular reference to some British populations. Ph.D.Thesis, Univ. of Keele.

Holling, C.S. 1973. Resilience and stability of ecological systems. Ann. Rev. Ecol. Syst. 4: 1–23.

Horn, H.S. 1976. Succession. In: R.M. May (ed.), Theoretical ecology, principles and applications. Blackwell, Oxford, pp. 187–204.

Hubbard, C.E. 1968. Grasses (2nd edn.). Penguin, Harmondsworth, 463 pp.

53

Leeuwen, C. G. van. 1966. A relation theoretical approach to pattern and process in vegetation. Wentia 15: 25–46.

Leeuwen, C. G. van. 1973. Ekologie. Collegedictaat HB 20A, Technische Hogeschool Delft, Afd. Bouwkunde. 79 pp.

Maarel, E. van der. 1976. On the establishment of plant community boundaries. Ber. Deutsch. Bot. Ges. 89: 415–443.

Ranwell, D.S. 1972. Ecology of salt marshes and sand dunes. Chapman & Hall, London, 258 pp.

Stewart, G.R., Lee, J.A. & Orebamjo, T.O. 1972. Nitrogen metabolism of halophytes. I. Nitrate reductase activity in Suaeda maritima. New Phytol. 71: 263–267.

Westhoff, V. 1947. The vegetation of dunes and salt marshes on the Dutch islands of Terschelling, Vlieland and Texel. Thesis Univ. of Utrecht. Mimeographed.

Wijnands, D.O. 1969. Een onderzoek naar de variabiliteit van Aster tripolium in Nederland. Report R.I.V.O.N. Zeist, 57 pp.

VEGETATIONAL SUCCESSION IN A SOUTH SWEDISH DECIDUOUS WOOD*

Nils MALMER[1], Lennart LINDGREN[2] & Stefan PERSSON[1]**

[1] Institute of Plant Ecology, Lund University, Östra Vallgatan 14, S-223 61 Lund, Sweden
[2] Ministry for Agriculture, S-103 20 Stockholm, Sweden

Keywords:

Anthropogenic influences, Diversity, Management, Nature reserves, Permanent sampling plots, Potential natural vegetation Species number, Stability, Succession, *Ulmo-Fraxinetum*, Vegetation analysis

Introduction

In 1918 the small area of woodland, Dalby Söderskog, in southwestern Skåne was designated as a national park. It was regarded as 'the only remaining representative of the type of deciduous forest, that during periods long ago commonly occurred in the southernmost province of our country' (translation from Lindquist 1938, p. 61).

During the years 1925–1935 extensive botanical and ecological investigations were carried out in order to establish a scientific basis for the management of the wood (Lindquist 1938). The results also included some conclusions concerning the further natural development of the vegetation in the forest.

During the period 1969–1976 parts of these investigations were repeated (Lindgren 1971, Persson, unpubl.). One reason for this was the need to revise the plans for the management of the wood. The investigations, however, have also yielded information on general trends in vegetational successions and have provided material suitable for testing methodological problems.

In this paper only the traditional methods in plant sociology will be applied, mainly to make clear the results thus obtained. However, together with other current projects on vegetational successions, the available data from this wood will also be treated by numerical methods, in order to determine whether the conclusions reached by means of traditional methods are confirmed or not and whether such methods yield more information.

In vegetation science there are, surprisingly enough, very few detailed investigations dealing with the year to year variation in abundance and cover of plants. Much more attention has been paid to the description of plant communities, their spatial variation and relationship to various environmental factors. This implies an important lack of knowledge for studies of plant succession.

All investigations dealing with the repetition of previous analyses or comparisons made over long periods of time will always be open to the criticism that the data obtained are not wholly comparable. It is hardly probable that any scientific method remains unchanged for a period of say 30–50 years. Nor is it probable that the same person is able to repeat an investigation after so many years. In the early days of vegetation science hardly anybody planned for long-term studies on vegetational succession. Therefore in such studies, which today are felt to be very urgent, we have to make the greatest possible use of the available data. At the same time we have to keep in mind the methodological weaknesses and hope that in the future. we will get something better to work with.

* Nomenclature of species follows Lid (1974). *Viola riviniana* includes *V. reichenbachiana* as it was impossible to keep the two species separated throughout.
** The investigation was sponsored by the Research Council of the Swedish Environmental Board. Most of the text in this paper was prepared by the senior author as a lecture presented at the Symposium on Plant species and plant communities held at Nijmegen on the occasion of the 60th birthday of Professor Victor Westhoff.

Lennart Lindgren and Stefan Persson were responsible for the project in 1969–1970 and 1975–1976, respectively. They compiled all the material from these investigations. Skilful assistance was given by Brita Billstein, Mimmi Varga and Tommy Wikberg.

The site

Dalby Söderskog is a woodland area of 36 ha situated 10 km east of Lund (Fig. 1) on the plain of Skåne at an altitude of 65 m. The climatic conditions are indicated in Fig. 2.

The mineral soil is a heavy, calcareous clay representing the Baltic moraine of SW Skåne. The organic soil layer is a mull, the character of which has been described in detail by Lindquist (1938, pp. 135–185) together with the soil profile and other soil conditions. A rivulet runs through the wood and there are several small damp depressions, often filled with water during the winter and indicating a rather high ground-water level in the wood.

Today this woodland area is surrounded by fields and cultivated pastures. In that way it is more or less isolated from other areas with a natural or semi-natural vegetation cover. Together with the surrounding areas it was utilized for grazing, especially horses, even during medieval times (Lindquist 1938, pp. 33–61). The vegetation may then have had the character of coppiced woodland with extensive hazel scrubs and scattered taller trees (oak, beech and elm). During periods of less intensive grazing a regeneration of the trees was possible. One such period of regeneration seems to have been the war years during the first decades of the 18th century. A second one occurred

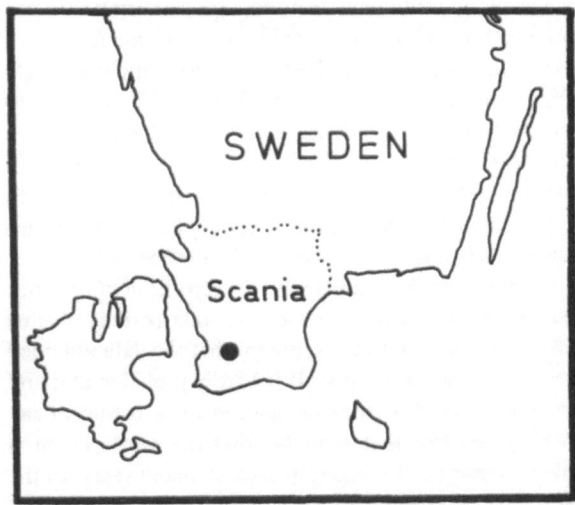

Fig. 1. Survey map of southern Sweden and Zealand (Denmark). The situation of Dalby Söderskog indicated.

in the beginning of the 19th century combined with intense cuttings in the shrub layer which seems to have favoured tree regeneration. At the end of the 19th century grazing

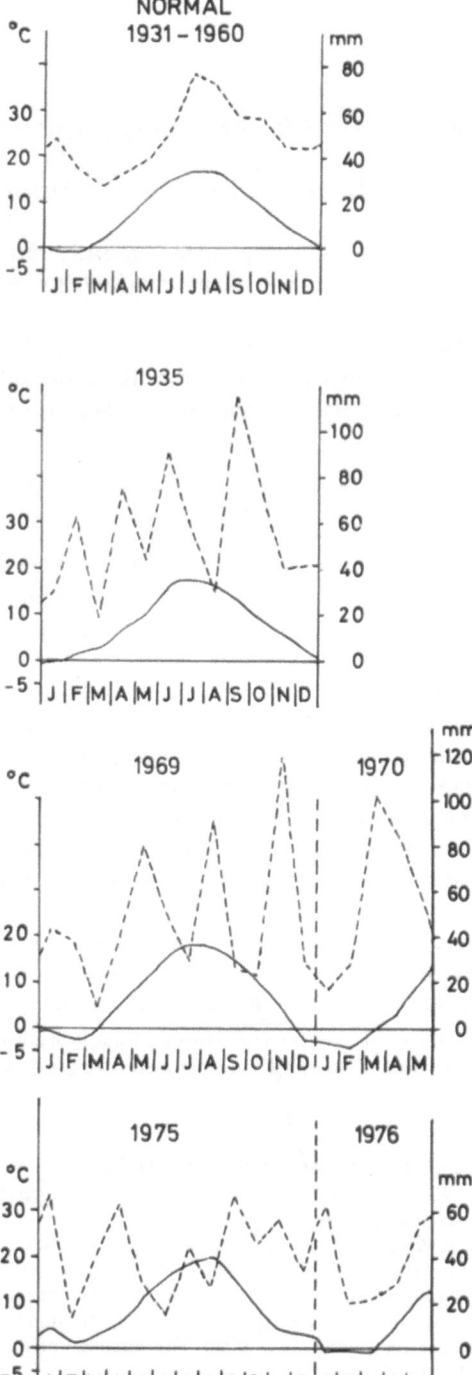

Fig. 2. Climate diagrams for Lund showing the monthly mean temperature (full drawn line) and precipitation (broken line) for the normal period 1931–1960 and for the three investigation periods.

56

ceased and the area was administratively transferred to the State Forest Commission. Heavy cuttings were undertaken in the beginning of the 20th century, the last ones during the years 1914–1916, just before the national park designation. Since then the area has been left to itself, except for some small cuttings, e.g. 1920–1924 and about 1960, when each time only a few trees were taken. Some trees have also been blown down during storms as in October 1967.

Material and methods

Complete floristic inventories from the area are available from the vegetation periods in 1925, 1935 and 1969.

The development of the tree layer has been investigated by the common forestry methods three times, viz. 1916, 1935 and 1969. Supplementary information is available from 1909. All trees of > 10 cm diameter at breast height are included in these inventories. The canopy cover of the tree and shrub layers has also been estimated in the vegetation analyses in 1935, 1969 and 1975.

Vegetation analyses of the field layer have been undertaken six times, viz. the periods April 24 – May 5 and August 15 – 25 in 1935, August 5 – 18 in 1969 and May 10 – 12 in 1970, July 21 – August 16 in 1975 and April 28 – May 10 in 1976. The spring analyses represent the vernal aspect of the vegetation, those from August the summer aspects. Some notes from supplementary analyses performed in 1925 are also available. Temperature and precipitation were closer to normal in 1935 than during the later investigations (Fig. 2). As the development of the vernal aspect ought to be highly dependent on the growth conditions the year before the analyses it is not inconsistent to carry out the investigations during two subsequent years as here has been done in 1969–1970 and in 1975–1976 respectively.

The data from the investigations in 1935 and earlier periods have been presented by Lindquist (1938). Most of the investigation in 1969–1970 has been accounted for in Lindgren 1971, while the results from the last period have not yet been published.

The method used for all vegetation analyses has been a small square analysis as carried out by Lindquist (1938, pp. 206–207). All vascular plants occurring within the sampling plots (squares) have been recorded with their cover according to the Hult-Sernander-Du Rietz scale (Table 1). In 1935 and 1975 the bryophytes were also included. For the high and low tree layers and the high

Table 1. Cover estimates according to the Hult-Sernander-Du Rietz scale

Part of the area covered (cover class)	Degree of cover	Middle of cover class
At most 1/16	1	1/32
1/16 - 1/8	2	3/32
1/8 - 1/4	3	6/32
1/4 - 1/2	4	12/32
More than 1/2	5	24/32

and low shrub layers (the A_1-, A_2-, B_1- and B_2-layers, height > 15, 15–8, 8–2 and 2–0.8 m respectively according to Lindquist) the size of the squares has been 16m², for the field and bottom layers 1 m². Each layer has been recorded separately, low-growing saplings of woody plants being included in the field layer.

In this study *frequency* (F) denotes the number of sampling plots in which a species occurs in per cent of the total number in the group referred to. In the same way the *characteristic degree of cover* (c) indicates the mean degree of cover for a species in the sampling plots where it has been noted. The method of calculation includes a transformation of the cover estimates to the middle values for each class (Table 1) and is presented in Malmer 1962, p. 49. In the tables the characteristic degree of cover is given as exponent together with the frequency (F^c). In some tables and figures a + or − is added to indicate a characteristic cover above or below the middle of each cover class. A complicated calculation is necessary in this case, as the Hult-Sernander-Du Rietz scale forms a geometric series and not an arithmetic one. This makes it difficult to transform the cover values to percentage of an area, especially when a cover of more than 50% is indicated. On the other hand we are dealing throughout with true cover estimates and not estimates which combine cover and some sort of abundance. (Cf. also Londo 1976).

The results of the contemporary spring and summer analyses have been put together in the tables. This is justified as the sampling plots have always been the same. For convenience the estimates of the summer species occurring in the vernal aspect have been excluded from the relevés.

All vegetation analyses dealt with in this paper refer to sampling plots in the fixed grid system established by Lindquist (1938, Fig. 35). As there are no other indications of the original positions of the sampling plots than those on the map there may be a slight uncertainty concerning the refinding of the same positions in 1969/70 and 1975/76. However, the difference in position of one plot between the investigation periods may nowhere exceed 10 m and is usually less than 4–5 m. The difference in positions between spring and summer analyses as well as between 1969/70 and 1975/76 is of course even less. In this very homogeneous type of vegetation such small differences in the positions of the sampling plots will hardly influence the results. It is also possible to regard the squares from each investigation period as a stratified sample of small areas for vegetational analyses.

Not only differences in the positions of the sampling plots but also personal differences between the field workers and annual differences in the growth conditions (cf., e.g., Ekstam & Sjögren 1973) will influence the results from investigations like this one. In this particular material the variation between the analyses of 1969/70 and of 1975/76 may yield some information on the combined variation between two years when two different persons have been responsible for the analyses.

Results

The flora

In 1925 the floristic inventory included 256 species of vascular plants in the wood (Lindquist 1938, pp. 190–194). In 1935 this number had been reduced to 204 and in 1970 to 149. This last figure includes 20 species introduced since 1925 and 129 that still occur, among them the 20 woody plants.

Most of the 127 species, which have disappeared since 1925 are those which mainly occur in open areas such as meadows, or which represent a distinct anthropogenic element of the flora. Therefore they may have been favoured by the open conditions resulting from the earlier grazing and the cuttings before 1916.

Vegetation

The high tree layer

The present tree population well reflects the history of

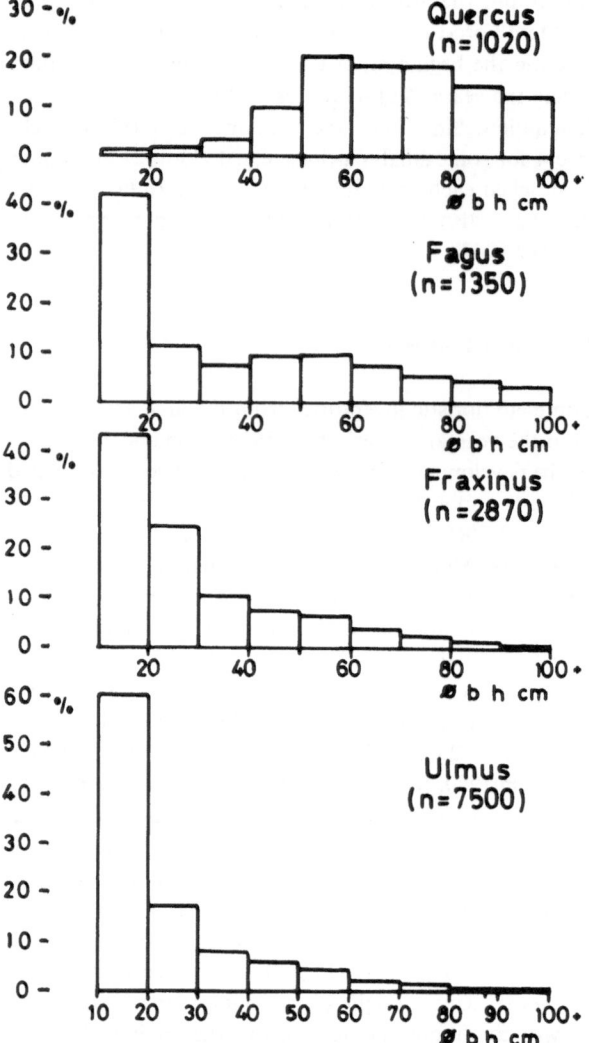

Fig. 3. The percentage distribution in stem diameter classes for oak, beech, ash and elm in Dalby Söderskog according to the measurements in 1970. – Only trees with stem diameter > 10 cm at breast height included.

the forest (Fig. 3). Among the oaks, *Quercus robur*, the greater diameter classes are strongly dominant. Only about 25 of the more than 1000 oaks are less than 25 cm at breast height which means that there are hardly any young oaks. The stem number remained fairly constant from 1916 to 1935 but has since then decreased by 26% (Fig. 4).

According to Lindquist (1938, p. 82) many of the existing big oaks must be about 250 years old, i.e. they grew up during the period of less intensive grazing at the

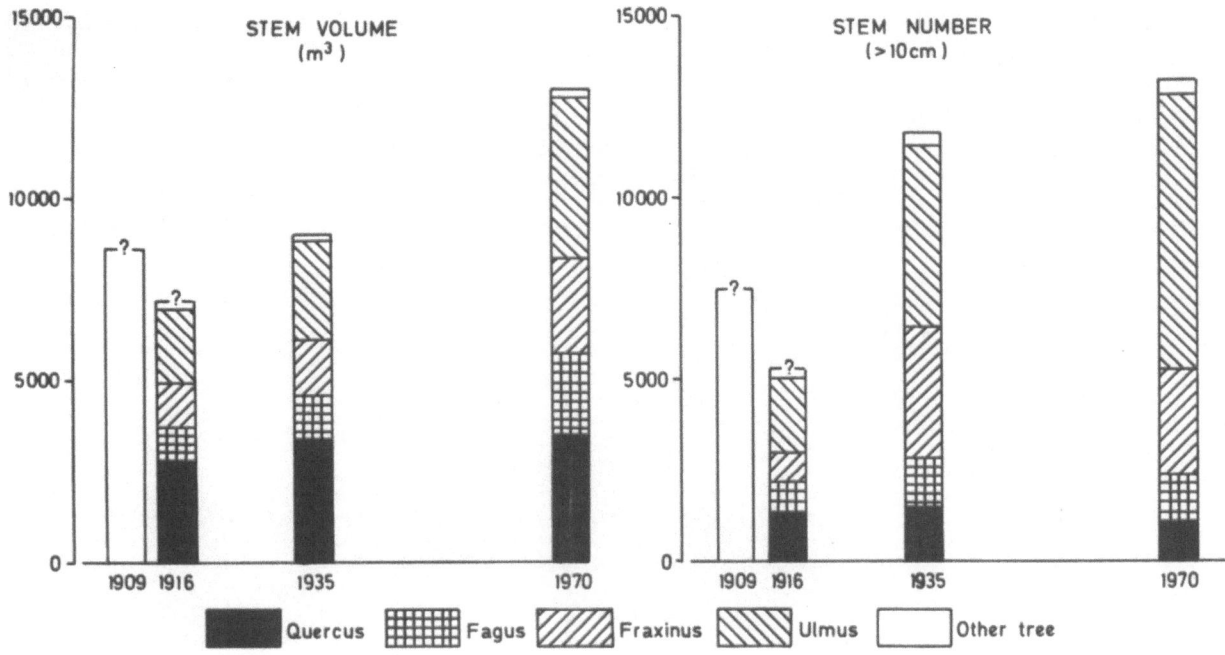

Fig. 4. Stem volume and stem number for the tree species in Dalby Söderskog 1909–1970. Trees with stem diameter > 10 cm at breast height included. The figures for 1909 are approximate.

first half of the 19th century. Most of the big beeches, *Fagus sylvatica*, ashes *Fraxinus excelsior*, and elms, *Ulmus glabra*, date from that period, too. For ash and elm and even beech there is a younger generation growing up (Fig. 3). Since 1916 the number of stems of ash and elm together has increased by four times (Fig. 4). Elm is strongly dominant in the smaller diameter classes while in fact a decrease of nearly 40 % in the number of ashes has been observed since 1935. Even for beech there has been an increase but much smaller and only during the first half of the period.

The total stem volume of the trees (Fig. 4) has doubled itself since 1916. All species are included in this increase but it is far greater for elm and ash than for oak and beech owing to the age structure and the increasing stem number. Elm and ash are also becoming several metres taller than the oaks (Fig. 5) in spite of the fact that they are younger. This development may have been influenced by the earlier cuttings as mainly low and less robust oaks were left.

The vegetation analyses show the same trends. In 1935 a high tree layer was lacking in 24 % of the sampling plots, in 1970 only in 7 %. The number of plots covered with either ash, elm or beech remained virtually constant (49 % and 56 % respectively) while those with oak alone decreased from 23 % to 16 %. Mixed stands increased from 7 % to 15 %.

The low tree and shrub layers

A low tree layer (A₂) occurs in about half of the sampling plots investigated while a B₁ shrub layer appears in most of them (Table 2). For both layers the occurrence seems to have become more widespread since 1935, but this does not prove that their total cover has increased, too, For the B₁ shrub layer it seems more probable that there has been a reduction of about 50 %. The low shrub layer (B₂) is less frequent than the others. It seems as if its total cover has been increasing during the last years.

As among the high trees the A₂-layer is dominated by *Ulmus glabra. Quercus* is lacking. *Fagus* and even *Fraxinus* are very poorly represented. This is the same development as among the high trees which showed an increasing dominance of elm.

In the B₁-layer hazel, *Corylus avellana*, and young elm have been the most frequent species during the whole period. In 1935 hazel was clearly dominant. The figures for 1969 and 1975 indicate a reduction in cover of about

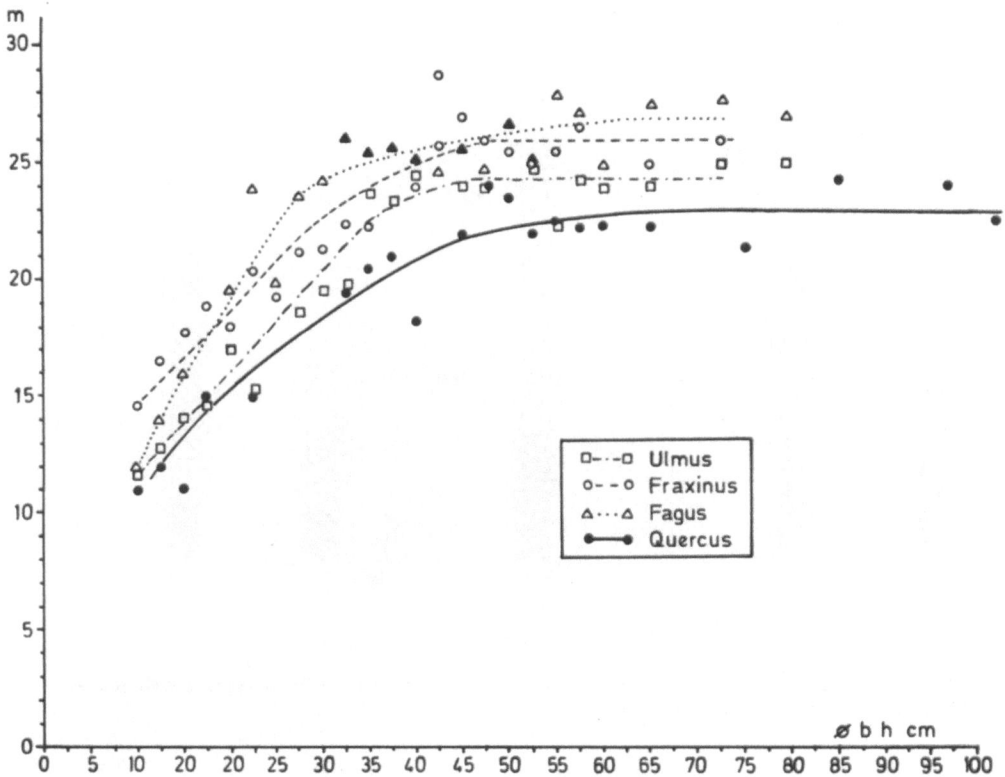

Fig. 5. Relationship between mean height and stem diameter at breast height for the main tree species in Dalby Söderskog according to the measurements in 1970.

50% since then. Among the other species hawthorn, *Crataegus laevigata*, seems to be disappearing from the B_1-layer while it remains in the B_2-layer. During the same period the frequency for *Ulmus* has doubled itself in the B_1-layer and increased still more in the B_2-layer, especially during the last years. The development for oak and ash in the B_1-layer is the same as in the higher layers while there is a weak tendency towards an increase in the number of beeches. However, because of the methodological approach all these figures indicate tendencies rather than give exact information.

The field layer

A total of 62 species has been noted in the small square analyses of the field layer during the period 1935–1976 (Table 3). Among them 12 are true vernal species in the sense that they occur only in the analyses from April and May. Eight are saplings of woody species, which also occur in one or more of the upper layers.

With regard not only to Dalby Söderskog but to 'all

plant communities of the eutrophic, deciduous forests of southern Scandinavia rich in herbs and grasses, but where dwarf scrubs are lacking', Lindquist (1938, p. 211, translation) made a first division of the deciduous forest vegetation. This was done according to the presence of efficiently dispersed, common species such as *Anemone nemorosa*, *Ranunculus ficaria* and *Oxalis acetosella*. Three main, one-layer (synusiae) plant communities, *unions*, were also distinguished in Dalby Söderskog in 1935 (Table 4). They were separated by the characteristic combination of species in the sampling plots (Trass & Malmer 1973, p. 550). Each such union includes a number of plant communities of lower rank, societies, distinguished by dominating as well as differential species. In all 23 such societies were separated, most of them within the *Anemone nemorosa-Ranunculus ficaria-union*. This union was and is still today the most widespread, covering nearly 90% of the total area of the wood. Of the 74 small squares analysed in 1935 only two were regarded as belonging to the *Anemone nemorosa-Oxalis acetosella-union* and five to the *Ranunculus ficaria-union*.

60

Table 2. The low tree and shrub layers. Frequency and as exponent characteristic degree of cover when higher than 1 according to the vegetation analyses (n = 74)

Year	1935	1969	1975
Low tree (A$_2$, 8-15 m)			
Fagus sylvatica	7^5	12^3	9^4
Fraxinus excelsior	18^5	5^2	-
Ulmus glabra	16^5	34^4	51^4
Lacking	59%	52%	43%
High shrub (B$_1$, 2-8 m)			
Acer platanoides	-	3	5^2
Corylus avellana	42^5	36^4	31^4
Crataegus spp	8^4	11^2	-
Euonymus europaeus	1^3	-	3^2
Fagus sylvatica	10^5	22^2	23^4
Fraxinus excelsior	12^4	1^2	4^3
Quercus robur	1	-	-
Ulmus glabra	34^4	64^3	68^4
Lacking	27%	16%	12%
Low shrub (B$_2$, 0.8-2 m)			
Corylus avellana	-	8^1	11^2
Crataegus spp.	22^3	23	24^2
Fagus sylvatica	4^3	4	8^3
Fraxinus excelsior	1^3	5	12^2
Lonicera xylosteum	-	3	3^3
Prunus padus	-	-	3^2
Ulmus glabra	11^3	14^2	46^3
Lacking	69%	61%	41%

Using the plant sociological principles commonly applied today, Lindgren (1971, p. 27) proposed a somewhat different classification of the field layer plant communities (Table 4). The *Filipendula ulmaria-community* well resembles the *Ranunculus ficaria-union* according to Lindquist (op. cit.) while the *Lamium galeobdolon-community* is a broader unit than the *Anemone nemorosa-Oxalis acetosella-union*. The *Mercurialis perennis-com-*

Table 3. Species recorded in the field layer in the vegetation analyses of the small squares. Figures indicate frequency (n = 74) and as exponent the characteristic degree of cover when higher than 1

Year	1935	1969/70	1975/76
Vernal species			
Adoxa moschatellina	28^4	4^4	4^3
Anemone nemorosa	94^4	91^3	83^3
- ranunculoides	68^3	90^3	91^4
Caltha palustris	-^{3}	1	-
Chrysosplenium alternifolium	25^2	8^3	1^3
Corydalis bulbosa	17^2	25^3	20^3
- intermedia	13	1	-
Gagea lutea	29^3	6	5
- minima	24^3	-	-
- spathacea	40	17	5
Ranunculus auricomus	18^3	?^{3}	8^4
- ficaria	77^3	66^3	68^4
Summer species			
Corylus avellana (j)	2	1	1
Crataegus laevigata (j)	10	2	1
Fagus sylvatica (j)	6	5^2	2
Fraxinus excelsior (j)	9^2	13^2	28
Lonicera xylosteum	2^2	-	-
Quercus robur (j)	4	-	-
Ulmus glabra (j)	27	5	5
Viburnum opulus (j)	1	-	-
Actaea spicata	2^2	-^{2}	-^{3}
Aegopodium podagraria	20^3	32^2	25^3
Angelica sylvestris	1^3	-	-
Anthriscus sylvestris	5^2	-	-
Campanula latifolia	2^2	2	1
- trachelium	1	-^{2}	-
Circaea lutetiana	10	14^2	9^3
Crepis paludosa	.2	-	-
Epilobium montanum	2^3	2^4	-^{4}
Filipendula ulmaria	18^3	8^4	5^4
Fragaria vesca	2^3	-	-
Galium aparine	1^3	-	-^{4}
Geranium robertianum	5^3	4^2	1^4
Geum rivale	29^2	10^2	4^3
- urbanum	32^2	13^2	6^3
Lamium galeobdolon	27^3	13^2	6^2
Lathyrus vernus	1^4	-^{4}	-^{5}
Mercurialis perennis	81^5	93^4	91^3
Oxalis acetosella	5^3	1	1^3
Paris quadrifolia	2^2	-	-
Polygonatum multiflorum	-^{2}	-	2
Pulmonaria officinalis	9^5	3	8
Ranunculus repens	1^5	-	-
Rumex sanguineus	4	2	-
Sanicula europaea	-^{3}	1	-^{2}
Stachys sylvatica	9^3	1	2^2
Stellaria nemorum	2^3	1^2	-^{4}
Urtica dioica	5^2	5^2	5^4
Veronica chamaedrys	6	-	-
- hederifolia	4	-	1
- montana	5	-	4
Vicia sepium	6	-	1
Viola riviniana	12	6	1
Taraxacum (coll)	1	1	-
Carex remota	1^2	-	-
- sylvatica	5	1	1
Dactylis glomerata	5	-	2
Equisetum arvense	-	-	1
- sylvaticum	1	-	-
Festuca gigantea	1	-	-
Poa nemoralis	5	-	-
- trivialis	4	-	-
Total species number	59	35	35

Table 4. Vegetation units distinguished in the field layer vegetation of Dalby Söderskog

Lindquist 1938	Lindgren 1971	Present paper
Anemone nemorosa--Oxalis acetosella-union	Lamium galeobdolon-com.	Group A
Anemone nemorosa--Ranunculus ficaria-union	Mercurialis perennis-com.	
	Typical-var	Group B
	Corydalis-var	Group C
	Aegopodium podagraria-var	
Ranunculus ficaria-union	Filipendula ulmaria-com.	Group D

munity is, like the *Anemone nemorosa-Ranunculus ficaria-union* much more widespread than the other two. Each of the three variants distinguished in this community corresponds to a group of societies according to Lindquist (1938, Table 26). (Cf. also the sociology of the South Swedish beech forest vegetation in Lindgren 1970).

For the problems dealt with in this paper another classification of the vegetation analyses from the field layer has been preferred (Table 4 and 5). Using the results of the analyses from 1935 the sampling plots have been divided into four groups. The development of the vegetation within each one of these groups is then followed up through the analyses of 1969/70 and 1975/76.

Group A comprises all sampling plots characterized by *Anemone nemorosa* and *Mercurialis perennis* and where *Lamium galeobdolon* and/or *Oxalis acetosella* were noted in 1935. This group corresponds mainly to the *Lamium galeobdolon-community*, according to Lindgren (1971, Appendix 9), but includes a few more sampling plots.

Group B comprises all sampling plots characterized by *Anemone nemorosa* and *Mercurialis perennis* but where neither *Lamium galeobdolon* and *Oxalis acetosella*, nor *Adoxa moschatellina* occurred in 1935. This group roughly corresponds to the typical variant of the *Mercurialis perennis-community* according to Lindgren (op. cit.).

Group C comprises all those sampling plots characterized by *Anemone nemorosa* and *Mercurialis perennis* where *Lamium galeobdolon* was lacking in 1935 but where *Adoxa moschatellina* occurred together with some other of the low frequent vernal herbs. (Cf. the *Corydalis cava-variant* of the *Mercurialis perennis-community* according to Lindgren (op. cit.).

Group D comprises those six sampling plots where *Filipendula ulmaria* dominated the summer aspect in 1935 and *Mercurialis perennis* was lacking. This very uniform group representing a tall-herb meadow vegetation in the damp depressions is clearly distinguished from the rest of the 68 sampling plots. This group corresponds to the *Filipendula ulmaria-community* separated by Lindgren (1971) but includes more sampling plots than the *Ranunculus ficaria-union* described by Lindquist (1938).

The distinct difference in vegetation between the groups A–C on the one hand and group D on the other has persisted throughout the investigation period (Table 5). The increase in frequency and cover for *Anemone ranunculoides* and *Mercurialis perennis* in group D somewhat diminishes this difference. In this special case, however, several of the sampling plots are situated very close to the border of the tall-herb areas. Therefore even small difference in their positions may give rise to distinct differences in species composition. The confidence of the vegetational differences found in this group between the investigations is therefore less than in the other ones. The number of sampling plots is smaller, too.

Doubtless the differences in the vegetation between the groups A, B and C have diminished since 1935. This development towards a more uniform vegetation all over the wood except in the damp areas is due partly to an increasing amount of *Mercurialis perennis*, partly to decreasing frequency and cover for those species which characterize groups A and C, e.g. *Lamium galeobdolon*, *Adoxa moschatellina*, and *Chrysosplenium alternifolium*.

During the two last investigations the total number of

62

Table 5. Frequency and as exponent the characteristic degree of cover when other than 1 for the most important field layer species in the groups of small squares according to the vegetation analyses 1935-1975/76

Group of sampling plots	Group A (n=21)			Group B (n=32)			Group C (n=15)			Group D (n=6)		
Investigation period	1935	1969/70	1975/76	1935	1969/70	1975/76	1935	1969/70	1975/76	1935	1969/70	1975/76
Vernal species												
Anemone nemorosa	100^{4+}	90^{4+}	90^{4-}	97^{4+}	100^{4-}	91^{3-}	100^{5-}	100^{4-}	87^{3+}	50	33^{2+}	17^{2+}
- ranunculoides	95^{3+}	86^{3-}	95^{4-}	91^{3+}	100^{4-}	97^{4+}	7	10^{3+}	93^{4+}	17	33^{3+}	50^{4-}
Corydalis bulbosa	10^{3-}	24	10^{3-}	31^{2-}	19^{3-}	13	7	53^{3+}	60^{3-}	-	-	-
Gagea lutea	38	14	5	38	6^{2-}	9	13	-	-	-	-	-
- spathacea	38	14	10	50	25	3	33	13	7	-	-	-
Ranunculus auricomus	33^{2-}	?	19^{3+}	16^{3+}	?	19^{4-}	13	$?^{2+}$	7^{3+}	-	17^{5-}	-
- ficaria	90^{3+}	67^{3+}	62^{3+}	91^{3+}	72^{2-}	75	20	47^{2+}	60^{3+}	100^{5+}	83^{5-}	83^{5-}
Adoxa moschatellina	24	5	10^{3-}	3	6	-	100^{4+}	-	7^{3+}	-	-	-
Chrysosplenium alternifolium	10	10	5	-	6	-	87^{3+}	-	-	67^{2+}	33	-
Corydalis intermedia	5	-	-	-	3	-	60^{3-}	-	-	-	-	-
Gagea minima	14	-	-	9	-	-	80^{3+}	-	-	-	-	-
Summer species												
Aegopodium podagraria	24^{3-}	14^{2+}	19^{3+}	22^{2+}	41	31^{3+}	13^{2-}	40^{2+}	33^{3+}	17	33^{3-}	-
Geum urbanum	38^{3-}	10^{4-}	5^{3+}	31^{4-}	16^{4+}	9^{2-}	40^{3-}	20^{4+}	7^{3+}	-	-	-
Mercurialis perennis	71^{3-}	90^{4-}	95^{5-}	97^{4-}	100^{4+}	97^{5-}	93^{4+}	100^{4+}	100^{5+}	-	50	33^{5+}
Pulmonaria officinalis	19	10	14	9^{2-}	-	3	-	-	-	-	-	-
Vicia sepium	19^{4+}	-	-	-	-	3	-	-	-	-	-	-
Stachys sylvatica	5^{4+}	-	5	6	3	-	20	-	-	17^{5+}	-	17^{3+}
Circaea lutetiana	10^{2-}	14^{2+}	10^{2+}	13^{2-}	9^{2-}	6^{3+}	13	13^{2-}	7	-	50^{2-}	33^{4+}
Lamium galeobdolon	95^{3+}	38^{2+}	19^{2+}	-	3	32^{2+}	-	-	-	-	17	-
Oxalis acetosella	19^{3}	-	-	-	-	33^{3+}	-	-	-	-	17	-
Viola riviniana	29	14	5	9	3	-	-	-	-	-	-	-
Filipendula ulmaria	10	5	5^{2+} 5^{4+}	3^{2+}	-	-	33^{2-}	-	-	100^{4+}	83^{5-}	50^{5-}
Geranium robertianum	-	$-^{2+}$	-	6^{2+}	3	-	-	-	-	33^{2+}	33^{3-}	-
Geum rivale	29	10^{2+}	5	25^{2+}	6	-	20	-	-	83^{4+}	67^{2-}	33^{3+}
Urtica dioica	-	10^{2+}	5	3	-	3	-	-	-	50^{3-}	33^{2+}	33^{5-}
Fagus sylvatica (j.)	14	19^{2-}	5^{2-}	-	-	3	7	-	-	17	-	-
Fraxinus excelsior (j.)	19	24^{2-}	43^{2-}	6	9	19	7	7^{4+}	27^{2+}	-	17	33
Ulmus glabra (j.)	33	5	-	22	6	6	33	-	13	17	17	-

Table 6. Number of field layer species in the small squares 1935 - 1975/76

	Group A (n=21)			Group B (n=32)			Group C (n=15)			Group D (n=6)		
	1935	1969/70	1975/76	1935	1969/70	1975/76	1935	1969/70	1975/76	1935	1969/70	1975/76
Total number of species	45	26	28	37	24	23	27	11	13	20	18	12
Among them vernal species	11	8	9	9	9	7	11	5	6	5	5	3
Mean number of species per m^2	10.7	6.0	5.9	7.5	5.6	5.1	8.4	5.0	5.1	7.3	6.7	4.2
Among them vernal species	4.7	3.1	3.1	4.3	3.4	3.0	5.2	3.1	3.1	2.5	2.2	1.5

species noted in the analyses was only 59% of that in 1935 (Table 3). A corresponding decrease is also observed for each one of the four groups of squares, both in the total number of species recorded and in the mean number per square (Table 6). Throughout the investigation period the total number of species has been highest in group A and lowest in groups C and D, even if group D is not completely comparable because of the small number of squares. These differences, however, refer to the summer aspect rather than to the spring aspect. Since 1935 the relative drop in species numbers has been somewhat greater in group C than in the others, mainly because of losses among the vernal species characterizing the group in 1935, e.g. *Adoxa moschatellina*, *Corydalis intermedia* and the *Gagea* spp (Table 5).

Anemone ranunculoides has increased in both frequency and cover, especially in the squares of group C where the number of vernal species was greatest in 1935 (Fig. 6). In this group of squares (C) *Corydalis bulbosa* and *Ranunculus ficaria* have also developed parallel to *Anemone ranunculoides*. Otherwise these two species regarded by Lindquist (1938) as especially characteristic of the field layer

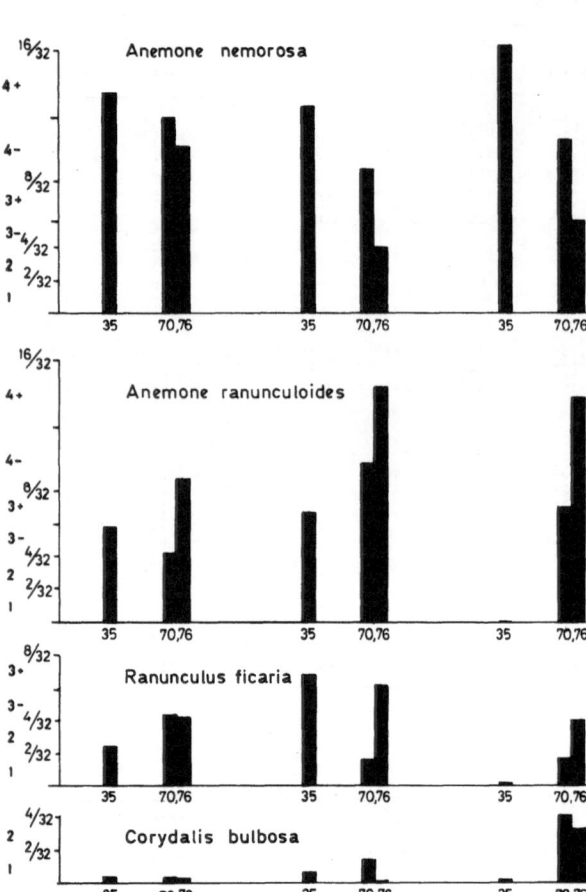

Fig. 6. Total cover of some species in the vernal aspect at the investigations in 1935, 1970 and 1976. – The calculations have been made separately for each group of sampling plots through multiplying the average of the middles for the cover classes with the frequency in per cent divided by 100. Cover (in parts of 32) and degree of cover indicated (Hult-Sernander-Du Rietz-scale, cf. Table 1).

community of the wood have remained fairly constant. For *Anemone nemorosa* the development is the reverse of that of *A. ranunculoides*. In 1935 *A. nemorosa* dominated over *A. ranunculoides* in most of the wood, while *A. ranunculoides* has now taken over in many areas (Tables 3 and 5, Fig. 6).

In the summer aspect *Mercurialis perennis* has increased considerably since 1935 (Fig. 7, Table 5). Today this species dominates all over the wood except in the wettest parts. Some increase is also observed for *Aegopodium podagraria*. In 1935 *Lamium galeobdolon* and *Geum urbanum* were both significant species but today they are of little impor-

tance (Fig. 7). For *Viola riviniana* (incl. *V. reichenbachiana*) the same development is observed (Table 3). In spite of the decreasing frequency for *Lamium galeobdolon* the species was in 1969 and 1976 noted in squares referred to group B in 1935. Even *Oxalis acetosella* was in 1969 recorded in this group and in 1975 in one of the squares of group D (Table 5). For *Oxalis* stumps and decaying wood provide suitable spots for an occasional colonization. A dispersal of *Lamium galeobdolon* to new spots may have occurred, too. However, it seems equally probable that such scattered records in new squares are only sampling errors due to small differences in the positions of the sampling plots.

Filipendula ulmaria and *Geum rivale* are two species in the summer aspect which have decreased since 1935. Their occurrence is now more restricted to the wettest areas (the squares of group D) than earlier. Even in such areas they seem to be of less importance. They seem to have been partly replaced by *Circaea lutetiana* and *Urtica dioica*.

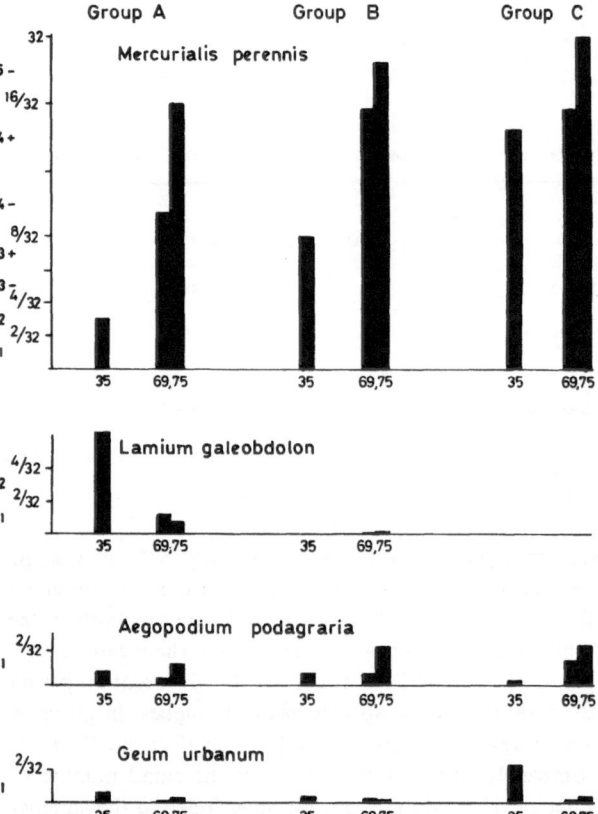

Fig. 7. Total cover of some species in the summer aspect at the investigations in 1935, 1969 and 1975. – Explanation, see Fig. 6.

Tree seedlings have been observed in all groups of squares, but are commonest in group A (Table 5). Corresponding to the development in the upper layers seedlings of *Fraxinus* and *Ulmus* are much more frequent than those of *Fagus* and *Quercus* (Table 3). In 1935 the frequency of *Ulmus* seedlings far exceeded the others. During the last two investigations *Fraxinus* seedlings have been the most widespread.

Bottom layer

The analyses presented by Lindquist (1938) indicate a well-developed bottom layer of bryophytes typical of woods like Dalby Söderskog. Originally a separate investigation of the whole bryophyte vegetation was planned for comparison with the work by Waldheim (1944). Therefore the bryophytes were not included in all these investigations. However, the observations made during the analyses clearly indicate that both the number of species and the total cover of the bottom layer must now be much less than it seems to have been in 1935

Discussion

Methodological remarks

So far the present material from Dalby Söderskog has been treated in a traditional manner only. Even if no treatments of vegetation analyses could improve on them, numerical methods should provide more information in this case. The classification of the groups can be made more objectively. Ordination techniques could help to distinguish vegetational gradients and to interpret them ecologically. In fact, there is not much new in this to vegetation science, but the amount of calculation constitutes a severe obstacle without modern date technology.

In a study such as this the presence of a species always implies a clearly defined vegetation structure. The importance of quantitative information such as cover is more difficult to evaluate and also more dependent on seasonal variation and annual climatic differences. Vegetational succession also includes both the colonization and the disappearance of species from an area. These processes are always delayed in relation to the environmental changes they depend on, which entails problems in the ecological evaluation of both the presence and the absence of a species.

Plant succession in Dalby Söderskog

The number of species in the wood is probably near that which can be expected. An examination of the species lists from 1969 and 1975/76 shows that most of the species noted are those which ecologically should be regarded as belonging to a deciduous woodland vegetation. Any further marked disappearance of plant species seems therefore improbable. This also indicates that somewhat stabler conditions than before may now have been reached in the wood.

Lindquist (1938, p. 265) concluded that in the future *Ulmus* would be the dominating tree species in Dalby Söderskog and that *Fraxinus* would be of rather scattered occurrence. So far the development of the woody plants since 1935 has proceeded in this direction. However, he also considered that *Corylus* could exist independent of the cover from the higher trees. The decrease in cover and of frequency for this species contradicts this and perhaps even his opinion that closed *Corylus* woodland would represent a final and stable stage in the succession of the South Scandinavian forests on clay soils (op. cit., p. 266).

For the field layer Lindquist (1938, p. 273) states that not only meadow and arable field species would disappear but also various species belonging to the normal wood flora. He forecasted that vernal species like *Corydalis bulbosa*, *Anemone nemorosa*, *A. ranunculoides*, and *Gagea lutea* would increase together with dominant *Mercurialis perennis* and *Aegopodium podagraria*. The main trends in the development of the vegetation since 1935 are in accordance with these assumptions. However, the expected luxuriance of the vernal species in the meaning of an increasing diversity has not been realized. Instead a number of vernal species seem to be disappearing. Only *Corydalis bulbosa* and *Anemone ranunculoides* have apparently increased.

The figures for stem volume indicate that the plant biomass in the tree layer and thus also in the whole of the wood today may be up to double that of 50 years ago. The reduction of the shrub layer seems to be much too small to compensate for that. All information from 1935 indicates that the field layer at that time almost completely covered the ground. There is no strong indication of a decrease or increase in total cover since then. Therefore neither for the field layer nor for the bottom layer can an increase in plant biomass like that in the tree layer be approved of.

The increase in stem number and plant biomass for

the trees since 1935 must certainly have given rise to a much more closed tree canopy. This does not only imply an increased shading of the lower layers but also a somewhat higher humidity and somewhat lower spring and summer temperatures at the soil surface today than existed 40 years ago. Litter fall must have increased and the relative parts of the different species in its composition must have changed during this period, too. The reduction among the shrubs, which in itself is a result of the closer tree canopy will perhaps to some extent counteract the effects of the tree layer on the field and bottom layers. In spite of that the driving force behind the changes observed in these two layers must be the successions going on in the tree layer.

Both in the tree, shrub and field layers the remnants in the vegetation from the former periods of grazing and cuttings for forestry purposes are disappearing. In the tree layer the oaks represent one such element which will continue to survive for a considerable time, perhaps longer than most of the other species of that element. Even if the number of species in the field layer remains fairly constant in the future, this does not imply that the final equilibrium has been reached yet between forest species such as *Mercurialis perennis*, *Lamium galeobdolon* and the *Anemone* spp. Further, without stable conditions in the upper layers the conditions in the lower layers will never become stabilized. It is also questionable whether stable conditions in the sense of a permanent equilibrium between the species will appear at any time.

Even if no cuttings or similar measures are undertaken within the wood the development there may be influenced by activities outside the area. For instance it seems as if the damp areas in the wood have been smaller since 1935. This may be an effect of increased interception from the tree canopies but it seems more probable that it is caused by a general lowering of the ground water table in the whole area resulting from a more intense draining of the surrounding agricultural land. Airborne pollutants and wind-transported material, esp. fertilizers, from the surrounding fields may also influence the wood. The expansion of nitrophilous species such as *Aegopodium podagraria* and *Urtica dioica* may be favoured in that way.

Theoretical remarks

The vegetational succession in Dalby Söderskog is distinctly directed towards what is looked upon as the potential natural vegetation of the area. This should also imply the development of a stabler vegetation type with a virtually permanent equilibrium between the plant species and thus also an ecosystem with stabler conditions, at least if no natural catastrophes such as storm fellings occur.

This development is accompanied by a considerable increase in plant biomass. However, at the same time the vegetation has become more uniform and with fewer species included both in the area as a whole and when calculated per unit area. In this case we can say with van Leeuwen (1966) that there is a temporal concentration and a disappearance of internal boundary conditions within the area reflecting a development towards less variation in the environmental conditions. However, the succession in the community does not follow a series 'from a convergent pioneer stage poor in species and characterized by a high degree of environmental instability to a divergent terminal community rich in species with a degree of stability as high as allowed by the local situation' (op. cit. p. 42). In this case and probably in many others a development of a vascular plant community regarded as stable is not combined with an increase in number of species in an area but with the reverse. For woods such as this one it may be an effect of the development of an even-aged stand, but, for instance, even the natural succession in peatlands ends up with a bog vegetation poorer in species than many of the preceding stages.

In this particular case the possibilities of new species reaching the area are slight. The surrounding farmlands form an obstacle for all sorts of seed dispersal. There are no other sites with a *Mercurialis*-vegetation in the neighbourhood (the nearest is at a distance of 25 km). The plant community (*Ulmo-Fraxinetum*, Sjögren n.p. in Kielland-Lund 1971 and Ekstam & Sjögren 1973) is near its north-eastern distribution limit in Skåne and even isolated from its main distribution area in Central Europe. However, these factors could only partly explain the decreasing species number of the community.

Conservation

The original aim of the national park Dalby Söderskog was to preserve the vegetation as it was in c. 1920. For this purpose it was a mistake to let the vegetation develop undisturbed as was done in the beginning (Lindquist 1938, p. 61). Minor annual cuttings combined with some grazing would probably have maintained the vegetation in this state, the richness of 'interior boundaries' and the variation in the environmental conditions.

Today there is hardly any choice for the management of the wood as succession has advanced too far. There is not even a theoretical possibility of reconstructing the vegetation of the period around 1920. The only way is to let the natural succession continue in the future. In parts of the area the development may be slowed down through measures to conserve the present oak generation which also favours the spring flora. This is also in accordance with the measures that are desirable to carry out with regard to the importance of this woodland area for recreation and landscape scenery in this, the most densely populated part of Sweden. This guiding principle of working along with the natural vegetational succession has also been followed in plans for the future management of the national park which have now been drawn up.

Summary

The study deals with a small South Swedish deciduous wood (*Ulmo-Fraxinetum*, Sjögren n.p. 1971), a national park which has been left almost undisturbed during the last 60 years. Since 1925 the number of vascular plants in the area has diminished by 42%. In the tree layer *Ulmus* and *Fraxinus* have increased since 1918 while *Fagus* has remained fairly constant and *Quercus*, the former dominant, has become much less important. In the shrub layer young *Ulmus* are predominant, while the former dominants *Corylus* and *Crataegus*, have diminished markedly. Vegetation analyses of the field layer laid out in a fixed grid system in the years 1935, 1969/70 and 1975/76 show a decrease in both the total number of species recorded and in the number of species per sq. m. *Anemone ranunculoides*, *Aegopodium podagraria* and *Mercurialis perennis* are species which have expanded while e.g. *Anemone nemorosa*, *Lamium galeobdolon* and *Geum urbanum* have diminished. Increased shading from the closer tree canopies, increased litter fall, and lower temperature at the soil surface are changes in environmental factors which probably have come about sine 1935 and may be especially important for the field and bottom layer species. The results are discussed in relation to the stability of an ecosystem and to the management of nature reserves.

References

Ekstam, U. & E. Sjögren. 1973. Studies on past and present changes in deciduous forest vegetation on Öland. Zoon, Suppl. 1: 123–135.

Kielland-Lund, J. 1971. A classification of Scandinavian forest vegetation for mapping purposes. IBP i Norden 7: 13–43.

Lid, J. 1974. Norsk og svensk flora. Det norske samlaget, Oslo, 808 pp. (In Norwegian).

Lindgren, L. 1970. Beech forest vegetation in Sweden – a survey. Botaniska notiser 123: 401–424.

Lindgren, L. 1971. Skötsel av lövskogsområden. Vegetationsförändringar i Dalby Söderskog. Meddelanden från forskargruppen för skötsel av naturreservat 11, 43 pp. + 12 app. (In Swedish).

Leeuwen, C. G. van. 1966. A relation theoretical approach to pattern and process in vegetation. Wentia 15: 25–46.

Lindquist, B. 1938. Dalby Söderskog, en skånsk lövskog i forntid och nutid. Acta Phytogeogr. Suec. 10, 273 pp. (In Swedish with summary in German).

Londo, G. 1976. The decimal scale for relevés of permanent quadrats. Vegetatio 33: 61–64.

Malmer, N. 1962. Studies on mire vegetation in the archaean area of southwestern Götaland (South Sweden). Opera Bot. 7: 1, 322 pp.

Trass, H. & N. Malmer. 1973. North European approach to vegetation science. In: R.H. Whittaker (ed.), Ordination and classification of communities. Handbook of vegetation science 5, 737 pp. Junk, The Hague.

Waldheim, S. 1944. Mossvegetationen i Dalby Söderskogs nationalpark. Kungl Svenska Vetenskapsakademiens avhandlingar i naturskyddsärenden 4, 142 pp. (In Swedish).

UNTERSUCHUNGEN ZUR ENTWICKLUNG VON GEHÖLZ-AUFFORSTUNGEN AUF BERGBAUKIPPEN IN DER DÜBENER HEIDE (DDR)*, **

RUDOLF HUNDT***

Sektion Biowissenschaften, Wissenschaftsbereich Biologiemethodik, Forschungsgruppe Umwelterziehung, Martin-Luther-Universität, Halle-Wittenburg, DDR

Keywords: Aufforstung, Landeskultur, Pflanzengeographie, Pflanzensoziologie, Syngenese

Aufgabenstellung

Im fruchtbaren Lößlehmgebiet zwischen Borna und Zeitz, im Geiseltal bei Mücheln sowie im Altpleistozän der nördlichen Dübener Heide um Bitterfeld, Gräfenhainichen und Bergwitz, aber auch in der Lausitz wird das Bild der intensiv genutzten Kulturlandschaft in starkem Maße durch den Braunkohlentagebau und seine Folgeerscheinungen geprägt. Neben sich noch in Betrieb befindlichen ausgedehnten Tagebauen, vereinzelten ausgekohlten Restlöchern und gehölzbestandenen Hochkippen der dreißiger Jahre trifft man allenthalben auf Maßnahmen zur Einbindung der ausgekohlten Braunkohlentagebaue in die Kulturlandschaft und zur Rückführung ausgedehnter Bodenflächen in die gesellschaftliche Nutzung.

Der anfallende Abraum wird in den ausgekohlten Teilen des Tagebaues sofort wieder verkippt, eingeebnet und mit Mutterboden versehen, so daß unmittelbar nach dem Auflassen des Tagebaus mit der Wiederurbarmachung und Rekultivierung begonnen werden kann. In Gebieten mit von Natur aus fruchtbaren Böden, wie im Lößlehmgebiet bei Borna, Zeitz, Mücheln und westlich von Bitterfeld steht bei der Rekultivierung die Rückführung der Böden in landwirtschaftliche Produktion im Vordergrund, während in der altpleistozänen Dübener Heide und in der Lausitz mit ihren vorherrschenden sandigen Böden die forstwirt-

schaftliche Nutzung der rekultivierten Kippenflächen von großer Bedeutung ist (Tabelle 1).

Die durch die Braunkohlenentnahme entstehenden Restlöcher werden in vielen Fällen mit ihrer Umgebung zu Erholungs- und Urlaubsgebieten entwickelt, wofür es im Lausitzer Knappensee, im Bergwitzsee und in dem zur Zeit neu entstehenden großen See im ehemaligen Tagebau bei Pouch am Westrand der Dübener Heide recht gute Beispiele gibt. Andere Tagebaurestlöcher werden von der Wasserwirtschaft genutzt oder in den großen Ballungsgebieten um Halle, Merseburg und Bitterfeld für eine geordnete Mülldeponie verwendet.

In den pleistozänen Landschaften der Dübener Heide und der Lausitz steht bei der Wiederurbarmachung und der Kippenrekultivierung die Aufforstung im Vordergrund. Durch Gehölzanpflanzungen von *Populus nigra* bzw. *canadensis* und *Robinia pseudacacia* auf den mit Mutterboden überdeckten vegetationslosen tertiären Erdmassen entstehen monokulturartige Baumbestände.

In dieser Untersuchung soll der Frage nachgegangen werden, in welcher Weise sich die syngenetische Entwicklung dieser Gehölzanpflanzungen vollzieht. Zu diesem Zweck wurden folgende Teilaufgaben bearbeitet:
– Untersuchung der Bestandsentwicklung der Kippenaufforstungen durch Analyse verschieden alter Gehölzbestände mit vergleichbaren Aufforstungs- und Standortsbedingungen

* Contribution to the Symposium on Plant species and plant communities, held at Nijmegen, 11–12 November 1976, on the occasion of the 60th birthday of Professor Victor Westhoff.
** Nomenklatur der Pflanzenarten nach Rothmaler (1976).
*** In der Untersuchung wurden Ergebnisse, darunter Vegetationstabellen der zu erwähnenden Vegetationstypen der Lehrer-Diplomarbeit von Herrn Frenzel (1976) mit einbezogen, dem wir an dieser Stelle für seine Mitarbeit recht herzlich danken möchten.

Tabelle 1. Durch Braunkohlenbergbau devastierte bzw. entzogene und durch Wiederurbarmachung und Rekultivierung zurückgegebene Bodenfläche (vgl. Waldmann 1975)

Nutzungsformen	Devastierte bzw. entzogene Flächen	Rückgabeflächen
Landwirtschaftliche Nutzfläche	35.547 ha	10.000 ha
Forstwirtschaftliche Nutzflächen	32.837 ha	20.112 ha
Sonstige Nutzungsformen	5.218 ha	5.337 ha

69

– Untersuchung der Entwicklung ökologisch-standortkundlicher Bedingungen in den verschieden alten Kippenaufforstungen

– Analyse der pflanzengeographischen Beziehungen der untersuchten Bestände

– Herausarbeitung landeskultureller Aspekte der Kippenaufforstungen in den Gebieten des Braunkohlenbergbaus.

Methodik

Die vegetationskundlichen Untersuchungen erfolgten nach der Methode von Braun-Blanquet in verschieden alten Gehölzanpflanzungen, wobei die Amplitude von jungen, nur wenige Jahre alten Monokulturen bis zu recht dichten etwa sechzigjährigen waldähnlichen Beständen reicht. Um die Wirkung unterschiedlicher Baumarten auf die Entwicklung der Strauch- und Baumschicht möglichst konstant zu halten, wurden in die Untersuchung nur Bestände mit *Populus nigra* bzw. *canadensis* und *Robinia pseudacacia* einbezogen.

Zur ökologischen Charakterisierung der untersuchten Bestände wurden verschiedene Bodenfaktoren, wie der Humusgehalt (Glühverlust), der Kalkgehalt (Passongerät), der pH-Wert (Czensny-Indikator), der K_2O-Gehalt und der P_2O_5-Gehalt (beide nach der Laktatmethode) ermittelt und die Faktorenzahlen für das Licht, die Temperatur, die Kontinentalität, die Feuchtigkeit, die Bodenreaktion und den Stickstoff nach der Methode von Ellenberg (1974) berechnet, die eine feinere Differenzierung gegenüber den früheren Verfahren ermöglicht. Der Herausarbeitung ökologischer Beziehungen zwischen den untersuchten Beständen diente auch die Berechnung von Spektren der Lebensformen, von Gruppen mit unterschiedlicher Blatausdauer und unterschiedlichem umweltbedingtem anatomisch-morphologischem Bau, wobei vor allem der Wasserfaktor im Zusammenwirken mit dem Temperaturfaktor in den Beständen eine determinierende Rolle spielen dürfte. Bei der Bestimmung der Gruppenzugehörigkeit der Pflanzenarten und der Berechnung ihres Anteiles an der Gesamtzusammensetzung folgten wir den Verfahren von Ellenberg (1974), die sich nach unserer Ansicht für eine derartige vergleichende Betrachtung syngenetischer Entwicklungsstadien sehr gut eignen.

Der Herausarbeitung der pflanzengeographischen Beziehungen legten wir die Arealdiagnosen der Pflanzenarten nach Meusel, Jäger & Weinert (1965) zugrunde. Die Gruppenbildung erfolgte auf der Grundlage der zonalen Verbreitung der Pflanzenarten insofern, als Artengruppen

mit meridional-submeridionaler bis arktischer Verbreitung, Arten mit meridional-submeridionaler bis borealer Verbreitung und Arten mit meridional-submeridionaler bis temperater Verbreitung zur Ausscheidung gelangten. Innerhalb dieser Gruppen ergibt sich eine Differenzierung der Pflanzenarten nach ihrer Kontinentalität, wobei Arten mit ozeanischer, subozeanischer, subkontinentaler und kontinentaler Verbreitung sowie Pflanzen ohne ausgesprochen ozeanisch-subozeanische bzw. kontinental-subkontinentale Verbreitung herausgestellt wurden. Um die pflanzengeographischen Beziehungen recht anschaulich darzustellen, erfolgte eine separate Darstellung nach der Ozeanität und Zonalität der Pflanzenverbreitung. Auch bei der pflanzengeographischen Auswertung wurden die Anteile der einzelnen Arten in den untersuchten Beständen nach der Methode von Ellenberg (1974) bestimmt.

Bei den Berechnungen zur pflanzensoziologischen, ökologischen und pflanzengeographischen Analyse wurden nur die Pflanzen der Feldschicht berücksichtigt, um eine direkte Beeinflussung der Spektren durch angepflanzte Gehölze zu vermeiden. Da sich *Populus nigra* und *Populus canadensis* bei der Feldarbeit nur sehr schwer unterscheiden lassen, erfolgte eine Zusammenfassung beider Arten in den Vegetationstabellen, und bei der Bezeichnung der Vegetationstypen wurde in Analogie zu Bruens et al. (1975) nur der Gattungsname *Populus* benutzt.

Das Untersuchungsgebiet

Die Untersuchungen erfolgten im Bitterfelder Raum beiderseits der Mulde am Ostrand des Altpleistozängebietes der Dübener Heide gegen das Lößlehmgebiet der Bördelandschaft des Köthener Ackerlandes (vgl. Meusel 1955). Alle untersuchten Kippen liegen im Sanderbereich des zentralen Grundmoränengebietes und Endmoränenzuges der Dübener Heide, der sich östlich von Söllichau und Gräfenhainichen in Südost-Nordwest-Richtung hinzieht. Das Sandergebiet hat im Raum von Bitterfeld-Gräfenhainichen durch den Braunkohlentagebau weitgehend seinen natürlichen Landschaftscharakter und seine naturnahe Vegetation verloren. An Stelle der armen sauren ehemals podsolartigen Böden tritt heute in weiten Gebieten aufgeschüttetes tertiäres Erdmaterial, das mehr oder weniger stark von Humuserde überdeckt ist.

Das Klima wird bestimmt durch die Lage des Untersuchungsgebietes an der Ostgrenze des mitteldeutschen Trockengebietes. Sein subkontinentaler Charakter wird geprägt durch relativ geringe Niederschläge, die nur

70

Tabelle 2. Klimamittelwerte von Bitterfeld
(50jährige Mittelwerte)

Jahresniederschläge	539	mm
Sommerniederschläge (Mai-Okt.)	325	mm
Winterniederschläge (Nov.-Apr.)	214	mm
Januarniederschläge	38	mm
Juliniederschläge	74	mm
mittlere Jahrestemperatur	9,3	°C
Januartemperatur	0,2	°C
Julitemperatur	18,8	°C
Temperaturschwankung	18,6	°C

unwesentlich über 500 mm liegen, und durch ein ausgesprochenes Sommermaximum derselben. Eine subkontinentale Tendenz weisen auch die Temperaturverhältnisse auf. Bei relativ niedrigen Januartemperaturen und hohen Julitemperaturen erreicht die mittlere Jahresschwankung der Temperatur mit 18,6 °C einen relativ hohen Wert. Bei einem Vergleich der Klimawerte mit denen von Halle und der östlich vom Untersuchungsgebiet liegenden Dübener Heide kommt deutlich die Zwischenstellung des Untersuchungsgebietes zwischen der kontinentaleren Börde-Ackerlandschaft um Halle und Köthen und der von buchenreichen Kiefernmischwäldern bestandenen altpleistozänen Dübener Heide mit einem mehr subatlantischen Klima zum Ausdruck.

Vegetationsdifferenzierung und Vegetationsentwicklung

Ursachen der Vegetationsdifferenzierung

Beträchtliche Unterschiede im Pflanzenkleid treten durch die zur Aufforstung gelangten Baumarten in Erscheinung. Dominieren auch *Populus nigra* bzw. *Populus canadensis* und *Robinia pseudacacia*, so gibt es Gehölzflächen auf den Kippen, in denen auch *Pinus silvestris* und *Betula pendula*, bedingt durch den Anbau, eine beträchtliche Rolle spielen. Besonders interessant sind Unterschiede der Anpflanzungen infolge ihres unterschiedlichen Alters. Hierbei handelt es sich nicht um unterschiedliche Entwicklungsstadien eines bestimmten Waldtypes bzw. einer Phytozönose. Es erfolgt vielmehr eine Syngenese, die bei Monokulturen von Gehölzen beginnt und in der Tendenz sich auf die Phytozönosen der potentiellen natürlichen Vegetation des Gebietes hinbewegt, ohne in der Struktur und in der floristischen Zusammensetzung diese nach einer mehr als sechzigjährigen Vegetationsentwicklung zu erreichen.

Es ist anzunehmen, daß bei der Vegetationsentwicklung,

die sich in der Ausprägung verschieden alter Vegetationstypen widerspiegelt, der Entwicklungszeitraum der Bestände eine wesentliche Bedingung darstellt. Daneben sind jedoch trotz der relativen Übereinstimmung ler Bodenverhältnisse und eines mittleren Bodenfeuchtezustandes (frisch), die durch das aufgefüllte Tertiärmaterial und den aufgebrachten Mutterboden gegeben sind, edaphische Unterschiede nicht auszuschließen. Schließlich besteht die Möglichkeit des Einflusses der zur Bestockung verwendeten Baumart auf die Entwicklung der Baum-, Strauch- und Feldschicht. Offensichtlich tritt das bei *Robinia pseudacacia* und bei *Pinus silvestris* besonders in Erscheinung. Nicht auszuschließen sind bei den ehemaligen vegetationslosen Standorten auch Wirkungen der Konkurrenz im Verlauf der Syngenese durch das zufällige Auftreten recht unterschiedlich konkurrenzstarker Pflanzenarten im gleichen Entwicklungsstadium der Vegetation auf den verschiedenen Teilarealen.

Diese angedeuteten Bedingungen für die Syngenese der aufgeforsteten Monokulturen führen letztlich dazu, daß nicht wie bei den Sukzessionen halbnatürlicher oder natürlicher Vegetationsformen mehr oder weniger gut mit Hilfe von Kenn- und Trennarten zu charakterisierende Phytozönosen einander ablösen. Für die Charakterisierung der herausgearbeiteten Vegetationstypen wurde deshalb die Kombination der soziologischen Artengruppen verwendet, die in ihrer Abwandlung recht gut die Entwicklungstendenz der Vegetation zum Ausdruck bringt.

In den Mittelpunkt der vorliegenden Untersuchung wurde die Wirkung des Zeitfaktors bei der Syngenese der aufgeforsteten Kippenvegetation gestellt. Aus diesem Grunde gelangten verschieden alte Bestände zur Analyse, in denen mit *Populus nigra* und *Robinia pseudacacia* die gleichen Baumarten aufgeforstet wurden.

Die untersuchten Vegetationstypen der aufgeforsteten Kippen

Nach der Struktur der Bestände lassen sich folgende Typen herausarbeiten, die sich in ihrer Artengruppenkombination deutlich unterscheiden und mehr oder weniger gut auch Entwicklungstypen entsprechten (siehe Frenzel 1976 für vollständige Vegetationstabellen).

Populus-Artemisia vulgaris-Typ
    Untertyp von *Tripleurospermum maritimum*
    Untertyp von *Poa nemoralis*
Populus-Taraxacum officinale-Typ
Populus-Calamagrostis epigeios-Typ
Populus-Carpinus betulus-Typ

Der Populus-Artemisia vulgaris-Typ

Die Bestände des *Populus-Artemisia vulgaris*-Typs besiedeln größere Kippenflächen bei Holzweißig unmittelbar südöstlich von Bitterfeld. Ihre Aufforstung mit Reinkulturen von *Populus* und *Robinia pseudacacia* erfolgte erst sieben Jahre vor der Vegetationsuntersuchung. Sie besitzen deshalb eine etwa 10 Meter hohe recht gleichförmige Baumschicht, bei der die Baumreihen des Pflanzverbandes noch deutlich zu erkennen sind. Die Baumschicht erreicht eine relativ lockere Deckung von 30 bis 60 %, während die Feldschicht bei völligem Fehlen einer Strauchschicht den Boden meist mit Werten zwischen 60 und 70 % deckt.

Da es sich beim Standort des Bestandstyps um frisch aufgeforstete Rohböden mit einer Mutterbodenauflage auf Tertiärmaterial handelt, nehmen Ruderalpflanzen eine dominierende Rolle in der Feldschicht ein. Eine hohe Konstanz erreichen neben *Artemisia vulgaris* vor allem *Cirsium arvense* und *Agropyron repens*. Daneben fällt besonders *Calamagrostis epigeios* mit mittleren Deckungswerten auf.

Es lassen sich deutlich zwei Untertypen unterscheiden, die ebenfalls in ihren Differentialartengruppen stark von Ruderalelementen geprägt werden. Im *Tripleurospermum maritimum*-Untertyp baut neben der namengebenden Art vor allem *Diplotaxis tenuifolius* die Bestände auf. Daneben finden sich bereits einige Wiesenpflanzen und Elemente, die in armen sauren Rasengesellschaften ihren Verbreitungsschwerpunkt besitzen. Zu nennen wären hier *Achillea millefolium*, *Plantago lanceolata*, *Taraxacum officinale* und *Daucus carota* bzw. Anspruchslosere, wie *Hypochoeris radicata*, *Agrostis tenuis* und *Festuca ovina*. Im Untertyp von *Poa nemoralis* zeigt sich schon eine leichte Wirkung der Aufforstung. Neben Ruderalarten, wie *Cirsium vulgare*, *Oenothera biennis*, *Conyza canadensis*, und *Poa pratensis* sowie *Festuca rubra* als Wiesenpflanzen, weisen *Poa nemoralis* auf Waldvegetation und *Epilobium angustifolium* auf Kahlschlagvegetation hin.

Der Populus-Taraxacum officinale-Typ

Die Aufforstung der Bestände des *Populus-Taraxacum officinale*-Typs erfolgte vor etwa 28 Jahren. Auch hier bilden *Populus* und *Robinia pseudacacia* als Aufforstungselemente eine geschlossene obere Baumschicht mit einer Deckung von 80 bis 90 %. In einigen Beständen wachsen in einer unteren Baumschicht *Alnus glutinosa*, *Acer platanoides* und *Populus nigra* sowie einige andere Baumarten, die, wie *Acer pseudoplatanus*, *Quercus rubra*, *Padus avium*

und *Sorbus aucuparia*, augenscheinlich künstlich eingebracht wurden. In einer schwach entwickelten Strauchschicht tritt sehr regelmäßig *Acer pseudoplatanus* auf.

In der Feldschicht fallen vor allem im Frühjahr die zahlreichen Rosetten von *Taraxacum officinale* physiognomisch besonders auf. Zu einigen Ruderalpflanzen mit hohen bzw. mittleren Stetigkeitswerten, wie *Agropyron repens*, *Cirsium vulgare* und *Convolvulus arvensis* treten Jungpflanzen einer ganzen Anzahl von Gehölzarten, von denen vor allem *Crataegus monogyna*, *Sorbus aucuparia*, *Rubus fruticosus*, *Padus avium* und *Quercus robur* zu nennen sind. Durch eine recht hohe Stetigkeit zeichnen sich darüber hinaus *Agrostis alba*, die auf feuchtere Böden hinweist, und *Festuca ovina* aus, die armes Substrat anzeigt.

Der Populus-Calamagrostis epigeios-Typ

In den etwa 25 Jahre alten Beständen des *Populus-Calamagrostis epigeios*-Typ bestimmen zwar auch *Populus* und *Robina pseudacacia* die Baumschicht, die eine Deckung zwischen 40 und 80 % aufweist. Mit mittleren Stetigkeitswerten mischen sich in die Baumbestände aber auch *Populus tremula* und *Betula pendula*, die als Vorwaldarten bei der Waldentwicklung nach Kahlschlägen oder Waldbränden eine große Rolle im Untersuchungsgebiet spielen.

Die Strauchschicht besteht aus einer großen Anzahl von Arten, die aber im Typ nur eine geringe bis mittlere Stetigkeit erreichen. Zu erwähnen wären vor allem *Robinia pseudacacia*, *Alnus glutinosa*, *Betula pendula* und *Populus nigra*.

Das physiognomische Bild der Feldschicht wird völlig von *Calamagrostis epigeios* bestimmt, die nicht nur eine hohe Stetigkeit, sondern meist auch Deckungswerte von drei und vier erreicht. Ruderalarten erlangen nur noch einen geringen Anteil (*Cirsium arvense*, *Hypericum perforatum*, *Rumex acetosella* und *Equisetum arvense*). Dagegen mischen sich mit größerer Regelmäßigkeit eine Reihe von Rasenpflanzen, wie *Achillea millefolium*, *Leucanthemum vulgare*, *Anthriscus silvestris*, *Festuca ovina*, *Hieracium pilosella* und *Agrostis tenuis* in die Feldschicht. Besonderen Anteil erlangen dagegen weiterhin Gehölzarten: *Rubus fruticosus*, *Crataegus monogyna*, *Cornus sanguinea*, *Rosa canina* und *Crataegus laevigata* mit Verbreitungsschwerpunkt in *Querco-Fagetea*-Ges., aber auch Waldarten mit weiterer soziologischer Amplitude (*Quercus robur*, *Padus avium* und *Sorbus aucuparia*).

Der Populus-Carpinus betulus-Typ

Die Standorte einer mehr als 60 jährigen Aufforstung auf alten Braunkohlenkippen etwa zwei Kilometer südwestlich von Gräfenhainichen werden von Beständen des *Populus-Carpinus betulus*-Typs besiedelt. Die geschlossene, meist 90 bis 100 % Deckung aufweisende Baumschicht wird in ihrer oberen Etage ebenfalls noch stark von *Populus* und *Robinia pseudacacia* aufgebaut. Hinzu gesellen sich jedoch neben *Betula pendula* Arten mit einem Verbreitungsschwerpunkt in *Querco-Fagetalia*-Ges., wie *Acer platanoides*, *Fraxinus excelsior* und *Carpinus betulus*. Diese Arten erlangen in der unteren Baumschicht eine besonders hohe Stetigkeit, was eher auf eine natürliche Ansamung als auf eine Anpflanzung dieser Pflanzen hinweist.

In dieser unteren Baumschicht treten daneben mit mittlerer Stetigkeit zusätzlich noch *Crataegus monogyna* und *Acer pseudoplatanus* als *Querco-Fagetea*-Arten sowie weiter verbreitete Waldpflanzen, wie *Quercus robur*, *Sorbus aucuparia* und *Malus sylvestris*, auf, während hier die aufgeforsteten Gehölze *Populus* und *Robinia pseudacacia* keine Rolle mehr spielen.

Unter der recht dichten Baumschicht kommt es zur Entwicklung einer Strauchschicht mit Deckungswerten von 30 bis 40 %, die sich ebenfalls aus den in der Baumschicht vorkommenden Gehölzen mit Verbreitungsschwerpunkt in *Querco-Fagetea*-Ges. und weiter verbreiteten Baum- und Straucharten zusammensetzt. Die Hauptbestandsbildner der Feldschicht sind wiederum Jungpflanzen der Gehölze, die die Baumschicht und Strauchschicht zusammensetzen. Besonders hinzuweisen ist hier wieder auf die Pflanzen mit einem deutlichen Verbreitungsschwerpunkt in den *Querco-Fagetea*-Ges., wie *Acer platanoides*, *Crataegus monogyna*, *Carpinus betulus*, *Rubus fruticosus* und *Rosa canina*. Krautartige und Gräser treten, wenn man von *Brachypodium silvaticum*, ebenfalls einer

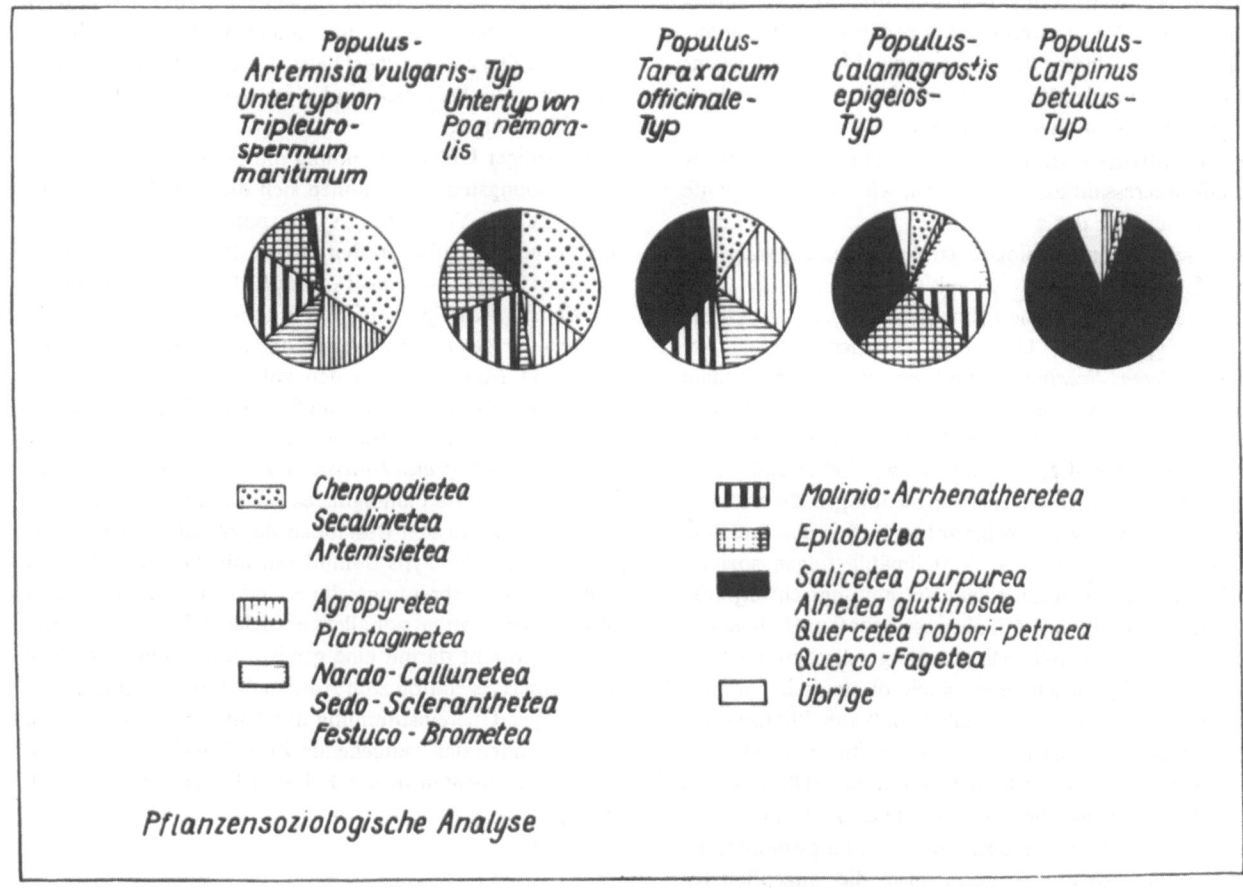

Abb. 1. Pflanzensoziologisches Artengruppen-Spektrum der untersuchten Phytozönosen.

Querco-Fagetea-Art, absieht, völlig in den Hintergrund.

Pflanzensoziologische Beziehungen

Die dargestellten Sukzessionsstadien der Aufforstungen auf dem frisch geschüttelten Substrat der Kippen des Braunkohlentagebaues zeigen deutliche physiognomische und floristische Unterschiede. Die Typen lassen sich jedoch nicht mit Hilfe von Charakterarten oder auch Differentialarten als scharf voneinander abgesetzte Assoziationen charakterisieren. Das gleiche gilt für die Abgrenzung der Typen von Assoziationen der naturnäheren oder halbnatürlichen Vegetation des Untersuchungsgebietes. Da die neu entstehenden, bzw. entstandenen Standorte meist eine mittlere Bodenwasserführung und mittlere bis gute Nährstoffverhältnisse aufweisen, aber auch keine Besonderheiten hinsichtlich der klimatischen Situation besitzen, erscheint es nur natürlich, daß je nach dem Entwicklungsstand der aufgeforsteten Bestände Pflanzen der unterschiedlichsten Vegetationsformationen bzw. aus den unterschiedlichsten Klassen des pflanzensoziologischen Systems aus der umgebenden Kulturlandschaft in den untersuchten Typen gemeinsam anzutreffen sind.

Unter pflanzensoziologischen Aspekten erscheint es deshalb interessant zu untersuchen, wie sich im Verlaufe der Entwicklung der aufgeforsteten Bestände die Anteile der verschiedenen soziologischen Artengruppen standortbedingt verändern (vgl. Abbildung 1).

In beiden Ausbildungsformen des *Populus-Artemisia vulgaris*-Typs bilden Unkräuter aus den Klassen der *Chenopodietea*, *Secalinietea* und *Artemisietea* mit einem Anteil von etwa 35 % die größte soziologische Komponente. Die Arten dieser Gruppe besiedeln bevorzugt stark anthropogen beeinflußte Standorte, wie Äcker und Ruderalstellen vor allem im Bereich von Siedlungen. Sie zeichnen sich durch ein gewisses Nährstoffbedürfnis aus und sind fast durchweg lichtliebend. Anteilmäßig folgen soziologische Artengruppen mit einem Verbreitungsschwerpunkt in Trittrasen und Feuchtpioniergesellschaften (*Plantaginetea*) bzw. in Quecken-Trockenpioniergesellschaften (*Agropyretea*). Syngenetisch, aber auch ökologisch-standortkundlich von Interesse ist dabei, daß die Pflanzen der Trockenpioniergesellschaften sowohl im *Tripleurospermum maritimum*, als auch im *Poa nemoralis*-Untertyp mit etwa 12 % den gleichen Anteil erreichen, während die Arten der Feuchtpionierrasen nur im Anfangsstadium der Sukzessionsentwicklung, wenn also die angepflanzten Bäume die Feuchtigkeit des Bodens noch nicht so intensiv ausnutzen, mit 5,2 % deutlich in Erscheinung treten.

Auf ähnliche ökologische Phänomene weist Ellenberg (1963) unter Bezugnahme auf Untersuchungen Jeniks (1957) hin. Der im Hochwald vorhandene, auf den Kahlschlägen aber dann plötzlich fehlende starke Wasserentzug durch die Waldbäume führt zum Auftreten von Nässezeigern in der jungen Kahlschlagvegetation.

Mit einem Anteil von etwas über 20 % beteiligen sich die soziologischen Artengruppen aus der *Molinio-Arrhenatheretea*-Klasse an der Bestandszusammensetzung. Eine weitere beachtenswerte Artengruppe bilden in beiden Untertypen die Pflanzen der Kahlschlagfluren (*Epilobietea*), die im *Tripleurospermum*-Untertyp einen Anteil von 13 % und im *Poa nemoralis*-Untertyp einen solchen von 16 % erreichen. Deutliche Unterschiede zwischen beiden Untertypen treten bei den Pflanzen mit einem Verbreitungsschwerpunkt in Mager- und Trockenrasen und in Gehölzgesellschaften auf. Der Anteil der zuerst genannten Arten geht vom *Tripleurospermum*-Untertyp zum *Poa nemoralis*-Untertyp von 12,5 % auf 1,7 % zurück, während die Werte der Pflanzen mit einem Verbreitungsschwerpunkt in *Quercetea robori-petraeae*- und *Querco-Fagetea*Ges. bzw. in *Salicetea*- und *Alnetea*-Ges. von 1,4 % auf 13,5 % ansteigen.

In völliger Übereinstimmung mit der oben angeführten Entwicklungstendenz erhöhen sich die Anteile der Pflanzen mit einem Verbreitungsschwerpunkt in sommergrünen Gehölzgesellschaften in den älteren Beständen des *Populus-Taraxacum-officinale*-Typs und des *Populus Calamagrostis epigeios*-Typs auf Werte von 34–37 %. Während *Molinio-Arrhenatheretea*-Arten und die Pflanzen der Trocken- und Magerrasen in den zuletzt genannten Gesellschaften etwa gleich stark mit Werten zwischen 12 % und 18 % an der Bestandszusammensetzung beteiligt sind, erreichen im *Populus-Taraxacum officinale*-Typ Arten der Agropyretea- und Plantaginetea-Klasse mit 24,4 % beachtliche Anteile. In den Beständen des 60 jährigen *Populus-Carpinus betulus*-Typs dominieren mit 90 % die Pflanzen mit einem Verbreitungsschwerpunkt in sommergrünen Waldgesellschaften, vor allem der Klasse der *Querco-Fagetea*. Es kommt darin eine gewisse Entwicklungstendenz zur Klimaxvegetation zum Ausdruck. Interessant ist daher in völliger Übereinstimmung mit Untersuchungen in den Niederlanden das weitgehende Zurücktreten von *Querco-Fagetea*-Elementen in der Feldschicht (vgl. Bruens et al. 1975).

74

Tabelle 3. Werte der Bodenuntersuchungen

	Populus-Artemisia Typ		Populus-Tarax. off.- Typ	Populus-Calamagr. epig.- Typ.	Populus-Carp. bet.- Typ
	Untertyp v.Tripl. sperm.	Untertyp v. Poa nem.			
Humusgehalt (%)	1,5	2,1	4,8	3,6	7,9
Kalkgehalt (%)	0,1	-	0,2	0,4	1,4
pH-Wert	7,3	5,2	7,1	6,7	7,2
P_2O_5-Gehalt (mg/100 g Boden)	1,5	0,7	0,9	1,0	1,8
K_2O-Gehalt (mg/100 g Boden)	14	7,3	10,3	7,2	12,2

Ökologische Beziehungen

Zur Herausarbeitung ökologischer Beziehungen zwischen den untersuchten Pflanzengesellschaften erfolgte die Untersuchung einiger Bodenfaktoren und die Berechnung der Faktorenzahlen nach Ellenberg (1974). Bei den Bodenuntersuchungen hebt sich vor allem beim Humusgehalt eine deutliche, sicher mit der Vegetationsentwicklung zusammenhängende Differenzierung ab (vgl. Tab. 3). Während der Humusgehalt im aufgebrachten Mutterboden des *Tripleurospermum*-Untertyps noch nicht einmal 2% beträgt, steigt dieser in den 'Sukzessionstypen' kontinuierlich bis zu einem Wert von nahezu 8% im *Populus-Carpinus betulus*-Typ an. Ein gewisser Anstieg zeichnet sich auch beim Kalkgehalt ab, wobei nicht auszuschließen ist, daß durch den zunehmenden Tiefgang des Wurzelsystems der Typen der Kalkgehalt des tertiären Erdmaterials in den Stoffkreislauf einbezogen wurde.

Die Böden sind mit Werten zwischen 0,7 und 1,8 mg extrem phosphatarm, wobei sich in den mittleren Entwicklungsstadien noch zusätzlich eine leichte Depression abzeichnet. Besonders deutlich wird das beim K_2O-Gehalt. Nach einem Wert von fast 14 mg im aufgebrachten Mutterboden des *Tripleurospermum*-Untertyps sinkt der Gehalt in den zeitlich folgenden Ausbildungsformen auf unter 8 mg ab und erreicht im *Populus-Carpinus betulus*-Typ wieder einen Wert von 14 mg. Da Phosphat und Kali relativ schnell von den Pflanzen aufgenommen werden, ist diese Entwicklung sicher im Zusammenhang mit dem Stoffentzug durch die sich entwickelnde Pflanzenmasse zu sehen. In den ältesten Beständen spielt augenscheinlich die Stoffrückführung aus der abgestorbenen Biomasse in den Boden durch die Destruenten eine gewisse Rolle.

In ökologischer Hinsicht nicht uninteressant ist die Differenzierung der Faktorenzahlen nach Ellenberg (1974) in den verschiedenen alten Bestandstypen. So geht die Lichtfaktorenzahl von den noch relativ wenig bedeckten Beständen des *Tripleurospermum*-Untertyps bis zum hochwaldähnlichen *Populus-Carpinus betulus*-Typ von 7,2 auf 5,3 zurück (vgl. Abb. 2). Während die Temperaturzahl wenig differenziert ist und zwischen 5,0 und 5,9 schwankt, weist die Feuchtezahl einen starken Unterschied zwischen den jungen Typen und dem *Populus-Carpinus betulus*-Typ auf. Der Reaktionsfaktor zeigt insofern deutliche Parallelen zu den gemessenen Werten für die Bodenreaktion und den Kalkgehalt, als in den ältesten Beständen des *Populus-Carpinus betulus*-Typs mit 6,2 der höchste Wert erreicht wird. Die Stickstoffaktorzahl spiegelt insofern Beziehungen wider, die auch beim K_2O- und beim P_2O_5-Gehalt in Erscheinung treten, als in den mittleren Typen eine Depression dieser Faktorenzahl zu verzeichnen ist.

Gewisse Schlüsse in ökologischer Hinsicht lassen sich auch aus den Spektren der Lebensformen, der Blattausdauer und des anatomischen Baues der Blätter ziehen. In den Beständen des *Populus-Artemisia vulgaris*-Typs dominieren die Hemikryptophyten und die Geophyten mit Werten von über 60 bzw. 25%, während die übrigen Lebensformen nur geringe Anteile erreichen (vgl. Abb. 3).

In den Beständen des *Populus-Taraxacum officinale*-Typs und des *Populus-Calamagrostis epigeios*-Typs gehen die Werte für die Hemikryptophyten auf etwas unter 50% zurück, und die Geophyten verringern ihren Anteil um etwa die Hälfte auf 12% bzw. 14%. Dafür nehmen die Anteile der Nanophanerophyten und der Phanerophyten beträchtlich zu. Sie erreichen in beiden Typen zusammengefaßt fast einen Wert von 40%. Ein völlig anderes Lebensformenspektrum weist der *Populus-Carpinus* betulus-Typ auf. Hier gehören die Pflanzen der Feldschicht mit einem Anteil von nahezu 75% den Phanerophyten an, gefolgt von den Nanophanerophyten mit etwa 15% und den Hemikryptophyten mit 7,5%.

In allen untersuchten Typen dominieren Pflanzen mit sommergrünen Blättern. Die Arten mit grün überwinternden Blattorganen erreichen etwa einen Anteil von 25% (vgl. Abb. 3). Deutliche Abweichungen unter diesem Aspekt weist der *Populus-Carpinus betulus*-Typ auf. Hier erlangen die Pflanzen mit sommergrünen Blättern einen Anteil von 84%, während die Arten mit grünen Blattorganen im Winter auf einen Anteil von 9% zurückgehen und zusätzlich immergrüne mit einem Wert von 7% auftreten. Der anatomische Bau der Blätter stellt eine Anpassungserscheinung an den Wasserfaktor dar. In voller Übereinstimmung mit der Faktorenzahl des Wasserhaushaltes nach Ellenberg (1974) ergibt sich eine Differenzierung der Anteile der verschiedenen Artengruppen nach der Blattanatomie (vgl. Abb. 3).

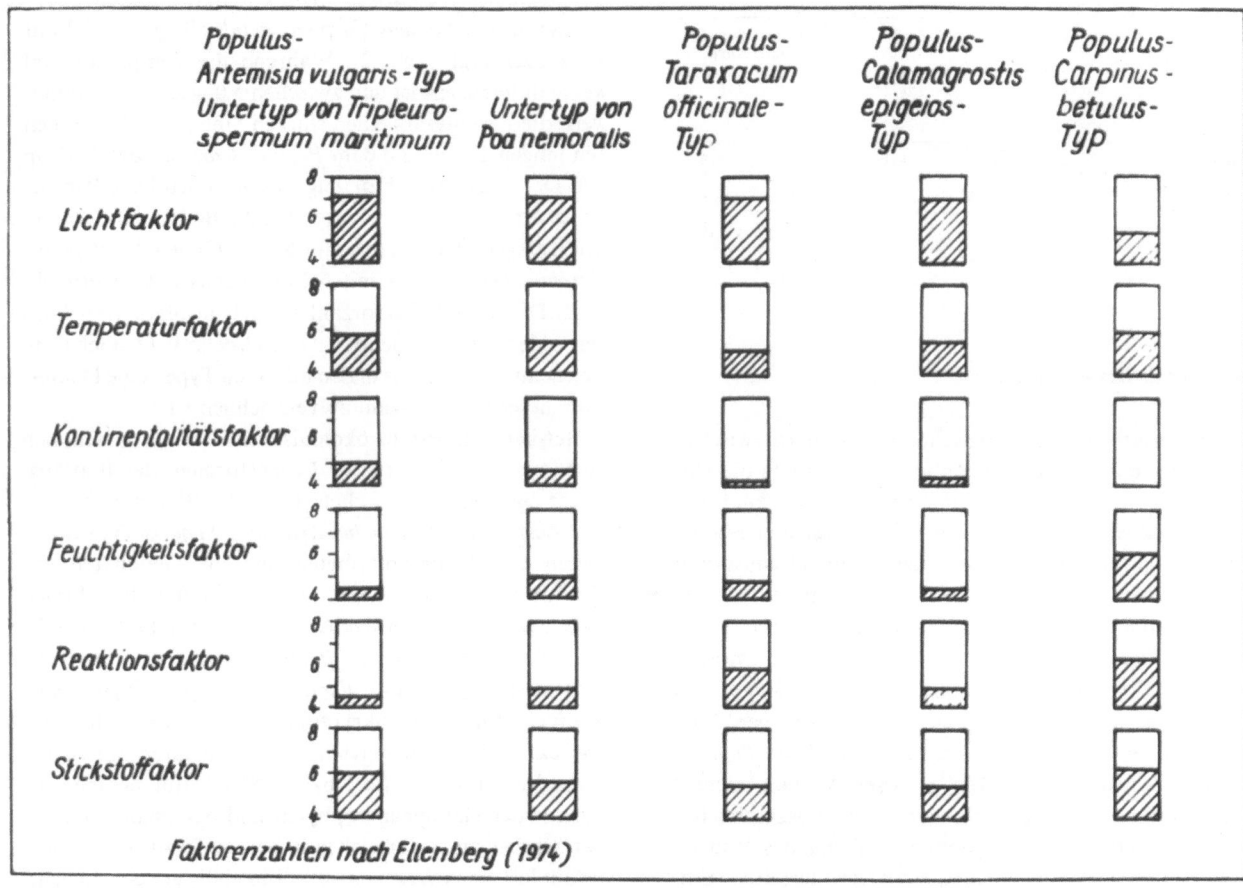

Abb. 2. Faktorenzehlen der untersuchten Phytozönosen nach Ellenberg (1974).

Im *Populus-Artemisia vulgaris*-Typ der infolge seiner geringen Wuchshöhe und Bestandsdichte eine recht starke Sonneneinstrahlung und damit verbundene Evaporation und Verdunstung aufweist, dominieren die Arten mit einem mesomorphen bis scleromorphen Blattbau und scleromorphen Blättern (jeweils etwa 22 %), gefolgt von Pflanzen mit mesomorphen bis hygromorphen Blättern. In den folgenden Typen nehmen die Anteile der Pflanzen mit mesomorphen bis scleromorphen Blättern zugunsten derjenigen mit mesomorphen Blättern kontinuierlich ab, bis schließlich im *Populus-Carpinus betulus*-Typ die mesomorphen Pflanzen einen Anteil von fast 70 % erreichen. An zweiter Stelle stehen in diesen dichten Beständen mit einem ausgeprägten Binnenklima, das sich durch eine relativ geringe Evaporation auszeichnet, die Arten mit mesomorph-hygromorphen Blattorganen.

Pflanzengeographische Beziehungen

Zur Charakterisierung der pflanzengeographischen Beziehungen zwischen den untersuchten Phytozönosen wurden die Artengruppen nach der Ozeanität und Zonalität (vgl. Meusel, Jäger & Weinert 1965, Rothmaler 1976, Ellenberg 1974) ihrer Verbreitung zu Gruppen zusammengefaßt und jeweils ein Ozeanitäts- und Zonalitätsspektrum errechnet. Im *Populus-Artemisia vulgaris*-Typ erreichen Arten mit einer atlantischen bzw. subatlantischen Verbreitungstendenz einen Anteil von etwa 45 %, gefolgt von Pflanzen ohne eine ausgesprochen atlantische bzw. kontinentale Verbreitung (vgl. Abb. 4).

Kontinental und subkontinental verbreitete Pflanzenarten nehmen einen Anteil von 13 % bzw. 19 % in den unkrautreichen Beständen des *Tripleurospermum maritimum*- bzw. im *Poa nemoralis*-Untertyp dieser Gesellschaft ein. Über den *Populus-Taraxacum officinale*-Typ und

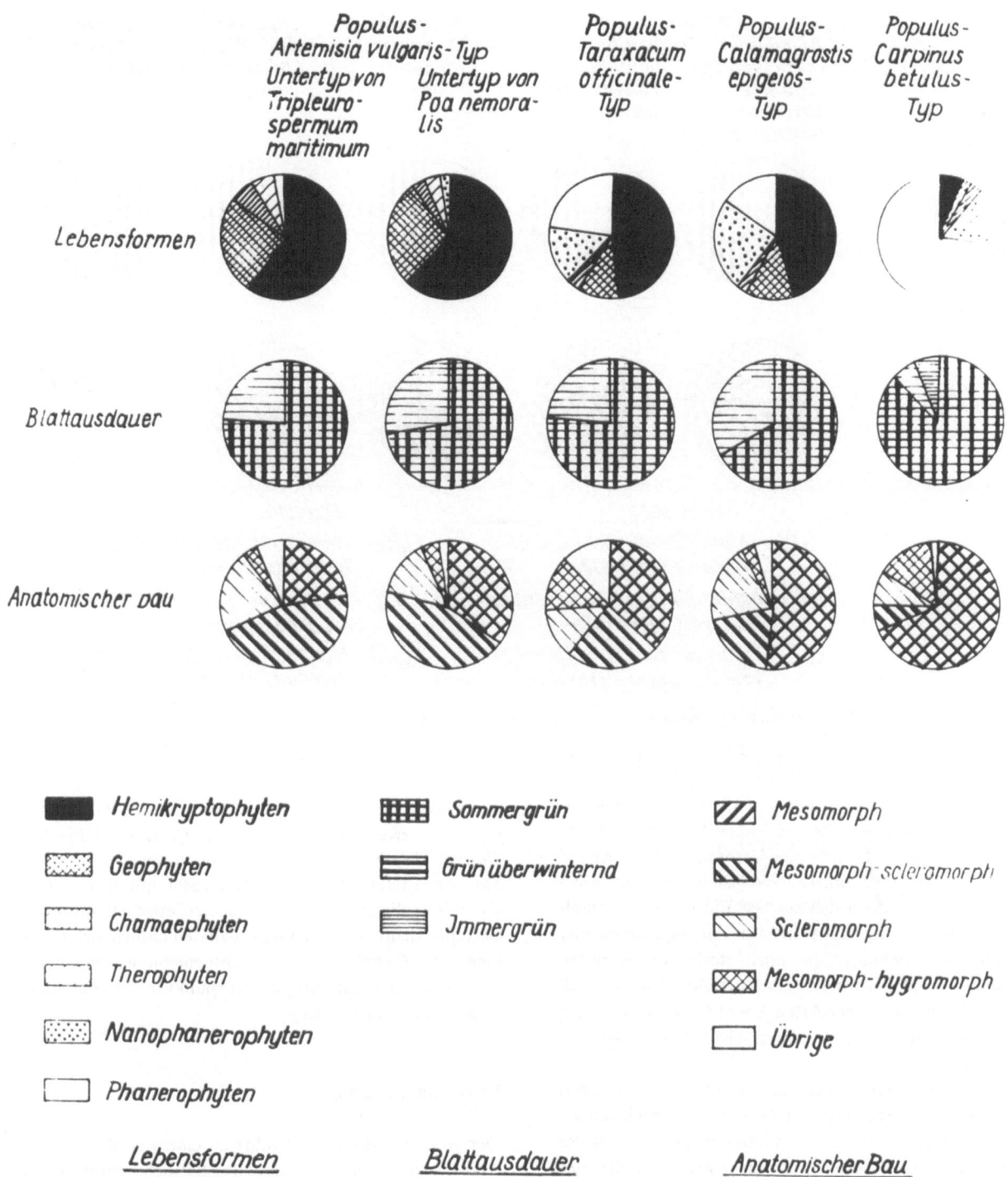

Abb. 3. Spektrum der Lebensformen, der Artengruppen mit unterschiedlicher Blattausdauer und der Artengruppen mit unterschiedlich anatomisch-morphologischem Bau.

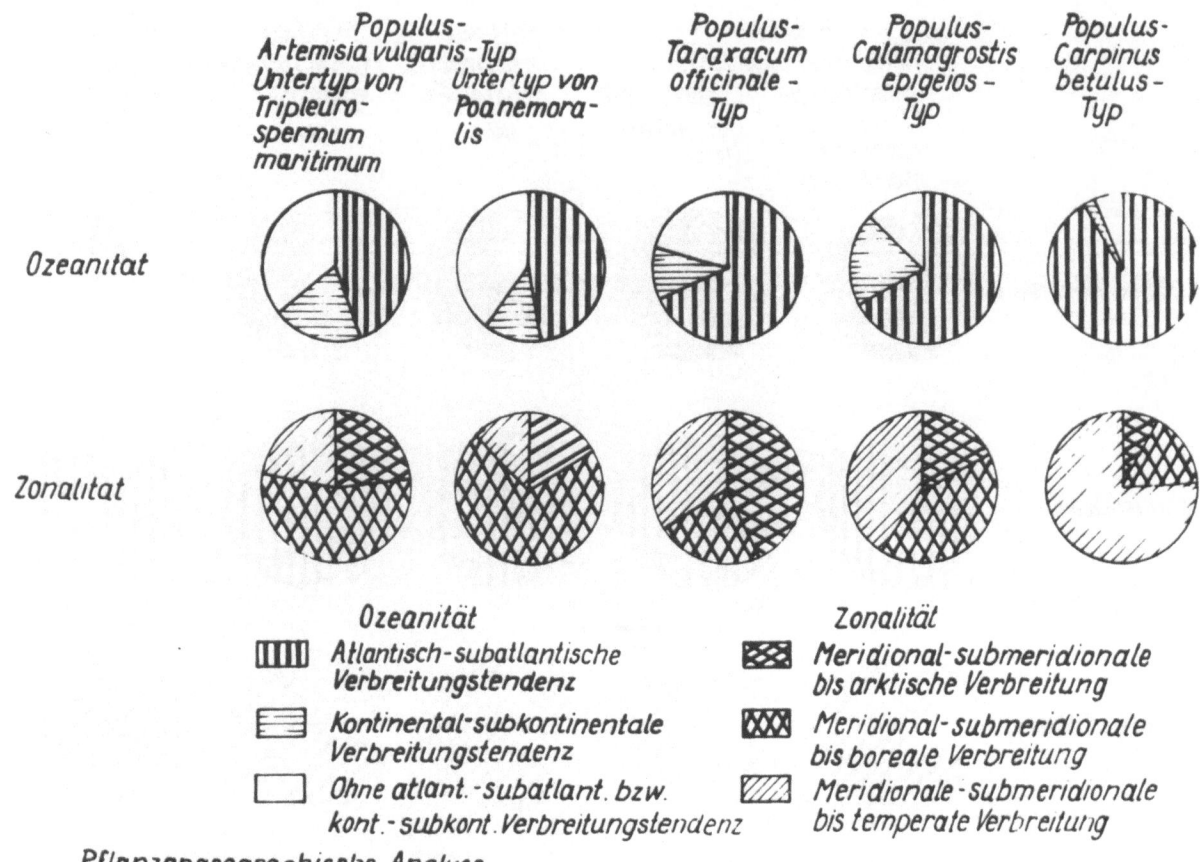

Ozeanität

Populus-Artemisia vulgaris-Typ
Untertyp von *Tripleurospermum maritimum*   Untertyp von *Poa nemoralis*

Populus-Taraxacum officinale-Typ

Populus-Calamagrostis epigeios-Typ

Populus-Carpinus betulus-Typ

Zonalität

Ozeanität

||||| *Atlantisch-subatlantische Verbreitungstendenz*

⊟ *Kontinental-subkontinentale Verbreitungstendenz*

□ *Ohne atlant.-subatlant. bzw. kont.-subkont. Verbreitungstendenz*

Zonalität

▨ *Meridional-submeridionale bis arktische Verbreitung*

▩ *Meridional-submeridionale bis boreale Verbreitung*

▨ *Meridionale-submeridionale bis temperate Verbreitung*

Pflanzengeographische Analyse

Abb. 4. Spektrum der Artengruppen unterschiedlicher Zonalität und Ozeanität.

Populus-Calamagrostis epigeios-Typ bis zum *Populus-Carpinus betulus-Typ* steigt der Anteil der Pflanzen mit atlantischer und subatlantischer Verbreitungstendenz auf Werte von über 90% an. Der Vergleich der Ozeanitätsspektren zeigt deutlich den starken Einfluß kontinental-subkontinentaler Elemente in den Anfangsstadien der Vegetationsentwicklung, die noch stark von Ruderalpflanzen geprägt werden und einen zunehmenden Einfluß ozeanisch-subozeanischer Arten je weiter die Entwicklung hin zu den naturnäheren Laubmischwäldern vonstatten gegangen ist.

Eine deutliche Differenzierung ergibt sich auch, wenn man die zonale Verbreitung der Arten einer vergleichenden Betrachtung zugrunde legt. Während sich die Typen der ersten Entwicklungsstadien durch starke Anteile der Pflanzen mit einer weiten zonalen Verbreitung von der meridional-submeridionalen bis zur arktischen Zone auszeichnen und die Pflanzen mit meridional-submeridionaler bis temperater Verbreitung in beiden Untertypen des

Populus-Artemisia vulgaris-Typs nur einen Wert von 22% bzw. 12% erreichen, nehmen diese Pflanzen im *Populus-Carpinus betulus*-Typ einen Anteil von mehr als 75% ein. Das Zonalitätsspektrum spiegelt recht gut den zonalen Charakter dieses Typs wider, dessen etwa 60 Jahre alte Bestände deutliche Beziehungen zu den naturnahen *Carpinion betuli*-Gesellschaften des Untersuchungsgebietes aufweisen, deren Verbreitungsschwerpunkt in der temperaten Zone Zentraleuropas liegt.

Landeskulturelle Aspekte

Der Braunkohlentagebau führt zu einem starken Verlust landwirtschaftlich bzw. forstwirtschaftlich genutzter Flächen und fügt der Kulturlandschaft beträchtliche Wunden zu, die ohne Wiederurbarmachung und Rekultivierung zu irreversiblen Schädigungen führen, wie sie uns noch heute in allen alten aufgelassenen Tagebauen ohne Folgemaß-

nahmen aus der Vorkriegs- und Kriegszeit in der Kulturlandschaft entgegentreten.

Die modernen Folgemaßnahmen des Braunkohlentage-
baus, unter denen die Wiederurbarmachung und Kippenrekultivierung eine zentrale Stellung einnehmen, führen zu
einer Einbindung der ehemaligen Tagebaue in die Kulturlandschaft und zu einer Rückführung beträchtlicher
Bodenflächen in die landwirtschaftliche und forstwirt-
schaftliche Pflanzenproduktion. Durch Gehölzanpflanzungen auf den Kippen wird die Produktivität der
Kulturlandschaft beträchtlich erhöht und auch ihr ästhe-
tisches Bild wesentlich verbessert. Die Aufforstungen
setzen aber auch die Denudation von Feinerdematerial
durch Wind und Wasser, die Rinnenerosion und Erdrut-
schungen wesentlich herab, die man an vegetationslosen
Kippenflächen, aber auch an den nicht behandelten
Rändern von Tagebaurestlöchern häufig beobachten kann.

Nicht übersehen werden darf die beträchtliche landeskulturelle Bedeutung der durch Aufforstung entstandenen
Gehölzflächen. Im fortgeschrittenen Alter, vor allem dann,
wenn sie einen waldähnlichen Charakter annehmen, üben
sie in der Ackerlandschaft einen nicht zu unterschätzenden
klimatischen Einfluß aus. Sie bieten gleichzeitig zahlreichen Pflanzenarten und Tierarten, die durch die modernen industriemäßigen Methoden in der Pflanzenproduktion stark verdrängt werden, z.T. recht günstige Biotope.
Dadurch kommt es durch diese zunächst recht einförmig
wirkenden Gehölzmonokulturen im Lauf der Entwicklung
der Bestände zu einer Bereicherung der Tier- und Pflanzenwelt in der intensiv genutzten Agrarlandschaft. Die damit
erzielte Vergrößerung der Mannigfaltigkeit der Biotope,
Biozönosen und Ökosysteme erhöht die biozönotische
Pufferwirkung und die Funktion des ökologischbiozöno-
tischen Regulationsmechanismus, was vor allem im Hin-
blick auf die biologische Schädlingsbekämpfung in der
durch Agrochemikalien stark belasteten, hochproduktiven
Kulturlandschaft von nicht zu unterschätzender Bedeutung
ist.

Schlußfolgerungen

Schließlich soll noch auf die Bedeutung der durch Aufforstung entstandenen Kippenvegetation für die Bio-
zönose- und Ökosystemforschung hingewiesen werden.
Da die Biotop-, Biozönose- und Ökosystementwicklung
auf dem vegetationslosen geschütteten Erd- und Mutter-
bodenmaterial mit der Aufforstung von Monokulturen
nahezu beim 'Punkt Null' beginnt, stellen diese Anpflanzungen sowohl unter dem Aspekt der Ökosystemdynamik
als auch unter dem Gesichtspunkt des Zusammenwirkens
der verschiedenen biotischen und abiotischen Elemente in
den Ökosystemen ein ausgezeichnetes Forschungsobjekt
dar. Es ließe sich ein umfangreiches Arbeitsprogramm der
Grundlagenforschung und der angewandten Forschung an
diesem Forschungsgegenstand entwickeln, von dem hier
nur einige besonders relevant erscheinende Aspekte ange-
führt werden sollen:
– Vergleichende Untersuchungen des auf den Kippen
aufgebrachten Mutterbodens mit entsprechenden intakten
A-Horizonten nicht beeinflußter Böden in der Umgebung
der Kippen auf Mikrobenbesatz, zellulolytische Aktivität
und Samenpotential.
– Eingehende Untersuchung der Entwicklung der Pflan-
zenbestände, des Tierbesatzes und der Bodenmikroben im
Rahmen einer umfassenden Sukzessionsforschung.
– Untersuchung der Veränderung des Nährsalzpotentials
der Böden durch den Aufschluß von Nährsalzen tieferer
Bodenschichten bis hin zu dem tertiären Erdmaterial unter
dem aufgebrachten Mutterboden durch das Zusammen-
wirken der Organismen in den Ernährungsstufen.
– Untersuchung der Entwicklung der zellulolytischen
Aktivität der sich entwickelnden Böden unter Labor- und
Standortbedingungen.
– Untersuchung der sich entwickelnden Nahrungsketten
unter besonderer Berücksichtigung von Organismen, die
die Funktion der biologischen Schädlingsbekämpfung in
den Ökosystemen besitzen.
– Untersuchung von Konkurrenzbeziehungen in den sich
entwickelnden Beständen.
– Untersuchung des ökologischen Verhaltens der Tier- und
Pflanzenarten unter den sich ständig verändernden abiotischen und biotischen Bedingungen der aufgeforsteten
Gehölzbestände zur besseren ökologischen Charakterisierung der verschiedenen Tier- und Pflanzenarten.
– Vergleichende Untersuchungen der Vegetationsentwicklung auf verschieden mächtig aufgebrachtem Mutterboden
und auf tertiärem Erdmaterial zur Erforschung des Verhaltens und der Entwicklungspotenz der verschiedenen
Pflanzenarten, was zu wichtigen Schlüssen hinsichtlich der
landeskulturellen Bedeutung der untersuchten Pflanzenarten bei der Entwicklung von Rohböden zu Kulturböden
führt.
– Untersuchungen zum Problem der Beziehungen zwischen
Forstgesellschaften, naturnahen Waldgesellschaften und
der potentiellen natürlichen Vegetation.
– Untersuchungen zum ökologischen Toleranzbereich der
Pflanzenarten, wobei sich für die Herausarbeitung der

ökologischen Amplituden der Unkrautarten unter besonderer Berücksichtigung des Licht- und Temperaturfaktors besonders gute Möglichkeiten bieten.

Die bisher durchgeführten syngenetischen Untersuchungen an Kippenaufforstungen zeigen, daß in dieser hochproduktiven Kulturlandschaft nicht nur eine starke Verarmung der Tier- und Pflanzenwelt und der Vielfalt der Biotope, Biozönosen und Ökosysteme vor allem unter dem Einfluß der industriemäßigen Pflanzenproduktion vonstatten geht, sondern, daß gerade in den Bergbaunachfolgegebieten durch wissenschaftlich fundierte landeskulturelle Maßnahmen die Entwicklung neuer Biotype gefördert wird, die zahlreichen Tieren und Pflanzen neue Lebensmöglichkeiten bieten. Dadurch wird das Bild der Kulturlandschaft bereichert, die landeskulturelle Situation verbessert und die Stabilität des Zusammenwirkens von abiotischen und biotischen aber auch gesellschaftlich-ökonomischen Komponenten erhöht.

Zusammenfassung

Der Braunkohlentagebau führt zu beträchtlichen Eingriffen in die Kulturlandschaft. Durch Wiederurbarmachung und Rekultivierung der Tagebauareale werden beträchtliche Bodenflächen der gesellschaftlichen Nutzung wieder zur Verfügung gestellt. In Landschaften mit von Natur aus fruchtbaren Böden steht dabei die Kippenrekultivierung für eine landwirtschaftliche Nutzung und in Gebieten mit von Natur aus weniger fruchtbaren Böden für eine forstliche Nutzung im Vordergrund. Die Tagebaurestlöcher werden häufig zu Naherholungsgebieten umgestaltet oder für wasserwirtschaftliche Zwecke bzw. für eine geordnete Mülldeponie verwandt. In der Arbeit wird die syngenetische Entwicklung von aufgeforsteten achtjärigen Monokulturen des *Populus-Artemisia vulgaris*-Typs über etwa 30 jährige Bestände des *Populus-Taraxacum officinale*-Typs und *Populus-Calamagrostis epigeios*-Typs bis zu etwa 60 jährigen Baumbeständen des *Populus-Carpinus betulus*-Typs unter pflanzensoziologischen, ökologischen und pflanzengeographischen Aspekten untersucht. Die Arbeit schließt mit einer Darstellung der landeskulturellen Bedeutung der Kippenaufforstung für die gesamte Kulturlandschaft.

Literatur

Barthel, H. 1962. Braunkohlenbergbau und Landschaftsdynamik. Gotha, 300 pp.

Bruens, J.E.M., Hendriks, J. L J., van der Putten, H.C.N. & Stortelder, A.H.F. 1975. Een kritisch onderzoek naar de botanische waarde van jonge en gestoorde bos-eco-systemen. Rapp. Afd. Geobotanie, Nijmegen, 91 pp.

Ellenberg, H. 1963. Vegetation Mitteleuropas mit den Alpen. (Bd. IV, Teil 2 der Einführung in die Phytologie von Heinrich Walter. Ulmer, Stuttgart, 943 pp.

Ellenberg, H. 1974. Zeigerwerte der Gefäßpflanzen Mitteleuropas. Script. Geobot. 9: 1–95.

Frenzel, S. 1976. Ökologisch-geobotanische Untersuchungen der Biozönosen von Kippenaufforstungen im Bitterfeld-Bergwitzer Braunkohlenrevier unter Berücksichtigung ihrer landeskulturellen Funktion. Diplomarbeit, Halle, 93 pp.

Hundt, R. 1963. Die Entwicklung der Grünlandwirtschaft und der Naturschutz. Arch. Naturschutz 3: 37–58.

Jenik, J. 1957. Das Wurzelsystem der Stiel- und Traubeneiche. Rozpravy Ceskosl. Akad. Ved. 67: 1–85.

Meusel, H. 1955. Entwurf einer Gliederung Mitteldeutschlands und seiner Umgebung in- pflanzengeographische Bezirke. Wiss. Z. Univ. Halle, Math.-Nat. R. 4: 637–642.

Meusel, H., Jäger, E. & Weinert, E. 1965. Vergleichende Chorologie der Zentraleuropäischen Flora. Fischer, Jena, 583 pp.

Rothmaler, W. 1976. Exkursionsflora. Berlin, 612 pp.

Waldmann, H. 1975. Die Wiederherstellung der landeskulturellen Funktion ehemals vom Braunkohlenbergbau genutzter Flächen. Neue Bergbautechn. 5: 770–774.

OPTIMAL WOODLAND DEVELOPMENT ON SANDY SOILS IN THE NETHERLANDS*·**

G. SISSINGH

Kemperbergerweg 601, 6816 RT Arnhem, the Netherlands

Keywords:

Multiple use, Optimum-community, *Quercion robori-petracae*, Regeneration, Woodland

Introduction

Little is left from the extensive forests that covered our country at the beginning of this era. By cutting, burning, grazing by cattle and by reclamation man has virtually extirpated our virginal forests. Only small fragments, the so called 'Malebossen' have been maintained in a nearly natural state, mainly because of hunting.

It was not before the middle of the 19th century that man became concerned about the need for timber and ever since he proceeded in planting trees. In the beginning planters thought in short term rotation in order to harvest by life. As a matter of fact this meant 'murder in childhood'; it was a policy still based on agricultural thinking. In 1899 the State Forest Service in the Netherlands was founded and during its development the idea gained ground that forestry is something quite different from agriculture and that the forest has other purposes than just timber production, e.g. recreation nature-study and nature-conservation.

In the middle of this century the idea of 'multiple use' became generally accepted. One realised that the forest had to be subserviant to different purposes. This finally resulted in the aim formulated recently by the Kampfraath-Committee: 'The maintenance and development of a natural environment, as varied as possible, in which biocoenoses of plants and animals can survive and through which a contribution to the material and spiritual fulfilment of human needs is realised'. How correctly this aim may have been formulated, one has to realise that carrying it into effect is

a long term undertaking. It does not only take much time, but we have hardly experience with it. The nearly 80 years old State Forest Service is, in terms of rotation of a tree or development of an ecosystem, only a child under age or even an infant!

The forest as an ecosystem

In our forest there is a cycle of organic matter, produced by the trees, shrubs, herbs and mosses. Part of it gets into the soil directly through dying roots. The major part of it accumulates on the surface of the soil as so-called 'raw humus' (Auflagehumus). This humus layer – the A_0 or litter – is intensively penetrated by roots. The dying off of the fine hair roots is a continuous process. A considerable part of the organic matter in the A_0-layer comes from these roots. The mosses are dying off continuously as well. According to Damman (1971) litter only partly contributes to the A_0, whilst the contribution of dying roots and mosses is not neglectible at all. The proportion may change strongly depending on the type of vegetation. Especially our needle-leaf woods are rich in mosses and ferns. In hardwood forests the annual fall of leaves covers the mosses and brings them to die (Ellenberg 1963). Only where leaves are blown away mosses have a chance.

Concerning the ferns: they are negatively influenced by the alternation of sunshine/shadow and changes in relative humidity occurring in deciduous woodlands but they are well adapted to the moderate and more constant climate of our coniferous woodlands.

The litter and the raw humus decay during the years. Thereby fauna and especially microfauna (insects, gacteria and fungi) play an important role. In most of the foresttypes an equilibrium is established in the course of time,

* Contribution to the Symposium on Plant species and plant communities, held at Nijmegen, 11–12 November 1976, on the occasion of the 60th birthday of Professor Victor Westhoff.
** Nomenclature follows Heukels-van Ooststroom, Flora van Nederland, 18e druk, 1975, Wolters-Noordhoff, Groningen.

where supply and decay are in balance. This equilibrium depends on the composition of the raw humus and therefore on the vegetation-type.

The Netherlands have a humid climate. Precipitation surpasses evaporation and thus there is a precipitation surplus, that disappears in the soil. The humus acids dissolved from the A_0 cause a stratification in the soil profile. There is a zone of leaching – the A-horizon – and a zone of infiltration – the B-horizon. The B-horizon in a forest-soil is still intensively penetrated by roots. The minerals precipitated in it are dissolved by these roots, assimilated by the trees and added back to the A_0 by way of litter. And so the cycle is closed.

In heathland, as compared with woodland, the soil is less deeply penetrated by roots. The moisture balance is different as well. As a result there is a strong podsolisation. Whilst we find a humus iron-podsol or 'holtpodsol' under woodland, under heathland we observe a humus podsol or 'haarpodsol', which is much poorer in nutrients. The heathland *(Genisto-Callunetum)* is a semi-natural ecosystem, developed from the wood through interference of man (burning, grazing by sheep and sodding). By overgrazing the soil can be completely denudated and wind may cause shifting of sand until large inland dunes arise, which may be stablished again by a *Spergulo-Corynephoretum.* There we have a 'vague' profile or even no profile at all ('vaaggronden').

When man withdraws, the wood returns. At first there is a *Pinus-Betula* phase in which the flamish jay sews his acorns. From this stade an oak phase develops with a gradualy growing humus-layer, in which shrubs like *Sorbus aucuparia, Frangula alnus* and recently also *Prunus serotina* establish. As a result of the changed conditions the heather disappears and shade-standing grasses *(Deschampsia flexuosa)* and dwarf shrubs *(Vaccinium myrtillus* and *V. vitis-idaea)* appear. Also *Fagus silvatica* returns and as a shade standing tree it reaches dominance after all.

Forestry and pioneer forests

This natural succession takes several hunderds of years. The forestry officer can accelerate this succession by planting trees. But he is bound to the laws of nature. In the beginning he has to use pioneer species. On the poor sandy soils in our country these are white birch *(Betula pendula),* aspen *(Populus tremula)* and scotch pine *(Pinus silvestris),* all species who show a better growth in a more continental climate (Ellenberg 1963, p. 68). From the possible coniferous trees *Pinus silvestris* has the preference as can be judged from the fact that 80 % of the first generation of our planted

woodlands consist of *Pinus silvestris.* Scotch pine – originally an indigenous species (Wolterson 1973) has been exterminated in our country by excessive cultivation. We should bear in mind that even *Fagus silvatica* almost disappeared from our landscape through exploitation as a coppice because of its low stooling capacity. It is also comprehensible that a coniferous tree like scotch pine disappeared long before. For afforestation we had to import seeds.

The proveniences of *Pinus silvestris* imported from elsewhere were not adapted to our climate and suffered much from diseases (e.g. *Lophoderminum pinastri).* In the more pronounced atlantic parts of our country (West-Frisian Islands and Drente) the culture even failed almost completely. Thus we had to look out for other tree-species and as such we adopted Corsican pine *(Pinus nigra* v. *corsicana)* in the southern parts of our country and in our dunes. For the northern part (Drente) *Larix leptolepis* has been a lucky hit. Nevertheless, Japanese larch grows only on soils with sufficient water-holding capacity, i.e. on loamy soils or on soils with ground-water in reach. For zonal soils we cannot use it so much.

Second generation

In the second tree-generation, after a woodland microclimate was created and a first layer of litter and raw humus was formed, so-called 'more demanding species' can be used. In addition to the indigenous hardwood species: pedunculate and sessil oak *(Quercus robur* and *Q. petraea)* and common beech *(Fagus silvatica)* on economic considerations also coniferous trees like Norway spruce *(Picea exelsa),* sitka spruce *(P. sitchensis)* douglas fir *(Pseudotsuga taxifolia)* and grand fir *(Abies grandis)* were used. It was especially the two last-mentioned species, which are native on the atlantic coast of North-America, that have proved their usefulness in our country. The production of timber was considerably higher than that of our native deciduous trees like *Quercus* and *Fagus* and even more than twice that of *Pinus silvestris.* Also the penetration of the roots in the soil-profile (Groszkopf 1954), as well as the decay were much more pronounced.

The minerals, brought into circulation, and the humification of the litter of *Pseudotsusa taxifolia* leads to an ecosystem, that seems to be richer than that of other conifers *(Pinus* and *Picea)* or even those of our indegenous hardwood-species *(Quercus* and *Fagus).* This finds its expression in an understorey rich in ferns and mesotrophic mosses. Nevertheless, the climate of the West coast of

North-America – as a result of the NNW-SSE-coursing chains of mountains like Rocky Mountains and Cascades, who catch off the east winds – is considerably more oceanic than ours. In Europe where the chains of mountains run in E.W.-direction, the atlantic character of the climate is strongly weakened, especially in the latter part of the winter and spring, when dry east winds blow, with an attendant cloudless sky that *Pseudotsuga* madly endures. It is for this reason that *Pseudotsuga* – having a pioneer character in its native country – is a more demanding species here which must be planted in the shelter of older trees.

Succession

If we consider the 'Malebossen' *(Ilici-Fagetum)* as the climax-association, there is a great number of derivated 'substitute communities' (Ersatz-Gesellschaften) who can be considered as associations and also are described like this. In a scheme of break down and building up we can represent this as follows *(Figure 1)*. Braun-Blanquet 1951, p. 468) distinguished an 'Optimum-community' (Optimal Gesellschaft) besides the climax. This optimum-community is defined as the association that produces the greatest

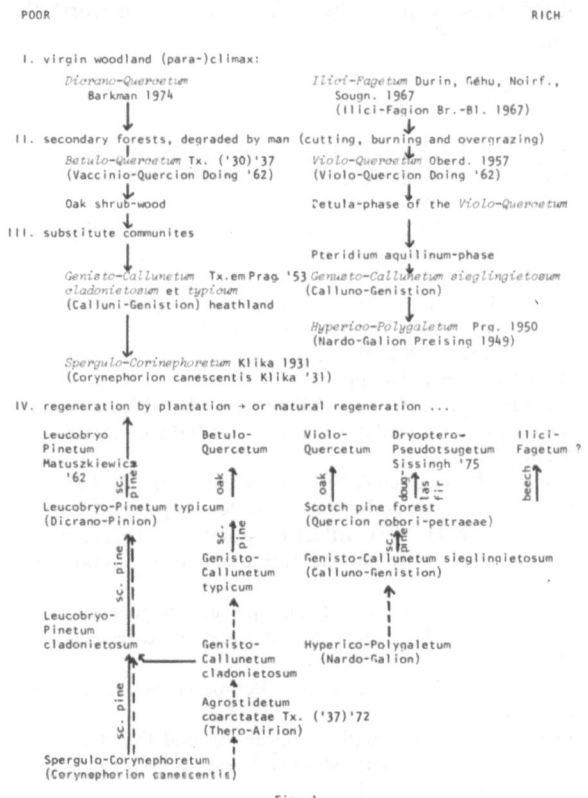

Fig. 1.

biomass. It may coincide with the climax but it need not. In the scheme presented above such is surely not the case. Whereas the *Ilici-Fagetum* is the climax, the coppice of sessile oak *(Violo-Quercetum petraeae)* is the secondary forest and the coppice of pedunculate oak *(Betulo-Quercetum roboris)* is a man-made substitute association (Ersatz-Gesellschaft). The equally man-made pine forest *(Leucobryo-Pinetum, Dicrano-Pinion)* is a pioneer forest and the douglas fir-forest *(Dryoptero-Pseudotsugetum)* approximates the optimum-community. This process is still going on. The *Dryoptero-Pseudotsugetum* is an association in statu nacendi. In the second generation its timber-production is already considerably greater than in the first generation (van Goor 1974). The floristic composition in this fern-douglas-fir-forest is in development as well (Sissingh 1976).

As a rule this development goes faster on soil rich in minerals (loamy soils or soils with a previous agricultural history) than on poor soils arisen from heathland-reclamation.

Development of the optimal forest

Most of our cultivated douglas-fir-forests are still even-aged pure stands. Therefore these stands – as a rule with horizontal density of canopy – are not stable and especially sensitive to winds and gales. According to the transformation of into uneven-aged stands by natural regeneration and their mixing with storm-proof species, stability will be built up. For instance we can give a structure to the forest by *Pinus silvestris* reserved from a preceeding generation, by coniferous species with a tap root as *Abies grandis* or shade-bearing hardwood species with a rapid youth-growth as common beech *(Fagus silvatica)* and American oak *(Quercus borealis)*.

Finally the durability has to be guaranteed by a form of group felling (Femelschlag). Regeneration over large areas has to be avoided because not only the micro-climate – a condition for douglas-fir – would be disturbed, but also the destruction of humus is strongly stimulated by access of light. A clearance plant-community like the *Corydalo-Epilobietum* appears and the development process to the optimal forest has been turned back. Removal of forest litter, which for some years was thought to be sufficiently compensated by adding an equal amount of minerals and organic matter, e.g. some tons of V.A.M.-compost, is now considered very unwise. Not only because it results in the destruction of an important part of the hair roots of the stand, but also because it obstructs the development of the ecosystem in the direction of the optimal forest.

Evaluation

The State Forest Service in the Netherlands consists of three divisions, each of which approaches the forest from its own point of view. Whilst the division of forestry considers the forest from the viewpoint of timber-production, the division of nature conservancy aims primarily at forest-protection and the division of landscape-architecture takes care of the scenery. We can show this clearly by representing the wood as an equilateral triangle, in which the three divisions operate from a different angular point *(Figure 2)*.

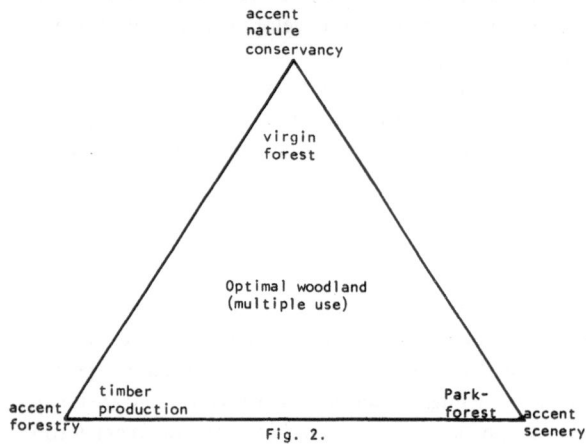

Fig. 2.

In the centre of the triangle we situate the optimal wood (the multiple use forest) where timber production, forest protection and scenery are balanced. It is an ecosystem of different tree-species (not entirely natives but also exotic species) from different ages and dimensions. It is a forest which not only has a high production of timber but is also sound, stable, durable and multilateral. From the point of view of vegetation science it is a 'substitute-community' since the climax, which just like other substitute-communities like heathland *(Genisto-Callunetum)* and semi-natural grass-communities *(Violion caninae, Cirsio-Molinietum* etc.) can regenerate naturally by purposive intervention of man and can be maintained permanently through rejuvenation. Provided it has the right composition of trees and is well managed, it also can deliver continously a much higher production than the virginal forest.

In our effort to establish the optimum forest we have to realise that time plays an important role. Our present day forests are hardly grown out of the stage of youth. They are still composed by pioneer species like *Pinus silvestris* or

Larix leptolepis. These monotonous forests are not in balance. They can be affected by insects, burned to death by fire or blown down by gales. In that case we have to take the consequences and start at the beginning.

To fortify our forests we have to mix them with other species and therefore bring in some so called 'more demanding' trees, i.e., trees that need a micro-climate and a litter-layer for good growth. We bring them in by planting three or four years old plants. It will take a generation of trees – that is to say more generations of man – before they are fertile, produce cones and seeds and rejuvenate. Meanwhile we have to learn about composition and management of this future forest, how to make it sound and stable. How does the mixture of tree-species behave in competition? Which blending makes our forest stable for gales? In which proportions do we have to plant or rejuvenate our tree-species? How do we have to tend thinnnings? What will our cutting- and rejuvenate-system be?

Of course we also have to avoid the mistakes we made in the past, like plundering the litter-layer or clear cutting on large scale. Even the setting in of heavy earning-machines, who invent the soil and wound the roots can nullificate much of the achievement. So we still have a lot to learn. But the ultimate aim – the optimum forest – is worth the trouble.

References

Adriani, N.J. & J. Vlieger. 1936. Plantensociologisch onderzoek, in het bijzonder van de midden Nederlandse bossen. Natuurwet. Tijdschr. 18: 123–139

Bannink, J.P., H.N. Leijs & I.S. Zonneveld. 1973. Vegetatie, groeiplaats en boniteit in Nederlandse naaldhoutbossen. Versl. Landbouwk. Onderz. PUDOC, Wageningen.

Barkman, J.J. & V. Westhoff. 1969. Botanical evaluation of the Drenthian district. Vegetatio 19: 330–388.

Braun-Blanquet. 1964. Pflanzensoziologie. 3e Aufl. Springer, Wien/New York. 865 pp.

Braun-Blanquet, J. 1967. La chênaie acidophile ibéro-atlantique (Quercion occidentale) en Sologne. SIGMA Comm. 175, Montpellier.

Broek, J.M.M. van den & W.H. Diemont. 1966. Het Savelbos. Bosgezelschappen en bodem. PUDOC, Wageningen.

Damman, A.W.H. 1971. Effect of vegetation changes on the fertility of a Newfoundland forest site. Ecol. Monogr. 41: 253–270.

Diemont, W.H. 1938. Zur Soziologie und Synoekologie der Buchen- und Buchenmischwälder der nordwestdeutschen Mittelgebirgte. Diss., Wageningen.

Diemont, W.H. 1942. Het wintereiken-berkenbosch in Nederland. Ned. Kruidk. Arch. 52: 309–310.

Doing, H. 1962. Systematische Ordnung und floristische Zusammensetzung niederländischer Wald- und Gebüschgesellschaften. Wentia 8: 1–85.

Doing, H. 1969. Assoziations Tabellen von Niederländischen Wäldern und Gebuschen. Lab. v. Plantensyst. en Geografie. Landbouwhogesch. Wageningen.

Doing-Kraft, H. & V. Westhoff. 1959. De plaats van de beuk (Fagus silvatica) in het midden- en westeuropese bos. Jaarb. Ned. Dendr. Ver. 21: 226–254.

Ellenberg, H. 1963. Vegeration Mitteleuropas mit den Alpen. Einführung in die Phytologie. IV. (H. Walter, ed.). Ulmer, Stuttgart.

Ellenberg, H. & F. Klotzli, 1972. Waldgesellschaften und Waldstandorten der Schweiz. Mitt. Schweiz. Anstalt Forstl. Versuchswesen 48 (4)

Frileux, P.-N. 1975. Contribution à l'étude des forêts acidiphiles de Haute Normandie. In: J.M. Géhu (ed.), La végétation des forêts caducifoliees acidiphiles, pp. 287–300. Cramer, Vaduz.

Géhu, J.M. 1973. Unites taxonomiques et végétation potentielle naturelle du nord de la France. Doc. Phytosoc. 4: 1–22. Lille Bailleul.

Goor, C.P. van. 1975. Een analyse van de keuzemogelijkheden by de bestemming van de stormvlakten in het Nederlandse bos. Ned. Bosb. Tijdschr. 47, 2.

Groszkopf, W. 1950. Bestimmung der charakteristischen Feinwurzel-Intensitäten in ungünstigen Waldbodenprofilen und ihre oecologische Auswartung. Mitt. Bundesanstalt f. Forst. Holzwirtschaft. 1950, 11.

Hoyos, G. 1953. L'Ardenne et l'Ardennais. Ed. Univers. Bruxelles-Paris.

Kelley, D. & J.J. Moore. 1975. A preliminary sketch of the Irish acidiphilous oakwoods. In: J.-N. Géhu (ed.), La végétation des fôrets caducifoliees acidiphiles, pp. 375–387. Cramer, Vaduz.

Lohmeyer, W. & W. Trautman. 1976. Zur Kenntnis der Waldgesellschaften des Schutzgebietes 'Taubergieszen'. In: Das Taubergieszengebiet. Die Natur- und Landschaftsschutzgebiete Baden-Württembergs, Band 7.

Meijer Drees, E. 1936. De bosvegetatie van de Achterhoek en enkele aangrenzende gebieden. Diss. Wageningen

Noirfalise, A. & N. Sougnez. 1956. Les chênaies de l'Ardenne verviétoise. Pédologie 6: 119–143. Centr. Cart. Phyt. Centr. Rech. Ecol. Phyt. Gembloux. Comm. 25.

Noirfalise, A. & A. Thill. 1958. Les chênaies de l'Ardenne centrale. Centre Cart. Phyt. Centr. Rech. Ecol. Phyt. Gembloux. Comm. 28.

Oberdorfer, E. 1957. Süddeutsche Pflanzengesellschaften. Fischer, Jena.

Sissingh, G. 1970. De plantengemeenschappen in onze naaldhoutbossen. Ned. Bosb. Tijdschr. 42: 157–162.

Sissingh, G. 1970. Dänische Buchenwälder. Vegetatio 21: 245–254.

Sissingh, G. 1975. Forêts caducifoliées acidiphiles dans les Pays-bas (Quercion robori-petraeae). In: J.-M. Géhu (ed.), La végétation des forêts caducifoliées acidiphiles, pp. 363–373. Cramer, Vaduz.

Sissingh, G. 1976. Niederländische Nadelforsten und ihr Humus als Substrat für ihre Vegetation. In: Vegetation und Substrat. Ber. Symposium 1969 der Int. Ver. f. Vegetationskunde. Cramer, Lehre, pp. 317–341.

Sissingh, G. 1976a. Eisen van onze houtsoorten aan het klimaat. Ms.

Trautman, W. 1972. Erlauterung zur Karte-Vegetation (Potentielle natürliche Vegetation). In: Deutscher Planungsatlas Band I. Nordrhein-Westfalen, Lieferung 3. Jäneke-Verlag, Hanover.

Trautman, W. et al. 1973. Vegetationskarte der Bundesrepublik Deutschland 1 : 200.000. Potentielle natörliche Vegetation Blatt CC 5502 Köln. Schriftenreihe f. Vegetationskunde, Heft 6. Bundesanstalt f. Vegetationskunde, Naturschutz u. Landschaftspflege. Bonn-Bad Godesberg.

Tüxen, R. 1937. Die Pflanzengesellschaften Nordwestdeutschlands. Mitt. Flor.-soz. Arbeitsgem. N.F. 3: 1–170.

Tüxen, R. 1955. Das System der Nordwestdeutschen Pflanzengesellschaften. Mitt. Flor.-soz. Arbeitsgem. N.F. 5: 155–176.

Tüxen, R. & W.H. Diemont. 1937. Klimaxgruppe und Klimaxschwarm. Jahresber. Naturhist. Ges. Hannover 88/89: 73–87.

Westhoff, V. 1958. Boden- und Vegetationskartierungen von Wakd- und Forst-gesellschaften im Quercion robori-petraeae Gebiet der Veluwe (Niederlande). Angew. Pflanzensoz. 15: 23–30.

Wolterson, J.F. 1976. Leven met bomen en bossen. Geschiedenis, huidige en toekomstige functie van het Nederlandse bos. Staatsuitgeverij, Den Haag.

85

DIE BEDEUTUNG DES KLIMAS FÜR DIE ZUSAMMENSETZUNG DER VEGETATION SW-UNGARNS, DES ELSASZ UND DER UMGEBUNG VON BRIANÇON, ALPES MARITIMES*

A.O. HORVAT**

Janus P.u. 8, H-7621 Pécs, Ungarn

Keywords:
Hungary, Pannonicum, Phytogeography, Submediterranean climate

Einleitung

Während das südliche Ungarn von der Tiefebene beherrscht wird, erstreckt sich im Südwesten des Landes eine Berggegend mit submontanem Charakter, die noch pflanzengeographischen Gesichtspunkten (Abb. 1) folgendermaßen unterteilt werden kann: Der höchste Abschnitt, das Mecsekgebirge (683 m), liegt nördlich von Pécs (Mecsekense). Etwas südlich davon in der Nähe der Drau befindet sich das Villányer Gebirge (422 m, Villányense). Das Klima, die Flora und die Vegetation beider Gebirge haben submediterranen Charakter, wobei die südliche Eigenart des zuletzt genannten höher ist. Westlich vom Mecsekgebirge liegt das unter dessen Einfluß stehende Zselicense, das aber eher ein Hügelgebiet, reich an Florenelementen des westlichen Transdanubiens (= Ungarn westlich der Donau), darstellt. Im Draugebiet (Dravense) finden wir eine reiche Flora mit submontanen Elementen. Das Mecsekense wird weiters im Nordosten von der Tiefebene (Colocense, d.h. die Donaugegend), im Westen vom Somogyicum interius mit Sandböden umgeben. Der teilweise submediterrane Charakter der Flora aller dieser Gebiete wird neben den pedologischen Gegebenheiten vor allem vom Klima und hier weniger von den Temperaturverhältnissen, sondern von der Menge und Verteilung der Niederschläge im Jahresablauf bestimmt.

Früher bezeichnete man viele mittel- und ost-submediterrane Florenelemente als 'illyrisch', obwohl sie auch in den submediterranen Teilen des Balkans und Italiens aufzufinden sind. Als echt subillyrisches Gebiet, mit Arten, die tatsächlich ihre Hauptverbreitung im NW

der Balkanhalbinsel haben, kann man nur einen kleinen Landstrich an der Drau südlich Nagykanizsa bezeichnen (Örtilosense), der in Niederschlagsverteilung und Flora dem benachbarten kroatischen Illyricum septentrionale orientale nahesteht. Die klimatische Ähnlichkeit beider Regionen ist aus den Gaussen- Walterschen Klimadiagrammen von Nagykanizsa und Lepoglava ersichtlich. Außerdem kommt nach Meusel und Mitarb. (1965) im nordöstlichen Kroatien in Klima und Vegetation evenfalls ein submontaner Einfluß zur Geltung. Das Klima des Illyricums unterscheidet sich von dem mehr kontinentaler submediterraner Gebiete des südöstlichen Europa vor

Abb. 1. Die pflanzengeographische Lage des südlichen Ungarns.

* Nomenclature follows Horvát (1972).
** Paper dedicated to Prof. Dr. V. Westhoff at the occasion of his 60th birthday.

allem durch seine hohen Niederschläge, die jedoch in submediterraner Verteilung, mit Frühjahrs- und Herbstmaxima, fallen.

Angaben zum Klima des Mecsekgebirges nach F. Simor.

Der berühmte Klimatologe dieser Region, F. Simor (1969), schreibt: 'Die seit vielen Jahrzehnten durchgeführten makroklimatischen Untersuchungen haben ermöglicht, das Klima dieser Landschaft mit den Klimaten anderer Regionen Ungarns zu vergleichen. Man muß nachdrücklich darauf hinweisen, daß hier ein von den übrigen Teilen Ungarns abweichendes spezifisches Makroklima herrscht.' Horvát & Papp (1962) und Horvát (1972) haben im Mecsekense und Villányense Mikroklima-Studien gemacht.

Lokale, spezifische, makroklimatische Eigenschaften sind folgende: früher Beginn des Frühlings, warmer, aber nicht sehr heißer Sommer, langer Herbst, milder, schneereicher Winter, typisch submediterraner Niederschlagsablauf (siehe Abb. 3). Dabei ist zu betonen, daß diese Kennzeichen sind auf das Mecsekgebirge, aber noch viel mehr auf das südlich davon gelegene Villányer-Gebirge beziehen.

Diese günstigen Klimamerkmale hängen teils mit der geographischen Lage, teils mit den spezifisch orographischen Verhältnissen zusammen. Unsere Landschaft ist der südlichst gelegene Teil Ungarns (45°45'–46°25' nördl. Breite) mit einer verhältnismäßig reichen Sonneneinstrahlung, hoher Mitteltemperatur und der längsten frostsicheren Periode von ganz Ungarn (mehr als 200 Tage). Dabei liegt unsere Region in Ungarn relativ dem mediterranen Raum am nächsten, somit ist dieses Gebiet das primäre Einfallgebiet der milden, feuchten mediterranen Luftmassen, wodurch submediterrane Züge, wie die Milde des Winters und die große Menge von Herbstniederschlag zur Geltung kommen können. Vor allem die Südhänge des Mecsekgebirges und des Villányer Berglandes, die beide aus Kalk aufgebaut sind, zeigen submediterrane Klimazüge, die sich in der Vegetation und dem Kulturpflanzensortiment bemerkbar machen. An diesen Südabhängen übersteigt die Versorgung mit Sonnenenergie die aller übrigen Gebiete Ungarns.

Flora und Vegetation des Mecsekgebirges als Ausdruck des Klimas

Die Bedeutung des Klimas ist aus der Florenanalyse gut

ersichtlich: Ein Viertel der Arten kann als submediterran im weiteren Sinne bezeichnet werden, wobei aber die submediterranmitteleuropäischen nicht eingerechnet sind. Davon sind 15,4 % als allgemein submediterrane, 2,2 % als ost-submediterrane, 4,3 % als pontisch-submediterrane und 2,2 % als subatlantisch-submediterrane Arten zu bezeichnen. Diese Zusammensetzung entspricht der früher beschriebenen klimatischen Eigenart der Region.

Eine Vegetationsanalyse zeigt uns, daß Pflanzengesellschaften auf südexponierten Kalkhängen am reichsten an submediterranen Elementen sind. Im Flaumeichenbuschwald des Mecsekgebirges (*Orchidi simiae – Quercetum pubescentis, Inulo spiraeifoliae – Quercetum pubescentis, Cotino – Quercetum mecsekense*) machen sie sogar 32,2 % aus, wozu noch balkanische Arten aus dem Südosten mit mit 7,5 % kommen. Da der Karst-Flaumeichenbuschwald im Mosaik mit *Festucion*-Rasen auftritt, ist auch die Zahl der kontinentalen Arten hoch.

Ebenfalls einen starken submediterranen Einfluß zeigt der Flaumeichen-Mittelwald (*Tamo – Orno – Quercetum pubescentis, Rusco – Orno – Quercetum*). Hier kommt *Tamus communis* massenhaft vor, aber auch *Ruscus aculeatus* ist, wenn auch spärlich, zu finden. In den Zerreichen- und Traubeneichenwäldern (*Potentillo micranthae – Quercetum dalechampii*) sind submediterrane Elemente mit 20 % vertreten. Sogar in dem klimazonalen Eichen-Hainbuchenwald, der die Hälfte des Mecsek-

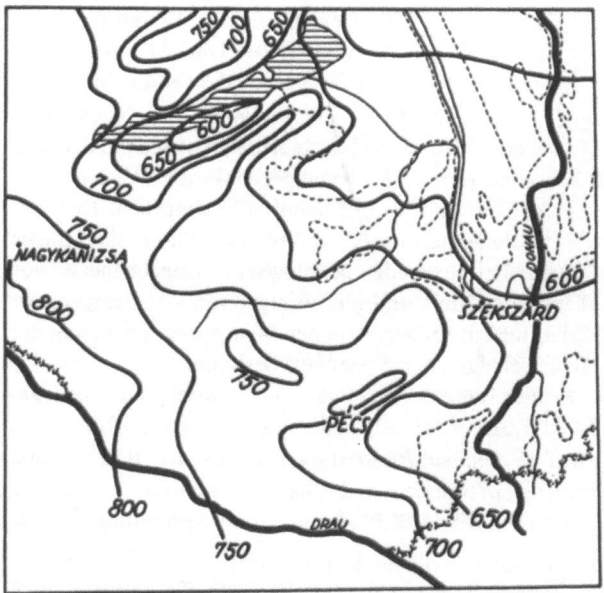

Abb. 2. **Regenfall Karte des Gebietes.**

88

gebirges bedeckt, und im extrazonalen Buchenwald (*Carici pilosae – Fagetum*) sind submediterrane und balkanische Arten zu finden (*Ruscus hypoglossum, R. aculeatus, Tamus, Helleborus odorus, Tilia argentea*).

Die Bedeutung der Niederschlagsmengen für die Zusammensetzung der Flora in unserem Gebiet wird dadurch anschaulich gemacht, daß für eine Reihe von Arten des Eichen-Hainbuchenwaldes die 650 mm Isohyete die Grenze darstellt und zwar vom Westen her z.B. für *Primula vulgaris, Knautia drymeia, Tamus* und *Helleborus odorus* (Abb. 2). Diese Linie markiert gleichzeitig die Ostgrenze der Verbreitung der Eichen-Hainbuchenwälder und der Braunerde in unserem Gebiet.

Dagegen erreichen *Asphodelus, Cyclamen, Erythronium* und *Vicia oroboides* nur den Westrand des Mecsekgebirges, dringen in dieses Gebiet aber nicht ein. Diese Arten kommen im Zselicense und im Somogycum interius vor. Das Bakonygebirge nordwestlich vom Mecsek und vom Plattensee (Balaton) weist mit unserem Klima, mit unserer Flora und Vegetation viele verwandte Züge auf (Balatonicum), während das an Niederschlag sehr reiche Örilosense südwestlich von Nagykanisza schon wahrhaft subillyrischen Charakter hat, mit einem dem noch regenreicheren Lepoglava in Jugoslawien ähnlichen Klimadiagramm und den illyrischen Arten *Anemone trifolia, Peucedanum verticillare* und *Lamium orvala*.

Vergleich von Klima, Flora und Vegetation im Elsaß und im südpannonischen Gebiet

Die Klimate des südpannonischen Gebietes (s.Abb. 3) und des elsässischen Raumes im Regenschatten der Vogesen (s. Abb. 4) zeigen Gemeinsamkeiten, auf die schon Issler (1951) hingewiesen hat. Die relativ hohen Sommer- und niedrigen Wintertemperaturen machen das Gebiet

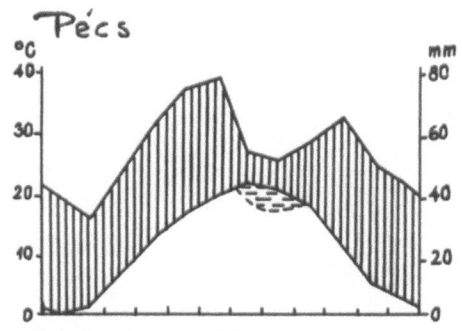

Abb. 3. Klimadiagramm von Pécs.

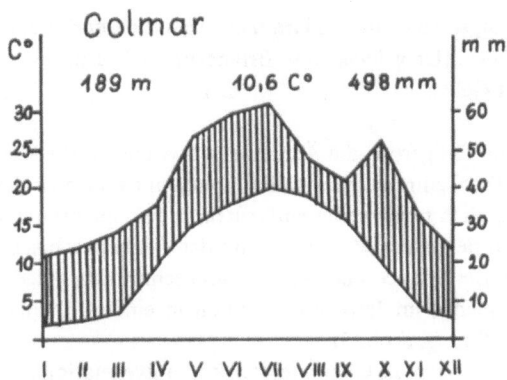

Abb. 4. Klimadiagramm von Colmar.

von Colmar und Straßburg in der Rheinebene zu einer subkontinentalen Klimainsel, die aber auch subatlantischen und submediterranen Klimaeinflüssen ausgesetzt ist. Besonders die geringen Sommerniederschläge sind ein submediterraner Zug des Klimas, der submediterranen Arten zugute kommt. Dies zeigt sich besonders im Gedeihen des *Quercus pubescens* – Waldes.

Ein subkontinentales Klima mit subatlantischen und besonders in seinem südwestlichen und südlichen Teil submediterranen Einflüssen charakterisiert auch den pannonischen Raum. Sowohl im Elsaß wie auch im pannonischen Gebiet werden diese klimatischen Einflüsse durch Flora und Vegetation widergespiegelt. Die natürliche Vegetation der elsässischen Trockeninsel ist durch das subkontinentalsubmediterran bestimmte *Potentillo-Quercetum* s.l. gekennzeichnet. In der Landwirtschaft deuten Wein-, Mais- und Obstkulturen auf die hohe Sommerwärme hin. Die klimatischen Verhältnisse erklären das Auftreten von *Festucion* und *Festucetalia* sowie *Quercion pubescentis* und *Quercetalia* Arten: *Potentilla alba, Anthericum liliago, Dictamnus albus, Anemone sylvestris, Melampyrum cristatum, Orchis simia, O. purpurea, Muscari botryoides, Lithospermum purpureo-coeruleum, Adonis vernalis, Aster amellus, Geranium sanguineum, Brachypodium pinnatum, Inula hirta, Achillea collina, Potentilla arenaria, Veronica spicata, V. prostrata, V. teucrium, Teucrium montanum, Verbascum lychnitis, Seseli annuum, Peucedanum cervaria, P. oreoselinum, Asperula glauca, A. cynanchica, Carex humilis, Thalictrum minus*.

89

Beziehungen von Klima, Flora und Vegetation in den Alpes Maritimes (Umgebung von Briançon) und dem pannonischen Gebiet

Noch auffälliger als die Ähnlichkeit zwischen der elsässischen Trockeninsel und dem Pannonicum ist die zwischen der räumlich noch weiter entfernten Ebene um Briançon, westlich der Alpes Maritimes und dem südpannonischen Gebiet. Bei Briançon beträgt die durchschnittliche Niederschlagsmenge im Jahr bloß 550 mm in einer Höhenlage von 1200 m (s. Abb. 5).

Der kontinentale Charakter der Ruderalvegetation ist so deutlich (*Salvia verticillata, Sisymbrium strictissimum, Diplotaxis muralis*), daß er sogar vom fahrenden Autobus aus auffällt. Besonders reich sind die Trockenrasen an kontinentalen Arten: *Stipa pennata, Lactuca viminea, Silene otites, Salvia aethiopis, Herniaria incana, Chondrilla juncea, Astragalus austriacus, Seseli annuum, Lactuca perennis, Potentilla argentea, Koeleria vallesiana, Astragalus alopecurioides.*

Ähnliches hat auch Braun-Blanquet (1961) in den inneralpinen Trockentälern konstatiert. Auch in anderen Trockengebieten von Mitteleuropa (BRD, DDR, CSSR) ist ein Klima 'pannonischen Charakters' von subkontinentaler Flora und Vegetation begleitet.

Zusammenfassung

Die Unterschiede in Flora und Vegetation verschiedener (submontaner) Gebiete SW-Ungarns werden durch Vergleich auf klimatische Unterschiede, vor allem unterschiedliche submediterrane Klimaausprägung zurückgeführt. Vergleiche mit klimatisch ähnlichen Bereichen in der Rheinebene (Elsaß) und den nordwestlichen Alpes Maritimes zeigen, daß dort entsprechende Klimate von einer subkontinentalsubmediterranen Flora und Vegetation begleitet werden.

Summary

The differences in flora and vegetation in various (submontane) areas in SW Hungary result from differences in the submediterranean climate. Comparison with climatologically similar areas in the Alsace and in the NW Alpes Maritimes show that in those areas these climates are also indicated by a subcontinental-submediterranean flora and vegetation.

Literatur

Braun-Blanquet, J. 1961. Die inneralpine Trockenvegetation. Fischer, Stuttgart.
Horvát, A.O. 1972. Vegetation des Mecsekgebirges und seiner Umgebung. Akadémiai Kiadó, Budapest.
Horvát, A.O. & L. Papp. 1962. Mikroklimavizsgálatok a pécsi Mecsek növénytársulásaiban (Untersuchungen des Mikrohlimas der Pflanzengesellschaften des Mecsekgebirges). Erd. Kut. 58 : 137–163.
Issler, E. 1951. Trockenrasen und Trockenwaldgesellschaften der oberelsässischen Niederterrassen und ihre Beziehungen zu denjenigen der Kalkhügel und der Silikatberge des Osthanges der Vogesen. Ber. Schweiz. Bot. Ges. 61 : 664–699.
Meusel, H., E. Jäger u. E. Weinert. 1965. Vergleichende Chorologie der zentraleuropäischen Flora. Fischer, Jena.
Simor, F. 1965. Adatok a Délkelet Dunántul éghajlatához (Angaben zum Klima von Südostdanubien). MTA Dunántuli Tudományoas Intézete Értekezések 1964–1965. 103–216/1–114.

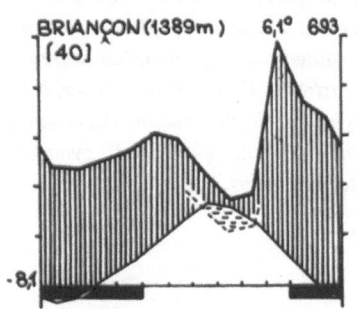

Abb. 5. Klimadiagramm von Briançon.

GEOGRAPHICAL CHANGES IN THE VEGETATION PATTERN OF RAISED BOGS IN THE BAY OF FUNDY REGION OF MAINE AND NEW BRUNSWICK*, **

A.W.H. DAMMAN***

Ecology Section, Biological Sciences Group, University of Connecticut, Storrs, Conn. 06268, USA

Keyword:

Fire, Plant communities, Raised bog, Regional differences, Snow cover, Vegetation classification, Water levels

Introduction

In eastern North America raised bogs occur roughly north and east of a line connecting stations with an average annual water deficiency (Thornthwaite 1948) based on a soil storage capacity of 10 cm (*Figure 1*). In the southern part of their range they are particularly abundant in an area near the Bay of Fundy which corresponds more or less with the Fundy Coast Section of the Acadian Forest Region (Rowe 1972) and the area mapped as spruce fir forest by Davis (1966).

Ganong (1891, 1897) was the first to recognize the different nature of the coastal bogs of New Brunswick which he considered to be the only raised bogs in the region. Several authors (Nichols 1919, Dachnowski-Stokes 1929, 1930, Osvald 1928, 1955) give a general description of the bogs of the area. Osvald (1928) described some of the general trends in the North American peatbog vegetation, and points out that species such as *Scirpus cespitosus*, *Rubus chamaemorus*, and *Empetrum nigrum* are common in

the bogs of the coastal area. However, he nor any of the others report on the strikingly different topography and vegetation cover of the bogs in a narrow coastal belt.

The purpose of this paper is (1) to describe the floristic composition of the plant communities of these ombrotrophic bogs, (2) to show the differences in topography and vegetation pattern between the coastal and inland bogs, and (3) to discuss the reasons for these differences. Some comparisons with European bogs will also be made.

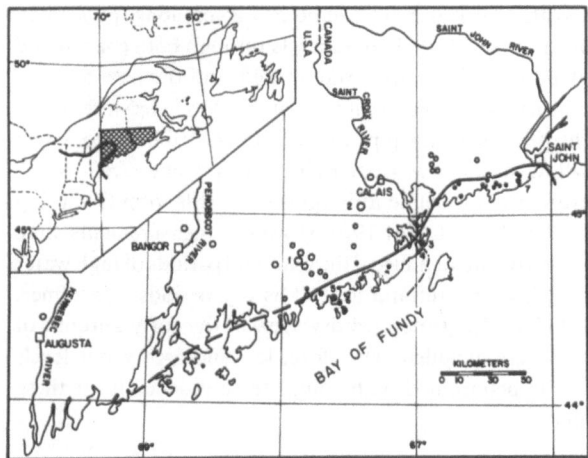

Fig. 1. The distribution of plateau and convex raised bogs along the west side of the Bay of Fundy. Numbers refer to bogs mentioned in the text: 1-Pleasant River bogs, 2-Meddybemps Heath, 3-Campobello Island, 4-Quoddy Neck bog, 5-Jonesport bog, 6-Kelley Point bog, 7-Spruce Lake bog. The location of the study area along the east coast of North America is shown in insert; the heavy line is the north-eastern limit of continuous area with an average annual water deficiency based on 10 cm storage.

* The nomenclature for vascular plants follows Fernald (1950), for lichens Hale & Culberson (1966), for hepaticae Schuster (1953), and for mosses Crum et al. (1965), except for *Sphagnum rubellum* Wils., *S. nemoreum* Scop., *S. apiculatum* Lindb., *S. parvifolium* Warnst., *Polytrichum strictum* Menz. ex Brid., and *Dicranum bergeri* Bland.
** Dedicated to Professor Victor Westhoff on the occasion of his 60th birthday.
*** This research was supported by NSF grant GB-8744 and grants from the University of Connecticut Research Foundation. I wish to thank K.J. Metzler for carefully tabulating the vegetation data, and A.E. Brower and W.J. Meades for their help in locating some bogs in Maine and New Brunswick, respectively.

Methods

The field data for this paper were collected during studies carried out in these bogs since 1969. The vegetation was described in plots large enough to include the normal species combination but care was taken to select plots homogeneous with respect to vegetation and soil conditions. To satisfy both criteria in bogs with a well-developed pattern of ridges and depressions, it was occasionally necessary to combine into one plot several small ridges or depressions. Abundance/cover and sociability of all species present were rated using the well-known scales of Braun-Blanquet (Mueller-Dombois & Ellenberg 1974) but no sociability values were assigned to species in the shrub stratum. Descriptions were tabulated, analyzed for similarities and differences in species composition, and then organized in vegetation synthesis tables using the standard analysis techniques of the Zürich-Montpellier School (Mueller-Dombois & Ellenberg 1974). Notes made during other field trips in the area were used to complete the data.

The topography of the bog surface was measured with an optical level; depth measurements were carried out simultaneously with a Hiller peat borer on the same transects so that level changes of bog surface and of underlying mineral soil were determined for the same points. In this way an accurate profile through the bogs could be constructed.

Water levels were measured in 1.25 cm long, perforated plastic pipes located on transects through both coastal and inland bogs. The pipes were vented, and the pipe top was used as a reference point for water level measurements. The position of the pipe top with respect to hummock and hollow surfaces was determined with a line level. Water levels were measured at irregular intervals from the spring of 1970 to the fall of 1976. However, measurements were carried out at all times of the year, and periods of high water in spring and autumn as well as dry periods in summer, including the prolonged dry spell in the early autumn of 1975 were included. Therefore, low and high water levels can de determined with complete confidence from these data.

The sucrose inversion method (Pallman 1940, Berthet 1960) was used to obtain average temperatures during the vegetative period. Sucrose ampoules were placed horizontally at 2.5 cm below the *Sphagnum* surface so that temperatures were measured at a depth from 2.5 to 4.5 cm. Sites were selected carefully to assure that temperatures applied to comparable conditions in different bogs.

Floristic composition of the vegetation

In order to discuss the differences between coastal and inland bogs it will be necessary to describe first the floristic composition of the plant communities found in the peat bogs. This discussion will be restricted to the vegetation of the ombrotrophic bog areas and the immediately adjacent extremely nutrient-poor fen vegetation. The remaining bog border vegetation is controlled primarily by the topography and parent material of the surrounding land, and therefore climatic differences are not so clearly expressed in this vegetation.

Four main groups of plant communities can be recognized: (1) dwarf shrub heaths, (2) *Scirpus cespitosus* lawns and solid *Sphagnum* carpets, (3) extremely nutrient-poor fens and (4) mud bottom communities. In spite of the physiognomic and floristic differences the first three share a large group of species (Table 1). The mud bottom communities occupy the wettest areas. Floristically they have little in common with the other communities, although some of the species can occur locally in the wettest parts of the *Scirpus* lawn and poor fen vegetation. Each of the communities will be very briefly discussed below. Tables 1 and 2 give detailed information on their floristic composition.

Dwarf shrub heaths

These occupy the driest parts of the bogs. They are dominated by ericaceous dwarf shrubs. The moss layer consists mainly of *Sphagnum* spp., especially *S. fuscum*; lichens are usually present and they can become dominant after fire on some sites.

As a group this type is differentiated by a large group of species (Table 1), primarily species occurring also on nutrient-poor mineral soils. *Sphagnum fuscum*, *Dicranum bergeri* and *Calopogon pulchellus* are the only true bog species among them. *Rubus chamaemorus* is a good differential species elsewhere, but in Maine and southeastern New Brunswick it is restricted to the coastal areas.

Gaylussacia baccata community

This occupies the driest areas. The dwarf shrub layer is 50 to 60 cm high, and completely dominated by *Gaylussacia baccata*. Some more nutrient demanding species (*Pogonia ophioglossoides*, *Coptis groenlandica*, etc.) occur sporadically, probably because of the nutrient accumulation near the surface resulting from the slow growth of the *Sphagnum* mat (Damman 1977). Trees occuur locally, either as isolated individuals or in patches. *Picea mariana* is most common, but *Pinus strobus* and *Larix laricina* occur also.

Additional material from *Plant Species and Plant Communities,*
ISBN 978-90-6193-591-9 (978-90-6193-591-9_OSFO3), is available at http://extras.springer.com

Kalmia angustifolia-Sphagnum fuscum community

This is the most common dwarf shrub heath community. *Kalmia angustifolia* forms a 30–40 cm high dwarf shrub layer which determines the physiognomy of the vegetation. A number of species occurring most commonly on nutrient-poor mineral soils are shared with the *Gaylussacia* community (Table 1). Several species such as *Empetrum nigrum**, *Sphagnum flavicomans*, *S. imbricatum*, *Icmadophila ericetorum* and *Rubus chamaemorus*, occur regularly in this community in the coastal areas but are absent or extremely rare further inland. Low *Picea mariana* and *Larix laricina* trees and shrubs occur locally.

This community occurs on relatively well-drained peat, and can cover the major part of bogs in both coastal and inland localities.

Empetrum nigrum-Sphagnum fuscum community

Physiognomically, this is very clearly differentiated from other bog communities by the continuous *Sphagnum fuscum* carpet, the low height (10 cm) of the dwarf shrubs, and the prominence of *Rubus chamaemorus* and *Empetrum nigrum*. *Scirpus cespitosus* occurs scattered throughout, and there are several other differences in floristic composition (Table 1). This community is restricted to the coastal bogs. In extreme coastal positions *Sphagnum imbricatum* can be very abundant (Table 1, nos. 28 and 29) and *Myrica gale* and *Juniperus communis* occur here on the ombrotrophic bog surface.

Scirpus cespitosus-Sphagnum fuscum community

These communities occupy the wet parts of ombrotrophic bogs with the water table above the surface in spring and after heavy rains. They cover extensive areas in the coastal bogs but are restricted to depressions and margins of pools in the inland bogs. The moss layer is dominated by *Sphagnum rubellum* and *S. flavicomans*, and large *Cladonia* patches often occur. The herbaceous layer is dominated by *S !irpus cespitosus* except in the inland areas where this species is only locally abundant. *Gaylussacia dumosa*, *Sphagnum flavicomans*, *S. tenellum*, and the abundance of *Sphagnum rubellum* and *Scirpus cespitosus* mostly clearly differentiate it from the dwarf shrub heath communities (Table 1).

Scirpus lawns and solid Sphagnum carpets

On wet bog flats and in depressions *Scirpus cespitosus* and *Sphagnum* spp. determine the physiognomy, with *Cladonia* spp. often covering over 25% of the surface. Floristically, this community

---

* Löve & Löve (1959) assert that the North American *Empetrum nigrum* sensu lato is actually *Empetrum hermaphroditum* Hagerup. This is clearly in error. *E. hermaphroditum* shows the same geographic pattern as in Europe; it is distinctly more arctic in its distribution, and in the southern part of its range it is restricted to alpine areas and cold coastal headlands. *Empetrum nigrum* L. occurs commonly throughout eastern North America, and is the only *Empetrum* in the bogs described here.

---

forms the transition from the lawn communities to the *Empetrum-Sphagnum fuscum* community (Table 1). *Rubus chamaemorus* and *Empetrum nigrum* occur regularly in this community. Although several other differential species of the dwarf shrub heaths occur sparsely, they are poorly developed, and do not affect the vegetation structure. The surface is somewhat uneven with the higher parts averaging 10–15 cm above the hollows. The differential species of the dwarf shrub heaths extending into this community occur mainly on these low hummocks.

Scirpus cespitosus-Sphagnum flavicomans community

On wetter sites than the *Scirpus-Sphagnum fuscum* community; usually in the well-developed depressions, 'Schlenken'. The bog surface is almost flat with few *Sphagnum clumps* rising more than 5–8 cm above the surface. It is differentiated from the *Scirpus-Sphagnum fuscum* community by the absence of *Rubus chamaemorus*, *Polytrichum strictum*, *Sphagnum fuscum*, and species of the dwarf shrub heaths (Table 1), and by the presence of *Andromeda glaucophylla* and often *Arethusa bulbosa*. A number of species of wet mud bottoms occur sporadically. *Sphagnum rubellum* and *S. flavicomans* are the dominant Sphagna.

Chamaedaphne-Sphagnum rubellum community

This community is distinguished most easily by the continuous carpet of *Sphagnum rubellum* with patches of moribund *Cladonia crispata* and sometimes also reindeer lichens, especially *Cladonia arbuscula*. *Chamaedaphne calyculata* is abundant but poorly developed and rarely over 10 cm high. *Gaylussacia dumosa* and *Andromeda glaucophylla* occur regularly. *Scirpus cespitosus* can be locally abundant. *Rhynchospora alba* and *Eriophorum virginicum* are found sporadically. This community occupies large depressions and filled-in pools in the ombrotrophic bog centers of the inland high moors, but it can also occur along some pool margins.

Extremely nutrient-poor fens

Included here are two floristically and physiognomically rather different communities which occur on slightly nutrient enriched parts of the bogs. They share some indicators of weakly minerotrophic conditions (Table 1).

Smilacina-Sphagnum pulchrum community

It is found on parts of the bogs affected by minerotrophic water. It is distinguished from the other communities discussed here by the occurrence of many more nutrient-demanding species (Table 1), of these *Eriophorum angustifolium*, *Smilacina trifolia*, and *Myrica gale* are almost always present. A *Sphagnum* carpet is still present but *Sphagnum pulchrum*, *S. papillosum*, *S. apiculatum* and *S. flavicomans* are most important, and *S. rubellum* and *S. fuscum* occur only as isolated cushions. This community is most extensive in the moats of the coastal bogs.

93

Rhododendron canadense-Chamaedaphne community

This is a wet dwarf shrub heath occupying the very slightly nutrient-enriched parts of the bogs. If it occurs as a zone, this community indicates the mineral soil water limit. However, it can also be found in clearly ombrotrophic areas where it occupies the intermittently wet channels draining the higher parts of the bogs. *Chamaedaphne calyculata* completely dominates this vegetation but some other dwarf shrubs, i.e. *Rhododendron canadense*, *Kalmia angustifolia*, *Ledum groenlandicum*, *Andromeda glaucophylla* and *Kalmia polifolia* are often present. Several weakly minerotrophic vascular plants occur, but *Eriophorum angustifolium* and *Rhododendron canadense* are the only ones occurring regularly. In addition, *Vaccinium angustifolium*, *Nemopanthus mucronatus*, *Viburnum cassinoides* and *Larix laricina* occur in this community (Table 1). The moss layer is poorly developed and variable in composition, *Sphagnum teres*, *S. robustum* and *S. cuspidatum* are locally abundant.

Mud bottoms

This includes wet depressions with much bare peat and areas with a very loose *Sphagnum* carpet. In this region *Sphagnum cuspidatum* is the most important species in these carpets. The physiognomy varies greatly depending on whether *Sphagnum* spp. or *Cladopodiella fluitans* are dominant, and whether *Utricularia cornuta* is in flower. The vegetation of the mud bottoms of the ombrotrophic bogs of this region can probably best be included in two communities: the *Rhynchospora alba-Sphagnum cuspidatum* community and the *Utricularia cornuta-Cladopodiella fluitans* community (Table 2). The first occurs primarily along the margins of pools and in depressions with small water level changes or where the *Sphagnum* mat can move with the water table. The second community occupies the peaty depressions in the bog. Both cover negligible areas in the bogs of this region.

Distribution of plant communities in the raised bogs

The coastal bogs occur in a very narrow zone along the coast (Fig. 1). The differences in the topography of the coastal and inland bogs is clearly shown in *Figure 2*. These differences, as well as those in the vegetation cover, are most clearly expressed in large peat bogs. Size affects peat depth, the relative importance of minerotrophic water from the surrounding uplands, as well as the direct effect of surrounding hills and forests on wind force and snow drifting. In small peat bogs these local environmental factors may overrule the regional factors or confound the the vegetation pattern. For this reason the following discussion concerns primarily the large peat bogs, although much of it will also apply to many of the smaller bogs. In general, the small bogs will deviate increasingly from the regional pattern the more rugged the topography, the

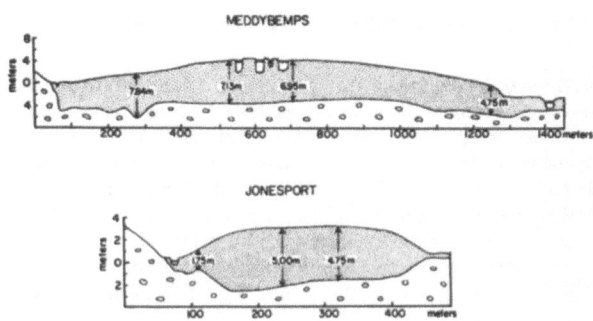

Fig. 2. Cross-section through an convex raised bog (Meddybemps Heath) and a coastal plateau bog (Jonesport bog). Note that the coastal bog has been drawn at twice the scale of the inland bog. The abrupt change in slope on the right side of the profile through Meddybemps Heath is a slump zone resulting from erosion by the brook.

nutrient-richer the surrounding mineral soil areas, and the higher and more uninterrupted the forest surrounding the bog.

Inland bogs

The inland bogs are clearly domed, and large bog complexes, such as those along the Pleasant River and at Meddybemps, consist of several raised areas separated by drainage channels. The centers of the domes of large bogs are frequently raised more than 4 m above the bog border. Bog pools occur usually in the central part of the bog.

The bog slope is very gentle and gradual (1 : 100–200), and consequently the minerotrophic bog border vegetation grades slowly into the ombrotrophic vegetation. A well-defined moat or lagg is usually absent.

Acer rubrum swamps and *Calamagrostis canadensis* meadows occur along the periphery of many of these raised bogs. The area covered by these vegetation types, as well as that of the poor fen vegetation, depends on the nature and abundance of seepage water from the surrounding upland areas, or the presence of flat areas near the bog border in which minerotrophic water accumulates. Without seepage, or if drainage of water from the bog border is rapid, an open, scrubby forest with high *Rhododendron canadense*, *Viburnum cassinoides*, *Nemopanthus mucronatus* and trees of *Picea mariana* and *Acer rubrum* may form the contact with the mineral soil (*Figure 3*). *Myrica gale*, *Spiraea latifolia*, *Sphagnum teres* and *S. apiculatum* usually occur here also.

The topographic sequence from the *Gaylussacia* community of the bog center to the bog border is essentially a

Table 2: Floristic composition of the mud bottom communities

	69	70	71	72	73	74	75	76	77	78	79	80	81
Description number	69	70	71	72	73	74	75	76	77	78	79	80	81
Location[1]	2	2	3	3	3	3	2	2	1	1	3	3	3
Plot size (m²)	6	24	15	20	15	10	15	16	15	16	15	12	9
Microtopography (cm)	0	0	0	9	0	0	0	0	5	5	8	0	0
Cladonia cover (%)	0	0	0	0	0	0	0	0	0	0	0	0	0
Sphagnum cover (%)	100	100	75	55	100	95	80	7	30	20	80	60	45
Number of vascular plants	7	3	7	9	5	4	9	6	12	8	10	9	7
Number of lichens and bryophytes	4	1	3	6	4	4	6	4	4	6	9	5	5
ASSOCIATION	Rhynchosporetum albae Koch 1926[2]												
VEGETATION UNIT	Rhynchospora - Sphagnum cuspidatum						Utricularia - Cladopodiella						
HERB LAYER cover (%)	40	40	25	20	20	20	35	35	40	40	25	25	30
height (cm)	15	15	10	15	25	15	12	15	13	12	25	20	20
MOSS LAYER cover (%)	100	100	95	90	100	100	100	25	80	100	100	100	100
Character and differential (*) species[3] of the Rhynchosporion and Rhynchosporetum													
Sphagnum cuspidatum	5.5	5.5	3.3	4.3	5.5	5.5	2.3	.	.	+.2	2.2	3.3	3.3
Cladopodiella fluitans	1.2	.	2.3	3.3	2.5	2.5	2.3	2.3	4.3	5.5	2.3	3.3	4.3
Rhynchospora alba	3.5	.	2.5	2.2	2.2	2.2	3.4	3.4	1.2	2.2	2.2	2.2	2.2
Sphagnum pulchrum	.	.	.	1.2	1.1	1.1	.	.	1.2	.	2.1	2.4	2.1
Eriophorum virginicum	.	.	.	.	+.2	.	1.1	.	+.1	+.1	1.1	+.2	.
* Drosera intermedia	+.1	+.1	2.1	2.1	2.1	1.1	+.1	1.1	1.1	.	2.1	1.1	2.1
* Xyris montana	.	.	.	1.2	+.2	1.2	1.2	.	.	.	2.2	1.2	1.2
* Utricularia cornuta	1.1	.	.	.	.	.	+.2	1.4	2.3	2.5	2.3	2.5	2.4
Transgressing species of the Oxycocco-Sphagnetea													
Chamaedaphne calyculata	1.1	.	1.1	+.2	.	.	2.2°	1.1	1.1	1.1	.	.	.
Andromeda glaucophylla	1.1	.	1.1	1.1	+.1	1.1	.	+.2	+.2	+.2	1.1	1.1	1.1
Vaccinium oxycoccos	1.1	.	1.1	.	.	.	+.1	2.4	1.1	1.1	1.1	1.1	1.1
Drosera rotundifolia	.	.	.	+.2	.	.	2.1	.	+.1	+.1	1.1	+.2	.
Kalmia polifolia	.	.	+.1	+.1	.	.	.	.	+.1	.	.	.	.
Sarracenia purpurea	.	.	.	.	.	.	r.1°	.	r.1°	.	.	.	.
Sphagnum magellanicum	+.1	.	.	+.2	1.1	1.1	1.1	+.2	.	+.2	2.1	1.1	+.2
Sphagnum rubellum	.	.	.	.	.	.	.	2.5	+.2	1.2	1.2	2.1	.
Cephalozia connivens	.	.	.	.	.	.	.	.	.	.	1.1	.	.
Microlepidozia setacea	.	.	.	.	.	.	.	.	.	.	r.1	.	.
Other species													
Sphagnum tenellum	1.1	.	3.3	+.2	.	.	+.1	1.1	2.3	2.2	.	.	.
Sphagnum flavicomans	.	.	.	2.2	.	.	4.4	.	.	.	3.3	+.2	+.2
Scirpus cespitosus	.	.	+.2	1.2	.	.	.	.	2.1	+.2	+.2	1.2	.
Vaccinium macrocarpon	2.5	3.5	.	2.1	.	.	.	.	.	.	.	.	2.5
Cetraria islandica	.	.	.	.	.	.	.	.	.	+.2	.	.	.
Pyrus melanocarpa	.	.	.	.	.	.	.	.	+.1	.	.	.	.
Eriophorum angustifolium	.	2.5	.	.	.	.	.	.	.	.	.	.	.
Eriophorum chamissonis	.	.	.	.	.	.	.	.	.	.	1.1	.	.
Drepanocladus fluitans	.	.	.	.	.	.	.	.	.	.	+.1	.	.

1. See Table 1 for further details on location
2. Or at least the North American counterpart of the European Rhynchosporetum albae; North American species are underlined.
3. Differential species with respect to the other Rhynchosporion communities; Rhynchospora alba itself is also most abundant in the Rhynchosporetum.

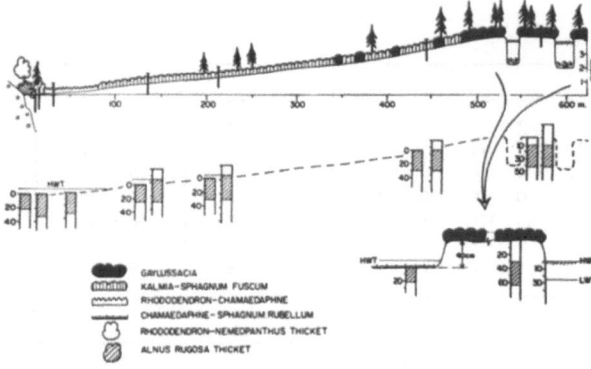

Fig. 3. Water level fluctuation and distribution of plant communities in a part of Meddybemps Heath, a large convex raised bog outside the coastal belt. For communities with an irregular surface the water level extremes are shown with respect to both average hollow and hummock surface.

moisture gradient (Fig. 3) but from the outside margin of the *Kalmia-Sphagnum fuscum* community increasing fertility also affects the floristic composition of the vegetation. The bog surface rises only very slightly from the contact with the mineral soil to the *Kalmia-Sphagnum fuscum* community. This part of the bog is flooded in early spring and after heavy rains in summer and fall. It shows the largest fluctuations in water level even though the minimum water level was never more than 33 cm below the surface.

A distinct change in the microtopography of the bog surface takes place at the boundary between the *Rhododendron-Chamaedaphne* and the *Kalmia-Sphagnum fuscum* community. The bog surface of the former is essentially smooth, although low *Eriophorum spissum* tussocks rise above the surface. However, from this boundary to the bog center the surface has a clear and regular hummock-hollow topography. The dwarf shrubs and most of the *Sphagnum* species, especially *Sphagnum fuscum*, form the hummocks whereas *Cladonia* is most abundant in the hollows.

The *Gaylussacia baccata* community is the best drained of the bog communities because it occupies the highest parts of the bog and due to the presence of pools which facilitate drainage after rain. The hollows of this community are never flooded, and the water level fluctuates from 30 to over 60 cm below the average hummock surface. In contrast, high water levels are at or slightly above the hollow surfaces throughout the *Kalmia-Sphagnum fuscum* zone, and water levels fluctuate from 14 to less than 50 cm below the average hummock surface.

The *Chamaedaphne-Sphagnum rubellum* community,

and the mud bottom communities described here, are restricted to the ombrotrophic part of the bog. They occur primarily in the central part dominated by the *Gaylussacia* community but can also be found in eroded parts of the *Kalmia-Sphagnum fuscum* zone or, fragmentarily developed, in some of the deeper hollows of the zone.

The distribution of these bog-center communities is entirely controlled by the depth of the water table (Fig. 3). The *Chamaedaphne-Sphagnum rubellum* community occupies large depressions, often 10–50 m in diameter, some of which represent filled-in bog pools. These depressions are flooded in spring and the water level drops to 25–30 cm below the surface during dry spells in summer. Water level changes of 27 cm were measured in the pools. Nevertheless, water levels vary much less in the *Rhynchospora-Sphagnum cuspidatum* community at the pool margins because the vegetation surface follows the water level fluctuations to some extent. Water levels in the mud bottoms with the *Utricularia cornuta-Cladopodiella* community ranged from 10–15 cm above the surface to 17–22 cm below the surface.

Scattered trees occur in the *Kalmia-Sphagnum fuscum* and *Gaylussacia* communities of some of the bog centers. They are most abundant in the *Gaylussacia* zone were patches of scrubby forest are most common near the bog pools. *Pinus strobus* occurs only in this part of the bogs.

Evidence of fire can be found throughout the bogs of Maine and New Brunswick. Osvald (1955) and Ganong (1897) also commented on the common occurrence of fires in these bogs. Recently burnt bogs were seen primarily in the commercially operated blueberry barrens, near settlements, and in bogs used for horticultural peatmoss production. In most other bogs fires had occurred before 1960. Fire frequency seems to have decreased during the last 20 years.

Burning primarily affects the dwarf shrub dominated parts of the bog. The Ericaceae regenerate easily from the rhizomes, and after 2 or 3 years a normal dwarf shrub layer has developed. The effect of fire on the *Sphagnum* carpet and tree layer is more long-lasting. Fire kills the hummock Sphagna, and they only very slowly regenerate from the hummock bases. *Cladonia* species, such as *Cladonia crispata, C. cristatella, C. pyxidata, C. chlorophaea* and *C. verticillata* can become important after fire. Especially, *C. crispata* is often dominant in burnt bogs (Table 1, nos. 9 and 10). The *Sphagnum* carpet of the *Rhododendron-Chamaedaphne* community is less affected by fire because of its much wetter surface. Repeated fires can eliminate trees from the bog surface. These ombrotrophic bog sites represent marginal habitats for trees, and regeneration is slow and difficult. The presence of isolated trees and stumps throughout all of the *Gaylussacia* and most of the *Kalmia-Sphagnum fuscum* communities suggest that the entire bog center was originally forested. The greater abundance of trees in the areas near the center pools seems to be only partly due to the better drainage, and primarily a result of their importance as a fire barrier. Tree-less raised bogs outside the coastal zone seem to be caused by repeated fires. Without fires these bogs would be covered with a very open, poorly growing forest, with the best-developed forests in the center of the bog domes. Judging from the height of trees present on these sites, the canopy height probably rarely exceeded 7–10 m.

Coastal bogs

The central parts of the larger coastal bogs are also frequently raised 3–4 m above the bog border but the topography is very different from those of the inland bogs. These bogs have a pronounced slope, often 1 : 20–30, which rises from a well-developed bog moat or lagg to an almost flat central bog plain (*Figure 4*). The actual slope and extent of the moat depends on the nature and slope of the surrounding uplands. Thus, these bogs are very clearly raised and have a plateau-like shape. Topographically, they are similar to the plateau bogs described from Europe (Eurola 1962).

Similar to the raised bogs outside the coastal zone, *Calamagrostis canadensis* meadows and swamp forests occur in the bog border when the supply of minerotrophic water is abundant. However, the *Acer rubrum* swamps of the inland bogs are replaced by a *Carex trisperma-Picea mariana* forest comparable to that found in boreal forest areas (Jurdant 1964, Damman 1964). Usually the *Smilacina-Sphagnum pulchrum* community occupies the major part of the well-developed lagg, but in contrast to the inland bogs the *Rhododendron-Chamaedaphne* community does not form a clear zone at the lower part of the bog slope. It is frequently absent or occurs only near drainage channels.

The bog slope is occupied by a *Kalmia-Sphagnum fuscum* community often with some low trees (8–10 m), and on the best drained slopes patches of an open forest may be present. If present, trees decrease rapidly in height toward the upper part of the slope. The boundary with the *Smilacina-Sphagnum pulchrum* community at the base of the slope is clear, and consequently the mineral soil water limit (Du Rietz 1954) is well-defined in these bogs.

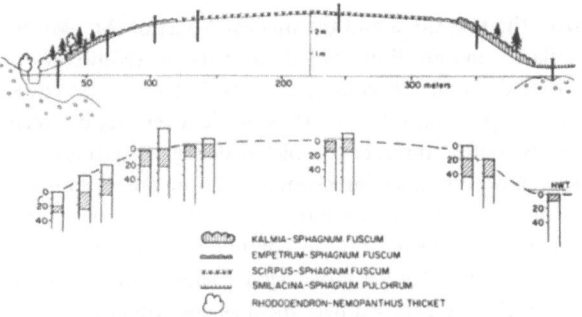

Fig. 4. Water level fluctuations and distribution of plant communities on a transect through Jonesport bog, a large plateau bog in the coastal belt. For communities with an irregular surface the water level extremes are shown with respect to both average hollow and hummock surface.

The almost flat plateau occupies the major part of these bogs (Fig. 4), and is covered almost entirely by the *Scirpus-Sphagnum* community. Hummocks are virtually absent on the plateau, and mud bottoms cover at most 5–10% of the surface. Bog pools occur rarely in these bogs. The *Scirpus-Sphagnum fuscum* covered areas lack a well-developed micro-topography. The *Sphagnum* cushions are small and distributed irregularly over the surface, and differences in microtopography average only about 15 cm. These areas are very wet most of the time. The surface is flooded in spring and after heavy rains, and even during prolonged dry periods the water levels remained within 16 cm of the hollow surface. Drainage does not improve until one approaches the margins of the plateau (Fig. 4) where the *Scirpus-Sphagnum fuscum* community is replaced by the *Empetrum-Sphagnum fuscum* community. Low water levels in these communities are similar to those in the *Kalmia-Sphagnum fuscum* communities of the inland bogs, but water levels fluctuate less, only 20–25 cm, and thus the high water levels remain farther below the surface.

Thus, in contrast to the inland bogs, the coastal bogs are wettest in the center and become progressively drier until one approaches the moat. However, the vegetation changes in the ombrotrophic part of the bog cannot be explained simply on the basis of moisture regime. This applies particularly to the *Empetrum-Sphagnum fuscum* community which occurs on sites which do not differ in moisture regime from many sites occupied by the *Kalmia-Sphagnum* community. All of these dwarf shrubs can grow vigorously on both much drier and much wetter sites. Therefore, moisture regime does not appear to be responsible for the extremely low growth of *Kalmia angustifolia* and the other dwarf shrubs. Temperature, precipitation and nutrient supply affect the bog slope and the marginal part of the plateau similarly, and are thus unlikely to cause these differences in growth habit. Of course, the possibility of differences in decomposition rate, and thus nutrient release, between these two sites cannot be completely disregarded. It is more likely, however, that winter snow conditions are responsible for these changes in vegetation. The sites supporting the *Empetrum-Sphagnum fuscum* community and the central part of the bog have a rather sparse snow cover, and the hummocks are exposed here most of the winter. In these exposed coastal bogs snow drifting from the open bog center accumulates on the slopes and in the moat. This snow drifting, combined with the occurrence of frequent winter thaws (Lautzenheiser 1959), exposes the hummocks of the *Empetrum-Sphagnum fuscum* zone, and at the same time provides a protecting snow cover on the slopes which allows the development of a *Kalmia-Sphagnum fuscum* vegetation. This also explains why the *Empetrum-Sphagnum fuscum* vegetation is absent in small and less exposed bogs.

Snowfall increases inland, in spite of the higher winter precipitation along the coast, and at the same time winter temperatures decrease (Fobes 1946), as a result winter snow cover changes rapidly when one moves away from the coast. At the end of February 1975 when the *Empetrum-Sphagnum fuscum* communities of the coastal bogs were exposed, the snow cover in bogs outside the coastal zone was continuous. In the Meddybemps Heath between 50–65 cm of snow covered the hummocks of the bog center, and the dwarf shrubs of the *Kalmia-Sphagnum fuscum* and *Gaylussacia* zone were completely snow covered. At the same time open bogs at Millinocket, in the interior of Maine, had a snow depth of 80–90 cm above the hummock surfaces. As a result of the thin snow cover in the coastal areas, frost penetrates much deeper in the peat, and ice persists longer in the spring. It is usually present in the hummocks of the coastal bogs until late May, and occasionally, as in 1970, until the middle of June. Thus, in spite of the milder winters much harsher conditions exist in parts of the coastal bogs. The conspicuous difference in the physiognomy of the *Empetrum-Sphagnum fuscum* community compared to the other dwarf shrub communities results mainly from the poor development of the dwarf shrub stratum. *Sphagnum fuscum. Rubus chamaemorus* and *Empetrum nigrum* are apparently best equipped to take advantage of the better light conditions.

Fire also occurs in the coastal bogs, and evidence of it is visible in most of them. However, it affects the vegetation much less than in the bogs outside the coastal zone. Fire damage occurs primarily on the bog slope. Under normal weather conditions the *Empetrum-Sphagnum fuscum* and *Scirpus-Sphagnum fuscum* communities lack the dry fuel to sustain a fire, and generally fire damage is limited to the sites adjacent the *Kalmia-Sphagnum fuscum* slopes. Trees are always absent in the central parts of the bogs but this is due to the wetness of the bog plateau, and is not a result of fire as it is in many bogs farther inland.

Causes of differences between raised bogs in coastal and inland areas

Most of the differences between bogs in the coastal zone and those farther inland (Table 3) seem to be associated with differences in moisture regime and topography of the bogs, or a combination of the two. The restriction of the *Empetrum-Sphagnum fuscum* community to the coastal bogs is an exception. As discussed earlier, this community occurs in the coastal bogs because of the intermittent snow cover and the resulting harsher conditions for plant growth in elevated parts of these bogs. The floristic differences between the coastal and inland bogs are another exception; this will be discussed later.

Even the restriction of the *Gaylussacia* community to the inland areas must be attributed to the absence of sufficiently well-drained sites in the coastal bogs of the Bay of Fundy region because this community can still be found abundantly in bogs in northern New Brunswick and in largely snow-free bogs in southernmost Nova Scotia where such sites do occur.

The coastal zone along the Bay of Fundy is characterized by frequent summer fogs and lower summer temperatures; precipitation is somewhat higher along the coast but mainly because of higher winter precipitation (Lautzenheiser 1959). The coastal bogs are located in a very narrow zone within 10 km from the open coast (Fig. 1). This includes only about half of the coastal climatic area recognized by Fobes (1946) and coincides roughly with the area in which fog horns operate for over 200 hours during July and August based on a 6-years average (Fobes 1946). Temperature data are available for some stations on the coast but all inland stations are located much farther from the coast. Data on temperature differences between the inland bogs, some located only 7 km from salt water and 25 km from the coast line, and coastal bogs were calculated from rotation changes in sucrose solutions (Berthet 1960, Damman 1975) placed in corresponding habitats of several study areas (Table 4). Average temperatures during the vegetative

Table 3. Principal differences between the bogs in and outside the coastal zone.

Inland	Coastal
Domed but with very gentle barely visible slopes	Clearly raised plateau with obvious bog slope and almost flat central part
Trees and patches of forest often present in the central, ombrotrophic parts of the bog (presumably trees covered most of this originally)	Trees restricted to the bog slope
Dwarf shrub communities cover almost the entire bog surface	Dwarf shrub communities only on slopes and edge of plateau; _Scirpus cespitosus_ lawn communities cover the plateau
Bog moat (lagg) poorly defined	Bog moat often well-developed
Pools common in bog center. Mud bottom vegetation unimportant and almost restricted to area with bog pools	Bog pools rare. Mud bottoms cover very little area but can occur anywhere on plateau; most common in disturbed areas
Gradual change from minerotrophic to ombrotrophic conditions; mineral soil water limit obscure, can only be recognized in floristic composition of vegetation	Mineral soil water limit clearly visible in vegetation and topography
Gaylussacia community restricted to these bogs	_Scirpus-Sphagnum fuscum_ and _Empetrum-Sphagnum fuscum_ communities restricted to these bogs
Empetrum nigrum and _Scirpus cespitosus_ occur very locally; _Rubus chamaemorus_ absent[*]	_Rubus chamaemorus_, _Empetrum nigrum_ and _Scirpus cespitosus_ extremely abundant

[*]For other floristic differences see section on distribution of individual species.

season in the inland bogs are 2.5–3 °C higher than in the coastal ones, and almost 4 °C higher than in the very exposed bogs of Campobello Island and Quoddy Neck. This is a considerable temperature difference, the magnitude of which is better appreciated if one realized that this corresponds to an altitudinal change of 500–600 m and 800 m, respectively.

Although the coldness of the coastal zone is responsible for the development of an almost boreal forest vegetation in this area (Davis 1966), its direct effect on the bog vegetation is minimal. The differences in bog development seem to be primarily due to the reduced evapotranspiration resulting from the lower temperatures and the high fog frequency in the coastal zone. Because of this, the zone experiences a lower excess of potential evapotranspiration over precipitation during the summer months and a higher annual moisture surplus. This causes the wetness of the coastal bogs which is further enhanced by their topography. Bog topography itself is also controlled by climate (Osvald 1925, Eurola 1962, Aletsee 1967). Paasio (1933) and Eurola (1962) report somewhat similar changes from convex highmoors, in this case the more northern Kermi-highmoors, in continental areas to plateau bogs along the coast of southern Finland. In eastern North America raised bogs with flat

centers also occur along the Nova Scotian coast and the west coast of Newfoundland. Although their development is obviously associated with an oceanic climate, thus far no adequate explanation has been given for the change from clearly domed bogs to plateau bogs with increasing oceanic conditions. It is suggested here that the inland raised bogs represent typical highmoors whose development is controlled by precipitation and evapotranspiration whereas the topography of the coastal raised bogs deviates because here other factors limit the growth of the _Sphagnum_ surface.

Raised bogs maintain their own water level (Moore & Bellamy 1974) which slopes with the bog surface and can be elevated far above the bog border. For instance, at Meddybemps (Fig. 3) the low water level on the margin of the bog dome is 33 cm below the surface, it remains at about 30 cm throughout the _Kalmia_ zone and is at 41 cm in the _Gaylussacia_ zone of the bog center where it is 3.85 m above the bog border (Fig. 2). Similarly, at Jonesport (Fig. 2) the bog water level is elevated about 3 m above the margin of the bog. The total rise of the water table of bogs depends on distance to the bog margin and climatic conditions.

The dependence of peat growth on an ample water supply limits the height to which the bog surface can grow above its water table. Although peat accumulation is due to slow

99

Table 4. Average temperature (°C) in coastal and inland bogs for the period June 8 - September 16 (100 days). Data based on rotation changes in sucrose solutions at 2.5 - 4.5 cm below surface of Sphagnum carpets.

Location		Coastal bogs				Inland bogs	
		Campobello Island	Quoddy Neck	Jonesport	Kelly Point	Meddybemps	Pleasant River
Sphagnum fuscum* hummock	1970	15.8 (n=3)	15.7 (n=3)	-	16.5 (n=3)	19.5 (n=3)	
	1971	-	17.3 (n=9)	18.2 (n=8)	18.1 (n=15)	21.5 (n=9)	20.5 (n=9)
Sphagnum rubellum carpet along pool	1970	-	15.8 (n=4)	-	16.9 (n=3)	-	-
	1971	-	17.5 (n=6)	-	18.7 (n=4)	20.3 (n=4)	-

*In Empetrum-Sphagnum fuscum vegetation of coastal bogs and in Gaylussacia community of inland bogs. In the latter the dwarf shrub cover was removed to create comparable light conditions.

decomposition rather than fast growth, it is obvious that eventually the maximum rise of the water table maintained by the bog will determine its topography, provided that growth is limited by water supply. In the inland bogs summer water levels appear to restrict growth of the Sphagnum carpet because the bog surface slopes gradually with the water level toward the bog center.

In the coastal areas moisture conditions are more favourable, and thus Sphagnum growth at the bog surface ought to be maintained to at least the same height above the water table as in the inland bogs. The fact that the water level on the plateau is much closer to the surface than in the inland bogs (Figs. 3 and 4) suggests that bog growth is limited by other factors than water supply. Length of the vegetative season, summer temperatures, or nutrient supply could be responsible. As limiting factors, all of these would affect peat growth equally throughout the bog and thus could account for the flat bog surface in the coastal areas. The relatively rapid drainage near the margin of the plateau lowers the water level in these parts so that once again water becomes limiting. This causes the development of the rather steep bog slope. Considering that the bogs discussed here occur in the southernmost part of the highmoor zone, and that plateau bogs also occur in southern Nova Scotia with a longer vegetative season, one would not expect that temperature or length of vegetative season are the limiting factors. It seems most likely that nutrient supply is limiting growth in these coastal ombrotrophic bogs. Although this supply increases toward the coast (Gorham 1961) it is still so small that it will soon limit moss growth if moisture supply is adequate.

In conclusion, the main differences between the bogs in the coastal zone and those farther inland appear to be a result of (1) the lower summer temperatures and greatly increased fog frequency in the coastal belt which cause the wetness as well as the topography of these bogs and (2) the difference in winter snow conditions between the two areas.

Distribution of individual species

A number of species do not show a regular distribution pattern within the plant communities. Two general patterns can be recognized: (1) the sporadic, although locally abundant, occurrence in ombrotrophic bog areas of species with higher nutrient demands and (2) the restriction or increased abundance of some species in the coastal areas. These will be discussed separately.

'Minerotrophic species' in ombrotrophic bogs

The presence of Gaultheria procumbens, Vaccinium vitisidaea and Rhododendron canadense in bogs is generally an indication of shallow peat or peat influenced by minerotrophic water. In contrast to most indicators of the mineral soil water limit (Du Rietz 1954) these species are not typical bog and fen species but they have their main distribution on mineral soils. In bogs these species occur on well-drained peat, although *Rhododendron canadense* can also be very abundant in wet oligotrophic bog borders.

Within ombrotrophic bog areas these species occur occasionally in forested parts where the stagnation of the peat growth and the more active nutrient cycle result in an enrichment of the surface peat (Damman 1977). Their occurrence in such areas could be attributed to this enrichment. However, these species can also be found locally in *Kalmia-Sphagnum fuscum* and *Gaylussacia* communities which show no clear evidence of stagnation. Invariably, such sites have been burnt rather recently, and the release of nutrients from the burnt peat temporarily increased their fertility.

The very local occurrence of a slightly more nutrient demanding species, such as *Eriophorum angustifolium*, within clearly ombrotrophic parts of a bog have been reported frequently (Malmer 1962, Svensson 1965). This is also the case in Maine and New Brunswick. It is possible that the occurrence of this species, as well as the local abundance of other weakly minerotrophic species such as *Andromeda glaucophylla* and *Scirpus cespitosus* in the *Chamaedaphne-Sphagnum rubellum* community, could also be due to fire. This requires further study, but a preliminary chemical analysis of peat cores indicated that their presence is not associated with 'fen windows' in the ombrotrophic peat.

Species restricted to coastal areas

In Maine and southeastern New Brunswick, *Myrica gale*, *M. pensylvanica* and *Juniperus communis* are restricted to mineral soils and minerotrophic bog borders. However, in the immediate vicinity of the coast they can be found also on ombrotrophic peat. *Sphagnum pulchrum* is very rare in the ombrotrophic bogs of the area, it was found in ombrotrophic mudbottoms of only one bog, Spruce Lake Bog, along the Bay of Fundy. However, in more oceanic areas it occurs commonly in such habitats, as it does in Europa. These distributions are related to the higher atmospheric nutrient input in coastal areas.

Sphagnum flavicomans, *S. tenellum* and *S. imbricatum* are abundant in the coastal zone but become much less common inland. The first two can still be found regularly in wet hollows in inland bogs but *S. imbricatum* is very rare outside the coastal belt. The increased continentality of the climate with distance from the coast appears to control their distribution. The decrease in abundance of *Solidago*

uliginosa with distance from the coast seems to be related to the same factors.

Scirpus cespitosus, *Rubus chamaemorus* and *Empetrum nigrum* are abundant and very prominent species in the coastal bogs. In bogs the last two are restricted to the coastal areas, although *Empetrum* can occur on an occasional hummock in some inland bogs. *Scirpus cespitosus* is also rare in the inland bogs but locally it is abundant in the wet *Chamaedaphne-Sphagnum rubellum* community (Table 1, 48–52). All of these species become more common to the north and, with the exception of *Rubus chamaemorus*, they are also most abundant in maritime regions. Although this geographic distribution suggests that their distribution may be controlled by the lower summer temperatures in maritime areas, their local distribution indicates that other factors may also be important. *Empetrum nigrum* occurs on mineral soils farther south along the coast, and is present in inland bogs only on some high hummocks with dead or dying *Sphagnum*, i.e. on stagnant surfaces which are nutrient enriched. *Scirpus cespitosus* occurs only very locally in ombrotrophic sites outside the coastal belt, but it can be found commonly on nutrient-richer fen sites throughout most of Maine and New Brunswick. *Rubus chamaemorus*, although generally considered a character species of the *Sphagnion fusci* (Moore 1968), occurs optimally in the minerotrophic bog border areas of the boreal zone. This suggests that nutrients may also play a role in determining the range of these species. Increased nutrient availability in the coastal bogs along the Bay of Fundy could result from (1) accelerated release of P and K from peat due to more severe freezing (Saebø 1969) in the absence of a reliable snow cover or (2) higher atmospheric nutrient input in the coastal areas (Gorham 1961). The factors controlling this species pattern ought to be investigated in more detail.

Syntaxonomic position of the bog communities

The plant communities of the peat bogs were described primarily to show the vegetation pattern in the bogs of the two zones. Although a discussion of the classification of the North American bog vegetation is outside the scope of this paper, it is worthwhile to discuss briefly the syntaxonomic position of these communities and to show their relationships to those described by others. Unfortunately, descriptions of the plant communities of the North American highmoors are sparse, and few (Pollett & Bridgewater 1973, Wells 1975, Gauthier & Grandtner 1975) provide enough floristic detail to permit a meaningful comparison.

Phytosociologically, the heath communities described here show enough similarity in their floristic composition to be included in one association. This applies also to the *Empetrum-Sphagnum fuscum* community. Although it is physiognomically very different, owing to the poorly developed ericaeous dwarf shrub stratum, it is floristically very closely related to the other dwarf shrub heaths.

These communities are similar in their overall floristic composition to the *Kalmio-Sphagnetum* described from Newfoundland (Pollett & Bridgewater 1973, Wells 1975), the *Sphagno-Chamaedaphnetum kalmietosum* described from the lower St. Lawrence River (Gauthier & Grandtner 1975), and possibly the *Rubo-Empetretum nigri* (Pollett 1972). The floristic differences between these plant communities are primarily due to the disappearance of oceanic and arctic species toward the interior of the continent. Thus, they reflect climatic changes in a vegetation type occupying the hummocks, slopes, and other rather well-drained parts of ombrotrophic bogs. These geographic changes do not affect the characteristic species composition nor the general physiognomy of the communities, and therefore they might be best combined into one association, the *Kalmio-Sphagnetum fusci* Pollett & Bridgewater 1973. Thus defined, the range of this association would cover the eastern part of Canada and the northeastern USA, and is replaced by a *Chamaedaphne* dominated hummock association northward and from central Quebec to the interior of the continent. Geographic forms of this association, or races in the sense of Oberdorfer (1968), could be recognized to express this climatic gradient. Sub-associations based on ecological differences can then be recognized regionally. The three ombrotrophic heath communities described here are at the level of such sub-associations and might be named: *Kalmio-Sphagnetum gaylussacietosum*, *typicum* and *empetretosum*, respectively (Table 1).

The *Scirpus cespitosus* lawn communities, including the *Scirpus-Sphagnum fuscum* and *Scirpus-Sphagnum flavicomans* communities, belong to one association which should probably also include the *Chamaedaphne-Sphagnum rubellum* community. The *Scirpo-Sphagnetum* (Wells 1975), described from eastern Newfoundland, occupies the same ecological position in these ombrotrophic bogs, and is also floristically and physiognomically related. This is probably also true for the *Scirpo-Gaylussacietum* (Pollett 1972) although no detailed floristic data are published for this association.

Using the same reasoning as above, these communities could best be grouped into one association, the *Scirpo-*

Sphagnetum tenelli Wells 1975, which occupies the wet, ombrotrophic bog surfaces and depressions. Its range includes the cold oceanic parts of eastern North America from the coast of Labrador and Newfoundland to Nova Scotia and the maritime parts of Maine and eastern and northern New Brunswick. It is dominated by *Scirpus cespitosus* except in the more continental eastern parts of its range where *Sphagnum rubellum* or *S. fuscum* determine its physiognomy. *Gaylussacia dumosa*, *Sphagnum tenellum* and *S. flavicomans* are optimally developed on these sites throughout the range of this association but these two *Sphagnum* species decrease rapidly in abundance with distance from the coast, and *Gaylussacia dumosa* does not reach the northern limit of the association. In oceanic areas of Nova Scotia and Newfoundland, *Sphagnum pulchrum* is often abundant.

The three communities included in Table 1 could be named *Scirpo-Sphagnetum empetretosum*, *sphagnetosum flavicomantis*, and *cladonietosum crispatae*. The second sub-association occurs primarily farther north; the descriptions included in Table 1 were made in northern New Brunswick. They are included here primarily to show the range in floristic composition within this association.

Tüxen et al. (1972) proposed to enlarge the *Oxycocco-Sphagnetea* to a class including the highmoor communities of the entire northern hemisphere. In such a classification the *Kalmio-Sphagnetum* belongs clearly to the *Sphagnetalia fusci* (Tx. 1955) em. 1970, and particularly the *Kalmio-Sphagnion fusci* Tx. 1970. The position of the *Scirpo-Sphagnetum tenelli* within such a scheme is less clear. It can probably be best included in a new alliance within the *Sphagnetalia fusci*, although this would require a redefinition of this order since at present it includes only hummock communities. Comparable communities, which do not fit the present classification, exist also in the European highmoors (Malmer 1962, 1968).

The variation in the floristic composition of the weakly minerotrophic plant communities is poorly known. They occur in wet oligotrophic habitats over a large area which extends also far south of the genuine highmoor zone.

The *Smilacina-Sphagnum pulchrum* community is phytosociologically related to the *Scirpeto-Sphagnum papillosi* (Pollett & Bridgewater 1973) and the *Scirpo-Sphagnetum niti* (Wells 1975). These are plant communjties of bog moats and other wet, slightly minerotrophic peats. They should probably be considered as one association, however, this requires further study. This plant community undoubtedly belongs to the *Oxycocco-Sphagnetea*. Judging by its floristic composition it appears to belong to the

North American equivalent of the *Erico-Sphagnion* Moore 1964 (1968). Although this is described as a hummock vegetation, which is undoubtedly true in the Atlantic parts of Europe, obviously, the plant communities of extremely poor fens also belong here (Jensen 1972).

The *Rhododendron-Chamaedaphne* community appears to be similar to the *Sphagno-Chamadaphnetum ledetosum* and *chamaedaphnetosum* (Gauthier & Grandtner 1975) but not the remainder of the *Sphagno-Chamaedaphnetum* which corresponds to the *Kalmio-Sphagnetum*. The presence of many minerotrophic species clearly differentiates it from the latter. Similar plant communities occur along the drainage channels of highmoors, on the lower slope and border areas of slightly convex highmoors as well as in many very weakly minerotrophic bog areas in southern New England and New York. Too few data are available to determine the variation in floristic composition within this community and to decide on its syntaxonomic status. However, it clearly belongs to the same alliance of the *Oxycocco-Sphagnetea* as the *Smilacina-Sphagnum papillosum* community. Corresponding European communities occur from easternmost Finland (Paasio 1933, Eurola 1962) into the USSR (Bogdanowskaya-Guihéneuf 1928).

The mudbottoms and soft *Sphagnum* carpets belong to the North American counterpart of the European *Rhynchosporetum albae* Koch 1926. The ones described here resemble in many ways the *Rhynchospora alba-Cladopodiella* association of Malmer (1968). It is differentiated by several North American endemics, *Utricularia cornuta*, *Xyris montana*, and in minerotrophic mudbottoms also *Sphagnum pylaesii* and *S. torreyanum*. Similar communities were described by Wells (1975) and Gauthier & Grandtner (1975). The classification problems in this association are comparable to those in Europe (Westhoff & den Held 1969); the dominance of one of the *Sphagnum* species, *Cladopodiella fluitans* or *Drepanocladus fluitans* results in a very different appearance (Table 2). These communities are relatively unimportant in the bogs of this area but become rapidly more abundant in the bogs to the north and east.

Conclusions

Bogs topographically similar to the coastal type have been described from Europe as plateau highmoors (Eurola 1962). The raised bogs outside the coastal zone are probably closest to the forest highmoors (Osvald 1925, Eurola 1962) but they are much more clearly domed and have a very poorly developed forest cover. In these respects they are more similar to the Kermi-highmoors but they lack the well-developed ridge and hollow pattern so characteristic for this bog type

There are, however, considerable differences in the zonation of bog types in Europe and eastern North America. Whereas plateau highmoors cover a considerable part of western Europe (Eurola 1962), they occur only as a narrow belt along the east coast of North America. This zone is particularly narrow along the Bay of Fundy coast. The reason for this is the thermally more continental climate of eastern North America, and it is only because of the much higher precipitation in this area that highmoors can develop. For instance, in the highmoor zone west of the Bay of Fundy the annual number of growing degree days, i.e. degree days above 42 °F, amounts to 2500–2750, and annual precipitation ranges from 1000–1300 mm (Chapman & Brown 1966, U.S. Dept. Commerce 1964). These values are about double those for southern Finland where the *Sphagnum fuscum*-plateau highmoors along the Baltic coast occupy a similar position with respect to the inland bogs (Eurola 1962, Bogdanowskaya-Guihéneuf 1928). Because of the warm summer weather in our area, plateau bogs can develop only in the cooler coastal zone. The fact that they are restricted to a very narrow belt with much higher fog frequency indicates that it is only because of this fog that conditions favourable for the development of plateau bogs occur here.

This difference in climatic conditions compared to Europe is also reflected in the floristic composition of the plant communities. Disregarding the differences due to the plant geographic history of the floras of the area, it is clear that the geographic distribution of species common to both regions is quite different. There is a peculiar mingling of species considered continental in Europe, such as *Chamaedaphne calyculata*, with species which have an obviously oceanic or sub-oceanic distribution in the ombrotrophic bogs of Europe, such as *Sphagnum imbricatum* and *Scirpus cespitosus*. This is primarily because temperature and humidity of the climate do not change in the same way in the two regions if one moves from oceanic to continental areas. Consequently, concepts such as oceanic and continental, as defined by species distributions in western Europe, cannot be applied to eastern North America. Obviously, it is necessary to define factors changing along a coast to inland gradient in more detail rather than to use vague concepts such as continental and oceanic climates. The effect of winter temperatures, summer temperatures, humidity of the climate, as well as other factors associated with distance to the coast, should be used individually in comparing species distribution or vegetation zonation on different continents.

103

Summary

The bogs of a narrow coastal belt along the Bay of Fundy are strikingly different in their topography and vegetation from those farther inland. The former are plateau bogs with an obvious bog slope with a scrubby forest cover, a wet treeless plain covered with *Scirpus cespitosus* lawn communities, and a well-developed lagg with a clearly defined mineral soil water limit. In contrast, the inland bogs are convex raised bogs sloping very gently to the high center; they are covered almost entirely with an ericaceous dwarf shrub vegetation with scattered trees in the central ombrotrophic part, and they have a very poorly defined nineral soil water limit. The high fog frquency in the coastal belt and the intermittent snow cover on the plateau of these exposed bogs appear to be responsible for the different topography and vegetation of the coastal bogs as compared with those further inland.

Water levels fluctuate most in the inland bogs. Within a plant community occurring in both bog types, maximum water levels are closer to the surface in the inland bogs but minimum water levels are at about the same level.

The absence of an open forest on the ombrotrophic central plain of some inland bogs seems to be caused by repeated fires. Fire also creates conditions favourable for the occurrence of some slightly more nutrient-demanding species, such as *Gaultheria procumbens*, *Rhododendron canadense*, and *Vaccinium vitis-idaea*, in the ombrotrophic parts of bogs.

The heath communities of the ombrotrophic parts of the bogs can be included in one association, the *Kalmio-Sphagnetum fusci*, which occupies the slopes, hummocks, and other well-drained parts. The *Scirpus* lawn communities make up another association, the *Scirpo-Sphagnetum tenelli*, which occurs on the wet, and occasionally flooded, ombrotrophic bog surfaces. The syntaxonomic position of these and the other plant communities is discussed briefly.

In contrast to Europe, the plateau bogs are restricted to a very narrow coastal zone along the east coast of North America because of the thermally continental climate of this region. This is also reflected in the floristic composition of the vegetation. It causes a peculiar mingling of 'continental' species such as *Chamaedaphne calyculata* with 'oceanic' species such as *Sphagnum imbricatum*.

References

Aletsee, L. 1967. Begriffliche und floristische Grundlagen zu einer pflanzengeographischen Analyse der europäischen Regenwassermoorstandorte. Beitr. Biol. Pflanzen 43: 117–160.

Berthet, P. 1960. La mesure écologique de la température par détermination de la vitesse d'inversion du saccharose. Vegetatio 9: 197–207.

Bogdanowskaya-Guihéneuf, Y. 1928. Die Vegetation der Hochmoore des russischen Ostbalticums (in Russian, German summary). Travaux Inst. Sci. Natur. Peterhof 5: 265–377.

Chapman, L.J. & D.M. Brown. 1966. The climates of Canada for agriculture. Can. Dept. Forestry and Rural Development, Ottawa, Canada Land Inventory Rept. 3. 24 pp.

Crum, H., W.C. Steere & L.E. Anderson. 1965. A list of mosses of North America. Bryologist 68: 377–432.

Dachnowski-Stokes, A.P. 1929. The botanical composition and morphological features of highmoor peat profiles in Maine. Soil Sci. 27: 379–388.

Dachnowski-Stokes, A.P. 1930. Peat profile studies in Maine. The south Lubec 'heath' in relation to sea level. J. Washingt. Acad. Sci. 20: 124–135.

Damman, A.W.H. 1964. Some forest types of central Newfoundland and their relation to environmental factors. Forest Science Monogr. 8: 1–62.

Damman, A.W.H. 1976. Plant distribution in Newfoundland especially in relation to summer temperatures measured with the sucrose inversion method. Can. J. Bot. 54: 1561–1585.

Damman, A.W.H. 1977. Distribution and movement of elements in ombrotrophic peat bogs. Ecology (in press).

Davis, R.B. 1966. Spruce-fir forests of the coast of Maine. Ecol. Monogr. 36: 79–94.

Du Rietz, G.E. 1954. Die Mineralbodenwasserzeigergrenze als Grundlage einer natürlichen Zweigliederung der nord und mitteleuropäischen Moore. Vegetatio 5–6: 571–585.

Eurola, S. 1962. Über die regionale Einteilung der Südfinnischen Moore. Ann. Bot. Soc. 'Vanamo' 33(2): 1–243.

Fernald, M.L. 1950. Gray's manual of botany. 8th ed. Amer. Book Co., New York. 1632 pp.

Fobes, C.B. 1946. Climatic divisions of Maine. Maine Techn. Exp. Sta., Univ. of Maine, Orono. Bull. 40, 44 pp.

Ganong, W.F. 1891. On raised peatbogs in New Brunswick. Bot. Gaz. 16: 123–126.

Ganong, W.F. 1897. Upon raised peat bogs in the province of New Brunswick. Trans. Roy. Soc. Can., Ser. 2(3): 131–163.

Gauthier, R. & M.M. Grandtner. 1975. Étude phytosociologique des tourbières du Bas Saint-Laurent, Québec. Naturaliste Canadienne 102: 109–153.

Gorham, E. 1961. Factors influencing supply of major ions to inland waters with special reference to the atmosphere. Geol. Soc. Amer., Bull. 72: 795–840.

Hale, M.E. & W.L. Culberson. 1966. A third checklist of the lichens of the continental United States and Canada. Bryologist 69: 141–182.

Jensen, U. 1972. Das System der europäischen Oxycocco-Sphagnetea. Ein Diskussionsbeitrag. pp. 491–496. In: van der Maarel, E. &. R. Tüxen (eds.), Grundfragen und Methoden in der Pflanzensoziologie. Junk, The Hague.

Jurdant, M. 1964. Carte phytosociologique et forestière de la

Forêt expérimentale de Montmorency. Can. Dept. Forestry, Ottawa, Publ. 1046F. 73 pp.

Lautzenheiser, R.E. 1959. Climates of the states: Maine. U.S. Dept. Commerce, Weather Bureau, Clim. U.S. 60–17. 16 pp.

Löve, A. & D. Löve. 1959. Biosystematics of the black crowberries of America. Can. J. Genetics and Cytology 1: 34–38.

Malmer, N. 1962. Studies on mire vegetation in the Archaean area of southwestern Götaland (South Sweden). I. Vegetation and habitat conditions on the Åkhult mire. Opera Bot. (Lund) 7(1): 1–322.

Malmer, N. 1968. Über die Gliederung der Oxycocco-Sphagnetea und Scheuchzerio-Caricetea fuscae. pp. 293–305. In: Tüxen, R. (ed), Pflanzensoziologische Systematik. Junk, The Hague.

Moore, J.J. 1968. Classification of the bogs and wet heaths of northern Europe. pp. 306–320. In: Tüxen, R. (ed.), Pflanzensoziologische Systematik. Junk, The Hague.

Moore, P.D. & D.J. Bellamy. 1974. Peatlands. Springer Verlag, New York. 221 pp.

Mueller-Dombois, D. & H. Ellenberg. 1974. Aims and methods of vegetation ecology. John Wiley, New York. 547 pp.

Nichols, G.E. 1919. Raised bogs in eastern Maine. Geogr. Rev. 7: 159–167.

Oberdorfer, E. 1968. Assoziation, Gebietsassoziation, Geographische Rasse. pp. 124–141. In: Tüxen, R. (ed.), Pflanzensoziologische Systematik. W. Junk, The Hague.

Osvald, H. 1925. Die Hochmoortypen Europas. Veröff. Geobot. Inst. Rubel. (Zürich) 3: 707–723.

Osvald, H. 1928. Nordamerikanska mosstyper. Svensk Bot. Tidskr. 22: 377–391.

Osvald, H. 1955. The vegetation of two raised bogs in northern Maine. Svensk Bot. Tidskr. 49: 110–118.

Paasio, I. 1933. Über die Vegetation der Hochmoore Finnlands. Acta For. Fenn. 39(3): 1–190.

Pallmann, H., E. Eichenberger & A. Hasler. 1940. Eine neue Methode der Temperaturmessung bei ökologischen und bodenkundlichen Untersuchungen. Ber. Schweiz. Bot. Gesellsch. 50: 337–362.

Pollett, F.C. 1972. Classification of peatlands in Newfoundland. Proc. 4th Int. Peat Congress, Helsinki, 1: 101–110.

Pollett, F.C. & P.B. Bridgewater.1973. A phytosociological classification of peat lands in central Newfoundland. Can. J. For. Res. 3: 433–442.

Rowe, J.S. 1972. Forest regions of Canada. Can. Dept. Environment, For. Serv., Ottawa, Publ. 1300. 172 pp.

Saebø, S. 1969. On the mechanism behind the effect of freezing and thawing on dissolved phosphorus in Sphagnum fuscum peat. Sci. Rep. Agr. Coll. Norway 48(14): 1–10.

Schuster, R.M. 1953. Boreal hepaticae. Amer. Midl. Naturalist 49: 257–684.

Svensson, G. 1965. Vegetationsundersökningar på Store Mosse. Bot. Not. 118: 49–86.

Thornthwaite, C.W. 1948. An approach toward a rational classification of climates. Geogr. Rev. 38: 55–94.

Tüxen, R., A. Miyawaki & K. Fujiwara. 1972. Eine erweiterte Gliederung der Oxycocco-Sphagnetea. pp. 500–520. In: Van der Maarel, E. & R. Tüxen (eds.), Grundfragen und Methoden in der Pflanzensoziologie. W. Junk, The Hague.

U.S. Dept. Commerce, Weather Bureau. 1964. Decennial census of United States Climate. Climatic summary of the United States 1951–1960. New England. Climatography of the U.S. 86–23. Washington, D.C. 142 pp.

Wells, E.D. 1975. A classification of peatlands in eastern Newfoundland. MSc. Thesis, Memorial Univ., St. John's Newfoundland. 194 pp.

Westhoff, V. & A.J. den Held. 1969. Plantengemeenschappen in Nederland. Thieme, Zutphen. 324 pp.

THE IMPORTANCE OF CLOSELY RELATED TAXA FOR THE DELIMITATION OF PHYTOSOCIOLOGICAL UNITS*

Elias LANDOLT**

Geobotanical Institute ETH, Rübel Foundation, Zürichbergstrasse 38, CH 8044 Zürich, Switzerland

Keywords:
Character-species, Differential species, Ecological races, Related taxa

Introduction

In the third edition of his 'Plant Sociology', page 91, Braun-Blanquet (1964) writes (translated from the German): 'In addition to the Linnean species, we can consider as character-species races whose morphology is difficult to distinguish, subspecies, and even varieties, all of which occur optimally in particular phytosociological units. Wide-spread species of expanding families often include parallel races fixed by heredity, i.e. genotypes which are attached to a particular area and often also to a clearly defined phytosociological unit (alliance, association). These parallel ecological races, usually subspecies, which are restricted to different climates, differ not only morphologically, but also physiologically and ecologically . . .'.

What Braun-Blanquet says about character-species naturally holds true for differential species as well. Most plant sociologists agree with Braun-Blanquet on this point; I shall only mention Westhoff's (1947) outstanding work on dune and salt marsh vegetation on West Frisian islands, in which he lists a large number of ecotypes (subspecies, varieties, and subvarieties) as characteristic taxa for particular phytosociological units (phytocoena). In 'Plantenge-meenschappen in Nederland' (Westhoff & den Held 1969), numerous microspecies, subspecies and varieties are also used to characterize phytocoena. More recently, Pignatti (1968) has pointed out the importance of closely related species for plant sociology.

Nevertheless, taxa within a species group have not always been used by phytosociologists for the following reasons:

* Dedicated to Professor V. Westhoff on the occasion of his 60th birthday.
** The author greatly appreciates valuable suggestions of Dr K. Urbanska-Worytkiewicz and her help in the translation of the manuscript.

1. Very few polymorphous species groups have been sufficiently investigated biosystematically to assure conclusive evidence on the value of the taxa involved. Often it is not even known whether described forms or varieties are really genetically fixed or represent only modifications of a wide-spread taxon. Thus Koch (1926) in his pioneer work on vegetation units in the Linth plain, used a series of forms and varieties which were probably only modifications (for example, his 'formae *submersae*') to characterize phytocoena.

2. Even when critical groups of species have been examined in detail, it is often almost impossible for a phytosociologist to identify them; the distinguishing features are often hard to recognize, applicable only on a statistical basis, or unequivocal solely when the conditions of cultivation were the same. Moreover, the borderlines between taxa are often not sharply drawn.

3. Taxa occurring within a given group often have a very limited distribution area and thus they are not identical with similar taxa found in other areas, which may be ecologically quite different. Thus, the results of experimental studies made in a limited area cannot, in many cases, be transposed to other areas.

Morphological and ecological variation

Before ecological races are dealt with in a more detailed way, a few words should be said on the meaning of phenotypic modifications. Modifications are non-hereditary reactions of plants to local ecological conditions and can be recognized in altered features of given individuals. In view of fact that ecological factors characterize some phytocoena, data on modifications are often very important. Variation in leaf size, recently noted in *Limonium vulgare*

107

Table 1. Leaf size variation of *Limonium vulgare* in different plant associations from Terschelling, Netherlands (Report on the excursion of Geobotanical Institute ETH, 1974, unpublished).

Puccinellietum maritimae	*Plantagini-Limonietum*	*Armerio-Festucetum*
1.6 – 2.4 (3)	1 – 2.2 (12)	2 – 6.2 (8)

1 = the smallest leaf surface = 4.4 cm^2.
The number of places (total of 23) on a transect where measurements of five leaves each were taken appears in parentheses.

(Table 1), is apparently due to modifications following a water and salt content gradient: the large leaves are characteristic for the *Armerio-Festucetum* whereas smaller ones occur in the other associations.

It should be noted, however, that a correct assessment of morphological variation in plants presents sometimes difficulties. On the one hand, morphological characters used for classification may correspond to modifications that under normal circumstances develop into a normal form. This might be the case of numerous 'taxa' described in various plant genera (e.g. formae *submersa*, *prostrata*, *microphylla* etc.). On the other hand, modifications often merely reinforce the expression of a hereditary character. Well known is the example of mountain plants that grow considerably taller than in their natural environment when they are transplanted into a lowland station. A corresponding lowland taxon, however, grows still taller than its mountain relative in the same conditions, differences between the two taxa being genetically fixed. Thus, it is sometimes not easy to decide whether one is dealing with ecological races, phenotypic modifications, or both.

Unfortunately, the term 'forma' is reserved for genetically fixed morphological variations within a population. If one wishes to use modifications for the ecological characterization of a plant community, it is possible to use the term 'phenodeme' proposed by Gilmour & Heslop-Harrison (1945), or one may alternatively give a description of the modified appearance of a taxon.

It is generally accepted today that species with a wide distribution possess a considerable amount of ecotypic, i.e. directly adaptive, genetic variation. Classical studies of Turesson (1922, 1925, 1930), Gregor (1930, 1938, 1946), Clausen, Keck & Hiesey (1940, 1948), Böcher (1944), and Böcher & Lewis (1962) show the relationships between particular races and various ecological factors. It has been found that the trends of ecotypical variation are either dis-

continuous or continous; in the former case one speaks about ecotypes, in the latter – an ecoclinal variation is described. Intensive studies in limited areas show that well-defined ecotypes are rare; they are most likely to be found if ecological factors change abruptly or if there are gaps in the geographical distribution of various ecotypes.

Bradshaw (1959, 1960) carried out extensive research on *Agrostis tenuis* in Wales. He found many locations with characteristic, morphologically distinct races which were, however, connected to one another through many intermediate types. He also found morphologically identical races which nevertheless differed from one another in their physiological tolerances (e.g. plants growing in soil that contained lead and those growing in soil free of lead). *Agrostis tenuis* is an example of a recent, multivarious ecological differentiation; the different ecological races of this group can be used only locally for delimitation of phytosociological units. In other groups it is possible to discern ecological races that have obviously undergone a more pronounced ecological and morphological differentiation, and which can thus be considered as subspecies or species.

The criteria for classifying a taxon as a species, a subspecies or a variety have long been the object of controversy. However, it is not relevant whether a good differential taxon of a given phytosociological unit is assigned to a variety or a species. We are inclined to think that the major criteria for taxonomic classification are morphological differentiation and independent geographical and ecological distribution; the ability to hybridize seems important only insofar as it can explain unclear or clear boundary lines. Following these criteria we propose to give the rank of species to taxa which inhabit ecologically well-defined, similar stations within a rather large area (in general, of more than 500 km in at least one direction). Such taxa are morphologically distinct, although often with only a few characters. A species can of course exist in a smaller area, but then its morphological difference from other related taxa should be correspondingly greater and its boundaries more sharply drawn. According to this definition, subspecies would be valid only regionally and would show small differences; the rank of subspecies might also be applicable to intermediary taxa of a recurrent appearance in places where two species meet. Varieties would, in turn, correspond to local populations with particular combinations of characters. Ecological races that are not distinguishable in the field can contribute to the ecological understanding of a phytocoenon when experimented with, but are of no practical value to the phytosociologist.

The afore-mentioned criteria might apply as well to the apomicts. Easily distinguishable taxa with independent ecological and geographical distribution, similar to that of sexual species, might be regarded as species, e.g. many *Alchemilla* taxa, main species of *Hieracium* etc. On the other hand, taxa occurring in geographically very limited areas and which geneticians call microspecies, agamospecies, or apospecies, might be regarded taxonomically as varieties. The same rank would be applicable to plants with similar combinations of characters which appear in locations that are widely apart and probably are of an independent origin, apomictic plants being only rarely fully asexual. A good example of this kind is given by most of the taxa in the *Vulgaria* section of the genus *Taraxacum*. Just as with sexual taxa, a classification can be made only after thorough, biosystematic and ecological investigation carried out on the material from a wide geographic range. However, apomictic taxa, even on the variety level, are often useful for characterizing different phytosociological units, because they are easily identifiable morphologically and well-defined ecologically owing to their limited genetic variability. A good example is presented by Westhoff & den Held (1969) who mention a whole series of *Taraxacum* taxa as character-species and differential species for certain phytocoena in the Netherlands.

The boundaries between morphologically isolated species are not always identical with those separating the phytosociological units involved, particularly when ecologically significant factors gradually change (Fig. 1). Should these factors change suddenly, there is often no overlap in the areas of ecologically different species as it has been found for meadows on calcareous and on siliceous soils in the Alps (Gigon 1971), or soils containing heavy metals (Ernst 1974). If a species with a broad ecological amplitude as the No. 9 in Fig. 1 can be divided into two or more ecological taxa, various types of behaviour can be found;

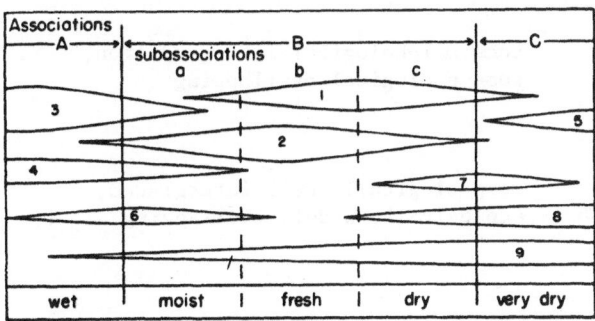

Fig. 1. Diagnostic species along a moisture gradient (from Westhoff & van der Maarel 1973).

they are presented in Fig. 2. A brief comment may follow:

No. 1 corresponds to No. 9 in Fig. 1, Nos. 2 and 3 occur only when the ecology of the two taxa is similar, the competition is not very strong (e.g. different growth forms, geographical separation) and the taxa are totally (No. 2) or partly (No. 3) isolated. Isolation can, in turn, be geographical, biological or genetical. Nos. 4 and 5 are analogous to Nos. 2 and 3, but the ecology of the two taxa is clearly different. In Nos. 6 and 7, competition between the given taxa is very strong, so that they are almost mutually exclusive. No. 6 might also occur when the differentiating ecological factor suddenly changes (e.g. calcareous – siliceous rocks, light – shadowy habitats, early – late mown meadows), even though there is no geographical, biological or genetical isolation. Hybrid plants, if they occur, are inferior to the parent plants in their own habitat, and are eliminated by selection. No. 8 corresponds to No. 6 if no morphological differentiation has gone parallel with the ecological differentiation. There is no effective isolation between the two taxa in Nos. 9, 10, and 11. No. 11 occurs primarily if a taxon differs only slightly from another one in its ecology and there are notable differences in size of the populations, one taxon being more agressive than another. It happens sometimes that morphologically and ecologically intermediate forms (as in No. 10), become geographically independent and cause the formation of additional taxa.

Examples

To illustrate various forms of plant behaviour and their usefulness for phytosociology, several examples will be discussed below. Numerous data on this subject are available (e.g. Bidault 1968); for the present purpose, we have chosen mostly groups that have been or are being studied at our Institute; the only exception represents *Molinia coerulea* s.l.

Molinia coerulea (L.) Moench s.l.

The group of *Molinia coerulea* is a wide-spread and ecologically many-sided species. Guinochet & Lemée (1950) were among the first to publish exact relevés and names of associations from which the plants were collected. Plants from eight different locations were cultivated under identical conditions; comparative morphological studies and transpiration measurements were carried out (Table 2). The plants of different origin differed primarily in quantitative

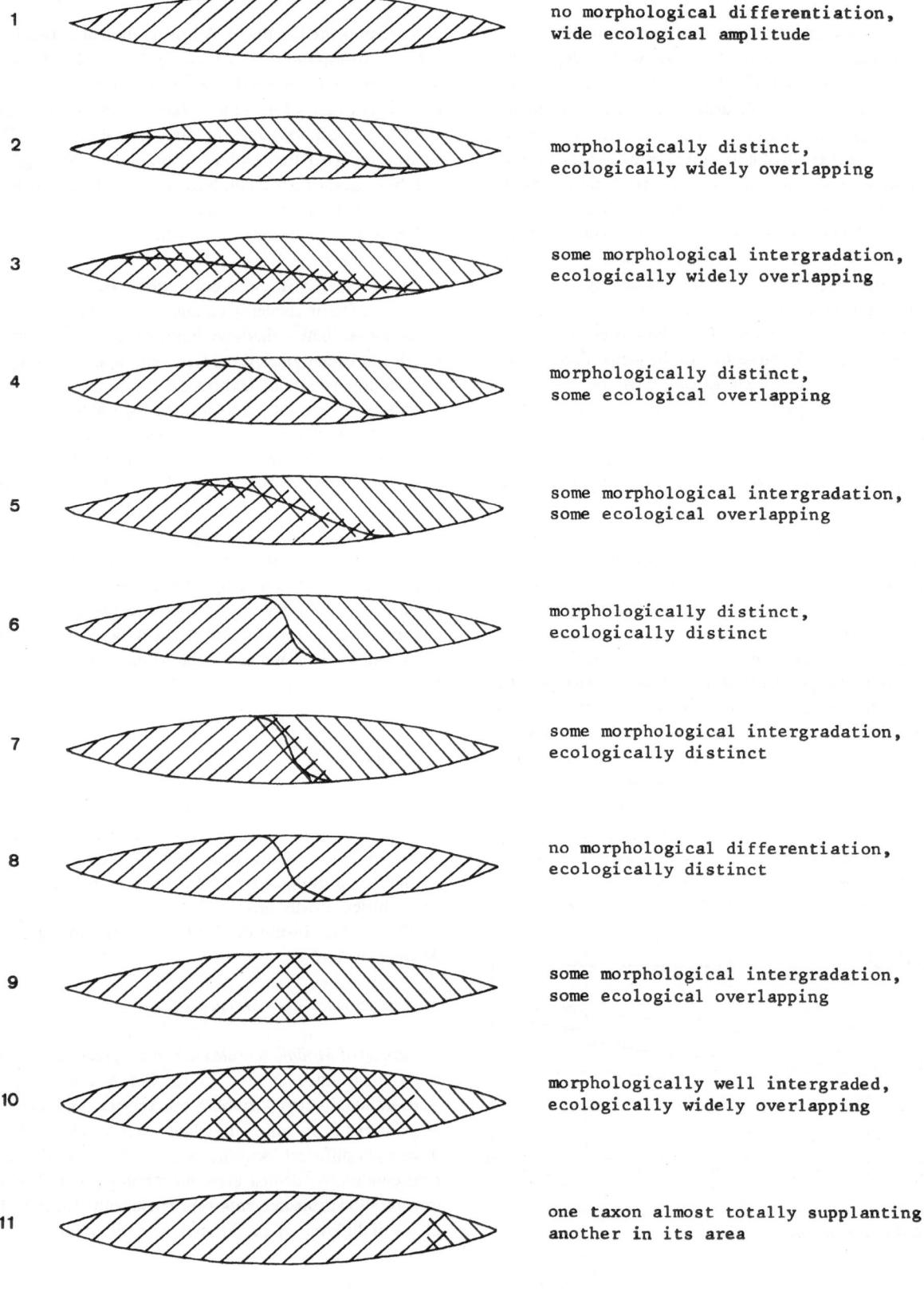

1   no morphological differentiation,
    wide ecological amplitude

2   morphologically distinct,
    ecologically widely overlapping

3   some morphological intergradation,
    ecologically widely overlapping

4   morphologically distinct,
    some ecological overlapping

5   some morphological intergradation,
    some ecological overlapping

6   morphologically distinct,
    ecologically distinct

7   some morphological intergradation,
    ecologically distinct

8   no morphological differentiation,
    ecologically distinct

9   some morphological intergradation,
    some ecological overlapping

10 morphologically well intergraded,
    ecologically widely overlapping

11 one taxon almost totally supplanting
    another in its area

Table 2. *Molinia coerulea* L. s.1.: Ecological and morphological differentiation (after Guinochet & Lemée 1950)

1	2	3	4	5	6	7	8	9	10	11	12
53	13	6.3	0.25	0.66	0.47	0.41	18.0	36	Molinion	6	var. *genuina*
46	10	6.4	0.22	0.87	0.60	–	19.5	36	. . . .		. . . . .
49	11	5.4	0.23	0.71	0.47	0.46	–	36	Quercion roboris sessiliflorae	5	var. *arundinacea*
68	14	7.0	0.20	0.59	0.50	0.46	23.0	36			
78	15	6.8	0.19	0.51	0.43	0.50	20.1	36	Oxycocco-Ericion	5	var. *depauperata*
57	10	5.6	0.18	0.52	0.56	0.45	16.8	36	. . . .		. . . . .
66	10	6.3	0.16	0.52	0.60	0.44	16.3	36	Mesobromion	7	var. *litoralis*
126	23	9.5	0.18	0.37	0.41	0.36	14.8	90			

1 = culm heigt in cm. 2 = leaf length in cm. 3 = leaf breadth in cm. $4 = \dfrac{\text{leaf length}}{\text{culm height}}$ $5 = \dfrac{\text{length of lower glumes} \times 100}{\text{culm height}}$

$6 = \dfrac{\text{leaf breadth} \times 10}{\text{leaf length}}$. $7 = \dfrac{\text{length of lower glumes}}{\text{breadth of lower glumes}}$. 8 = average transpiration in mg/g freshweight and minute.

9 = chromosome number (2n). 10 = phytosociological unit at the place of origin. 11 = pH of the soil. 12 = taxon

characters. The authors described four varieties, each originating from a different alliance. The variety *litoralis* was further divided into the subvariety *normalis* and the subvariety *gigantea*, the latter differing quite clearly from all the former taxa. Whether these varieties are of use for the plant sociologist in the field will only become clear after much more material has been examined, especially at the location itself. Many samples should be investigated from each alliance as well. In my opinion, plants from the two *Mesobromion*-origins given in Table 2 should not be classified together, even though their transpiration behaviour is similar. The pentaploid (2n = 90) plants classified as subvariety *gigantea* should be considered as a species, although they differ only in quantitative characters. This becomes clear when the ecological distribution of this taxon is being studied in a wider geographical area; it has an entirely different center as compared with *M. coerulea* s.str. Whereas the latter occurs alone in acidic straw-meadows and bogs, both taxa are to be found in base rich straw-meadows like the intermittently waterlogged *Mesobromion* stands mentioned by Guinochet & Leméé, (cf Klötzli 1969). On the other hand, only *Molinia litoralis* Host (= subvar. *gigantea* = *M. arundinacea* Schrank) occurs in pine forests of steep intermittently waterlogged slopes (e.g. in the *Molinio-Pinetum*), (cf also investigations of Prey 1975 in Poland).

The *Molinia coerulea* example shows us:
– ecological races can be closely bound to phytosociological units; however, long and difficult research is necessary before one can use one of these races as character- or differential taxon.
– to establish the taxonomic rank of the taxa, a monographic treatment of the species group over a wider geographical area is required.
– the two species *M. coerulea* and *M. litoralis* (perhaps other species can be defined, as well), are more or less isolated genetically owing to the difference in the number of their chromosomes and have different ecological centers. However, there is an overlap in their ecological distribution, and they therefore behave as No. 4 or No. 5, Fig. 2.

Centaurea Jacea L. s.l.

This group was studied cytologically and biosystematically by Marsden-Jones & Turril (1954) in England, Saarisalo-Taubert (1966) in Finland and Baltardive-Gardou (1970) in Western Europe (especially France). Hybridization and gene introgression were frequent between taxa with the same number of chromosomes within this group and many hybrid combinations have been given Latin names. The taxa given on Table 3 occur in Switzerland. *C. Jacea* and *C. dubia* are tetraploid (2n = 44) and have similar ecological distribution. However, they are separated geographically by the Alpine ridge. *C. dubia* is therefore characteristic for insubric vicariants of the same phytosociological units as *C. Jacea*. In the region of upper Ticino populations of *C. dubia* with gene introgressions from

Fig. 2. Possible behaviour of two closely related vicariant taxa along an ecological gradient. Differently shadowed parts correspond to a morphological differentiation.

Table 3. *Centaurea Jacea* L. s.1.: Ecological and geographical differentiation in Switzerland.

taxon	distribution in Switzerland	phytosociological center of distribution	chromosome number 2n
C. Jacea L.	Alps and regions situated north of the Alps	*Arrhenatheretalia*	44
C. dubia Suter	Southern Alps	*Arrhenatheretalia* *Molinietalia*	44
C. angustifolia Schrank	Alpine valleys and areas situated north of the Alps	*Molinion* *Mesobromion* *Origanetalia*	44
C. nemoralis Jordan	Northern Switzerland and northern Alpine foothills	*Trifolion medii* *Sarothamnion*	44
C. bracteata Scop.	Southern Alps	*Diplachnion* *Xerobromion*	22

C. Jacea have been found. Such populations can at best be classified as subspecies. However, this gives us little information from an ecological and phytosociological point of view. *C. nemoralis* occurs in the same area as *C. Jacea*, but grows in acidic soils in meadows and on the edge of forests. Intermediate forms are occasionally classified as *C. pratensis* Thuill. Since acidic meadows at lower altitudes in Switzerland are always local and small in size, *C. nemoralis* is very rare and only to be found in small populations; it is continually losing ground to *C. Jacea* or being integrated into larger populations of this species.

C. angustifolia also grows in the same area as *C. Jacea*, but in meagre meadows which are either not mowed at all, or not until autumn. These two species could also hybridize freely, but intermediary forms are rare, and if they do occur they are restricted to disturbed areas. The two species nearly exclude each other from their respective stations; the decisive factor here is probably not the difference in the nutrition status of the meadows, but rather the early mowing or grazing which prevents *C. angustifolia* from appearing on fertilized meadows and pastures. When this factor does not play a role, *C. angustifolia*, independently of good or bad nutritive conditions, is noticeably taller and probably more competitive than *C. Jacea* (Gebert 1972). In the southern Alps, the two species *C. dubia* and *C. bracteata* behave as two well-defined species since the difference in their number of chromosomes hardly permits them to hybridize (Baltardive-Gardou 1970). According to Meyer (1976), they can both be used as differential species in the *Carici humilis-Chrysopogonetum*. *C. dubia* is a good differential species of the subassociation with *Galium verum* against the drier subassociation with *Fumana ericoides*; *C. bracteata* grows in both subassociations and

is sharply delimiting from the *Arrhenatheretalia*-units where *C. dubia* has its center of distribution.

The *Centaurea Jacea* example shows us that a given species complex may show various types of relations between its taxa:
– genetically isolated taxa whose phytosociological distribution overlaps (No. 4 in Fig. 2: *C. dubia* - *C. bracteata*);
– taxa that are not isolated genetically and overlap a little phytosociologically (No. 7 in Fig. 2: *C. Jacea* - *C. angustifolia*);
– taxa which are isolated geographically, but not genetically (with limited introgression), and have similar phytosociological behaviour except for small climatic differences (No. 3 in Fig. 1: *C. Jacea* - *C. dubia*);
– taxa which are not genetically isolated and which are ecologically and sociologically insufficiently isolated, so that the survival of one of the species in the area under study is doubtful (No. 11 in Fig. 2: *C. Jacea* - *C. nemoralis*).

Cardamine pratensis L. s.l.

C. pratensis is one of the most common meadow plants in Central Europe. Already in 1946, Guinochet called attention to a cytological and ecological differentiation in the Jura mountains. Lövkvist (1956) carried out detailed cyto-taxonomical studies. Approximately 1000 populations have been studied biosystematically at our Institute (Landolt & Urbanska-Worytkiewicz 1971, Urbanska-Worytkiewicz & Landolt 1974). It was possible to distinguish taxa occurring in forests, fertilized meadows, reed meadows and mountains. Each of these ecological groups comprised various chromosomic races (Table 4). *C. nemorosa* is well-defined both ecologically and geograph-

Table 4. *Cardamine pratensis* L. s.l.: Ecological, geographical and cytological differentiation in Central Europe north of the Alps.

taxon	geographical distribution	phytosociological center of distribution	chromosome number 2n
C. nemorosa Lejeune	Central-european mountains (colline-montane)	*Fagion*	16, 20
C. pratensis L.	Central Europe north of the Alps (colline-montane)	*Arrhenatheretalia* *Alno-Padion*	16, 30, 40, 48
C. udicola Jordan	Central Europe (colline-montane)	*Caricion elatae* *Caricion gracilis* *Filipendulion*	16, 32, 40
C. palustris Peterm.	Central- and Northern Europe (colline)	*Caricion elatae* *Phragmition*	56 - 96
C. rivularis Schur	Central- and South-european mountains (subalpine)	*Caricion fuscae-canescenti* *Caricion rostratae*	16, 32

ically; it occurs in beech groves, primarily on base rich soils. A chromosome race of *C. pratensis* with 2n = 30 grows in the same area in acidic soil, primarily in the *Alno-Padion*. However, *C. nemorosa* is morphologically not distinct from the neighbouring *C. pratensis* taxa; it must thus be considered a crypto-species and as such it is of no interest in practical phytosociology and floristics. The same holds true for the numerous chromosome races of the other species which are partly distinct ecologically and geographically, but not morphologically. The morphological distinction between *C. udicola* and *C. pratensis* is more easily recognizable, but not always sharply drawn. *C. udicola* indicates wet soils with a rather low nutrient content, primarily in the *Caricion elatae*, and avoids the *Arrhenatheretalia* communities for which *C. pratensis* is characteristic. *C. pratensis* can also be found in *Phragmitetalia* communities where *C. udicola* does not grow. *C. palustris* occurs only in a very few locations in southern Central Europe, but is much more frequent in northern Europe. Its ecological distribution can vary according to the number of chromosomes and to the geographical area. Locally, *C. palustris* can be very characteristic for certain phytocoena. *C. rivularis*, finally, differs ecologically from *C. udicola* primarily in the altitude of its distribution.

The *Cardamine pratensis* example shows us:
– ecological differentiation without corresponding morphological diversification leads to the formation of taxa which are of no practical use in phytosociology (No. 8, Fig. 2: *C. nemorosa - C. pratensis*);
– the formation of chromosome races does not always lead to total genetic isolation, and morphological differentiation does not always, therefore, undergo an independent development.

– with *Cardamine pratensis* s.l., most of the ecologically vicariant taxa behave as in No. 7 or 9 in Fig. 2;
– discontinuous geographical distribution may result in locally distinct ecological races (*Cardamine palustris*).

Scabiosa columbaria L. s.l.

Only one number of chromosomes, i.e. 2n = 16 has hitherto been found in this wide-spread group. No barriers to hybridization occur between the taxa (Grossmann 1975). The six species listed in Table 5 can be found in the Alpine region. As Grossmann (1975) has shown, the morphological distinction between the species is not always a sharp one, especially when ecological variation is due almost exclusively to changes in altitude. Several quantitative characters run parallel between vicariant mountain and lowland taxa (Fig. 3). The morphological characters which were examined correlated very closely with the altitude when the degrees of exposure and slope were approximately equal. (This is not always the case for the stations appearing in Fig. 3.) Thus large homogeneous populations that have often been given Latin names can be found in certain wide, flat valleys of the Alps. Such populations could be classified as varieties. It seems advisable, however, to give them the names of the species between which they occur, and to add a special sign indicating their approximate location between the two species. The following possibilities occur, e.g. between *S. lucida* and *S. gramuntia*:

S. lucida
S. lucida　>　*S. gramuntia*
S. lucida　–　*S. gramuntia*
S. lucida　<　*S. gramuntia*
S. gramuntia

Table 5. *Scabiosa columbaria* L. s.1.: Ecological and geographical differentiation in the Alpine area.

taxon	Alpine distribution	phytosociological center of distribution
S. gramuntia L.	Valleys of central and southern Alps (colline-montane)	*Festucion valesiacae* and other dry grassland and dry forests
S. columbaria L.	Area north of the Alps (colline-montane)	*Mesobromion*
S. portae A. Kerner	Southside of the Alps (colline-montane)	*Arrhenatherion* *Mesobromion*
S. lucida Vill.	Alps (subalpine-alpine)	*Seslerion* *Caricion ferrugineae*
S. vestita Jordan	Southwestern Alps (montane-alpine)	*Seslerion* and other dry grassland
S. candicans Jordan	The Maritime Alps (planar-colline)	*Quercion ilicis*

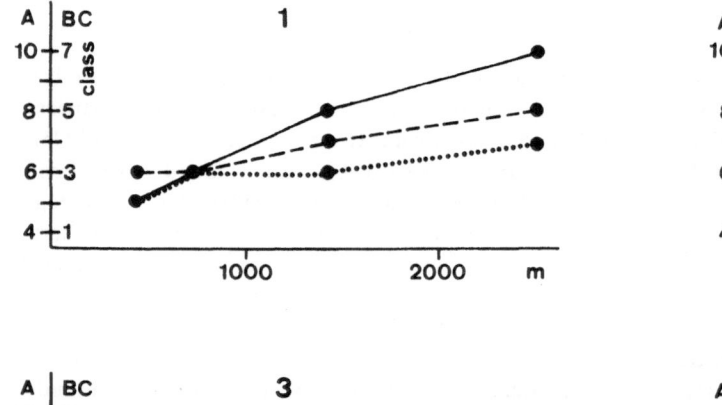
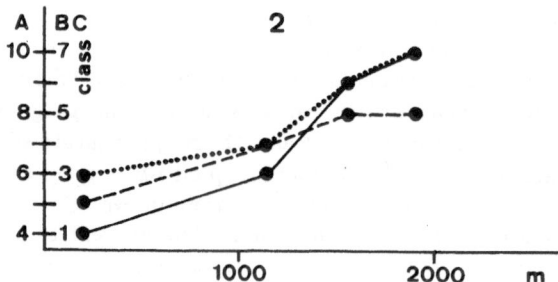
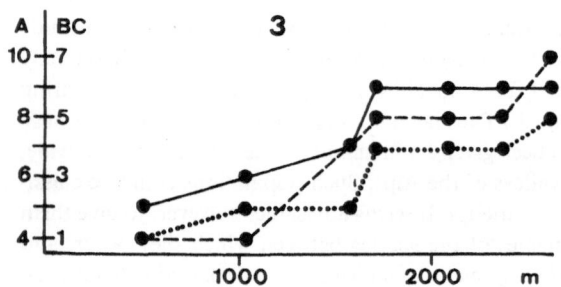
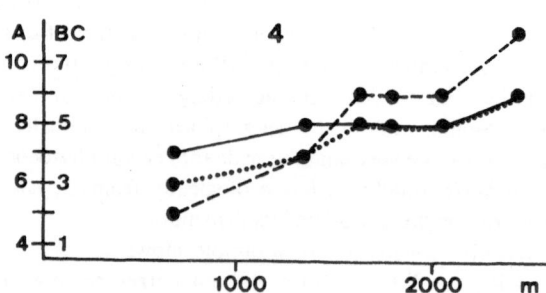

Fig. 3. *Scabiosa columbaria* L.s.l.: Correlation between three morphological characters and the altitudes a.s.l. (after Grossmann 1975, slightly modified).

1-4 investigated area (lowland taxon — mountain taxon)
1 Maritime Alps (*S. candicans* Jord. – *S. vestita* Jord.)
2 Alps of Bergamo (*S. Portae* A. Kerner – *S. dubia* Vel.)
3 Alps of Valais (*S. gramuntia* L. – *S. lucida* Vill.)
4 Alps of Grisons (*S. columbaria* L. – *S. lucida* Vill.)

A ●—● length of calyx setae (1 = short, 10 = long)
B ●····● relative length of leaf segment (1 = large, 7 = small)
C ●---● plant height (1 = large, 8 = small)

If ecological factors which determine the occurrence of two species undergo a relatively sudden change, then there are narrow morphological transitions, or none at all. In Ticino, *S. Portae* is found primarily on fertilized meadows and *S. gramuntia* on dry meadows. According to Meyer (1976) *S. gramuntia* is a differential species of the *Fumana ericoides* subassociation of the *Carici humilis - Chrysopogonetum Grylli* against the more mesophilic *Galium verum* subassociation. *S. Portae*, on the other hand, is typical for the *Galium verum* subassociation, and does not grow in the other subassociation. *S. gramuntia*, which is rather rare in that area, does occasionally show certain characters of *S. Portae*. *S. lucida* and *S. vestita* can be differentiated above all geographically; *S. lucida*, in contrast with *S. vestita*, does not grow in very dry and acidic soil. On Mount Cenis, where they both occur, they are morphologically distinct. *S. lucida* grows in the *Caricion ferrugineae* on calcareous schist, *S. vestita* in dry meadows on quartzit and other siliceous rock that contains little Ca. It seems clear that hybrids between *S. lucida* and *S. vestita* are not as competitive as their parents. Perhaps the difference in blossom colour (violet for *S. lucida*, red for *S. vestita*) causes a certain degree of biological isolation. It should be emphasized that the broad transitional areas between mountain and lowland species are due to the human influence. In nature, there would be a forested zone between the lowland species growing in steppe-like meadows, and the mountain species growing in alpine meadows; no *Scabiosa* could grow in this dark forest zone, so that there would be almost no contact between the two species.

The *S. columbaria* example shows us:
– two taxa which are not genetically isolated and which differentiate along ecological gradients, produce broad morphological transitions (No. 10 in Fig. 2). If they are studied in detail, transitional populations can nevertheless be used to characterize vegetation units. An analysis of characters can furnish data on temperature conditions of a habitat for transitional populations of mountain and lowland species of *Scabiosa*;
– even when they are not genetically isolated, two taxa can be distinguished rather sharply if the ecological factors change rapidly and/or if external isolation mechanisms are present (No. 7 in Fig. 2: *S. lucida - S. vestita*).

Edaphic vicariant species in the Alps
In order to determine the exact ecological boundaries between edaphic vicariant species, investigations were carried out in the Davos and Rätikon areas (Ruggli-Walser 1976). Special attention has been paid to the

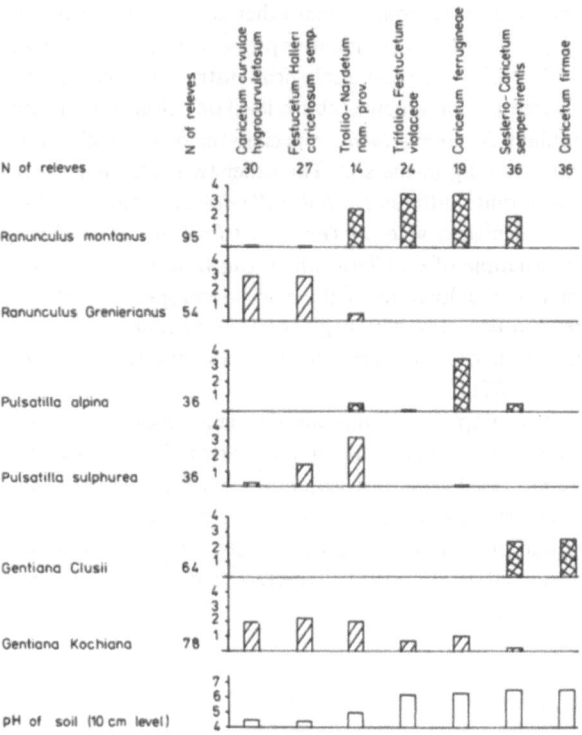

Fig. 4. Phytosociological distribution of edaphic vicariant species near Davos and in the Raetikon area (Grisons). The scale number on the ordinate indicates the average occurrence and frequency of the resp. species. 0 = r (scale of Braun-Blanquet); 2 = + ; 3 = 1; 4 = 2; 5 = 3.

following pairs of species: *Ranunculus montanus - R. Grenierianus*; *Pulsatilla alpina - P. sulphurea*, *Gentiana Clusii - G. Kochiana*. 200 relevés were taken between 1800 and 2570 m. These relevés were classified with a computer programme, and partly by hand according to affinities in species composition, and were then divided into 23 vegetation units. Those were in turn grouped into 6 associations (Fig. 4). The distinction between various vicariant taxa is quite well-marked, although some of them (e.g. *Pulsatilla*) are not genetically isolated. The studied vicariant taxa occur together only rarely, and then primarily in transitional vegetation units (7 times the *Ranunculus* species, 3 times the *Pulsatilla* species and not at all the *Gentiana* species). In a unit which was typical for the vicariant species, *Ranunculus montanus* occurred 5 times, *R. Grenierianus* once, *Gentiana Clusii* was never found whereas *G. Kochiana* occurred 3 times. *Pulsatilla alpina* and *P. sulphurea* were never observed in their resp. vicariant units. The individual species therefore behave like very good character-species and differential species. Only *Ranunculus Grenierianus* is

115

somewhat more limited than other siliceous taxa; this is probably due to the competitive pressure from *R. montanus*, which has the largest ecological distribution of all the calcareous taxa. It can establish itself on siliceous rock and supplant *R. Grenierianus*, especially in places with some water running in the soil. The other two calcareous taxa behave quite differently: *Pulsatilla alpina* avoids dry and exposed places, whereas *Gentiana Clusii* thrives in them. The example of *G. Clusii*, which can flourish in siliceous soil under cultivation if there is no competition, shows that mutual competition plays an important role in the formation of boundaries between the vicariant species (Gigon 1971).

The example of edaphically vicariant species shows us:
– in places where crucial ecological factors (in this case, the parent rocky material) undergo a sudden change, even closely related taxa, not isolated genetically, mutually exclude one another to a large degree; they can thus be considered as very good character- or differential species (No. 6 and 7 in Fig. 2).

Conclusion

The foregoing discussion shows that closely related taxa do not furnish new criteria for the separation of vegetation units in phytosociology. However, the sub-classifications of a species can provide new character-species and differential species, which is of particular importance for phytocoena that contain few species. In addition, the occurrence of closely related taxa, even if they do not differ sharply from one another, can furnish a great deal of ecological information on the given vegetation unit. Thus, phytosociologists are very much interested in biosystematic studies carried out on as many taxonomic groups as possible. It will take decades to do this; however, pertinent contributions of phytosociologists can encourage the biosystematists' understanding of the groups under study. Phytosociology and biosystematics are thus mutually complementary fields of endeavour and it would, in fact, be difficult to imagine the one without the other.

Summary

Phytosociological importance of closely related taxa has already been stressed by numerous authors. Difficulties in using closely related taxa as characteristics of a given plant association are mainly due to our insufficient knowledge of their biosystematics; the real status of many proposed units (e.g. microspecies, subspecies, variety, forma) is often not precised and it is often even not known whether the diagnostic characters are hereditary or merely represent some modifications. Furthermore, the ecological differentiation may show regional or local differences. Thorough experimental studies in many critical plant groups are required.

The behaviour of closely related taxa under different ecological conditions is shown in Fig. 2. Various trends are exemplified by several groups of species: *Molinia coerulea*, *Centaurea Jacea*, *Cardamine pratensis*, *Scabiosa columbaria*, also some Alpine vicariants on siliceous and calcareous soils.

Closely related taxa used in the phytosociology will not show any new criteria for delimitation of vegetation units, but often increase the number of characteristic components. Accordingly, actual limits of a given unit become better documented and more ecological information will be gathered.

Zusammenfassung

Die pflanzensoziologische Bedeutung von nah verwandten Sippen ist schon oft betont worden. Die Schwierigkeit ihrer Verwendung zur Charakterisierung von Pflanzengesellschaften liegt meist darin, dass sie biosystematisch nur schlecht untersucht und morphologisch schlecht fassbar sind. Die Wertigkeit von vielen unterschiedenen Taxa ist oft nicht bekannt, und gelegentlich ist es auch nicht bekannt, ob Unterscheidungsmerkmale zwischen Sippen wirklich genetisch fixiert sind oder nur Modifikationen darstellen. Dazu kommt, dass sehr oft ökologische Differenzierungen sich regional oder lokal unterscheiden. Nur experimentelle Untersuchungen in kritischen Pflanzengruppen können hier Aufklärung bringen.

Das Verhalten von nah verwandten Sippen unter verschiedenen ökologischen Bedingungen ist aus Fig. 2 ersichtlich. Einzelne Verhaltensmuster werden anhand von einigen Artengruppen gezeigt: *Molinia coerulea*, *Centaurea Jacea*, *Cardamine pratensis*, *Scabiosa columbaria*. Ebenso wird das Verhalten von vikariierenden Arten auf Karbonat- und auf Silikatgestein in den Alpen gezeigt.

Nah verwandte Sippen bringen keine neuen Kriterien für die Abgrenzung von Pflanzengesellschaftseinheiten, sondern vermehren lediglich die Zahl der charakteristischen Sippen für einzelne Einheiten. Dadurch können die Grenzen besser definiert werden, und viel zusätzliche ökologische Information wird gewonnen.

References

Baltardive-Gardou, Ch. 1970. Recherches biosystématiques sur la section Jacea Cass. et quelques sections voisines du genre Centaurea L. en France et dans les régions limitrophes. Thèse Orsay, 238 pp.

Bidault, M. 1968. Essai de taxonomie expérimentale et numérique sur Festuca ovina L.s.l. dans le sud-est de la France. Rev. Cyt. Biol. Vég. 31: 217–356.

Böcher, T. W. 1944. The leaf size of Veronica officinalis in relation to genetic and environmental factors. Dansk Bot. Ark. 11 (7): 1–20.

Böcher, T. W. & M. C. Lewis. 1962. Experimental and cytological studies on plant species 7, Geranium sanguineum. Biol. Skr. 11: 1–25.

Bradshaw, A. D. 1959. Population differentiation in Agrostis tenuis Sibth. 1. Morphological differentiation. New Phytol. 58: 208–227.

Bradshaw, A. D. 1960. Population differentiation in Agrostis tenuis Sibth. 3. Populations in varied environments. New Phytol. 59: 92–103.

Braun-Blanquet, J. 1964. Pflanzensoziologie. 3. Aufl. Springer, Wien. 865 pp.

Clausen, J., D. D. Keck & W. M. Hiesey. 1940. Experimental studies on the nature of species. 1. The effect of varied environments on western North American plants. Carnegie Inst. Wash. Publ. 520.

Clausen, J., D. D. Keck & W. M. Hiesey. 1948. Experimental studies on the nature of species. 3. Environmental responses of climatic races of Achillea. Carnegie Inst. Wash. Publ. 581.

Ernst, W. 1974. Schwermetallvegetation der Erde. Fischer, Stuttgart. 194 pp.

Frey, L. 1975. Taxonomical studies on the genus Molinia Schrank in Poland. Frag. Flor. Geobot. 21: 17–20.

Gebert, R. 1972. Konkurrenzversuche mit Centaurea Jacea L. und C. angustifolia Schrank. Ber. Geobot. Inst. ETH, Stiftung Rübel 41: 25–34.

Gigon, A. 1971. Vergleich alpiner Rasen auf Silikat- und Karbonatboden, Veröff. Geobot. Inst. ETH, Stiftung Rübel 48: 159 pp.

Gilmour, J. S. L. & J. Heslop-Harrison. 1954. The deme terminology and the units of micro-evolutionary change. Genetica 27: 147–161.

Gregor, J. W. 1938. Experimental taxonomy. 2. Initial population differentiation in Plantago maritima in Britain. New Phytol. 37: 15–49.

Gregor, J. W. 1946. Ecotypic differentiation. New Phytol. 45: 254–270.

Grossmann, F. 1975. Morphologisch-ökologische Untersuchungen an Scabiosa columbaria L.s.l. im mittleren und westlichen Alpengebiet. Veröff. Geobot. Inst. ETH, Stiftung Rübel 52: 122 pp.

Guinochet, M. 1938. Etude sur la végétation de l'étage alpin dans le bassin supérieur de la Tinée (Alpes-Maritimes). Thèse Lyon, 458 pp.

Guinochet, M. 1946. Sur l'existence, dans le Jura central, des races écologiques aneuploïdes et polyploïdes chez Cardamine pratensis L. C. R. Acad. Sci. 222: 1131–1133.

Guinochet, M. & G. Lemée. 1950. Contribution à la connaissance des races biologiques de Molinia coerulea (L.) Moench. Rev. Gén. Bot. 57: 565–594.

Huxley, J. S. 1938. Clines: an auxiliary taxonomic principle. Nature 142: 219.

Huxley, J. S. 1939. Clines: an auxiliary method in taxonomy. Bijdr. Dierkunde 27: 491–520.

Klötzli, F. 1969. Die Grundwasserbeziehungen der Streu- und Moorwiesen im nördlichen Schweizer Mittelland. Beitr. Geobot. Landesaufn. Schweiz 52: 296 pp.

Koch, W. 1926. Die Vegetationseinheiten der Linthebene. Jahrb. St. Gall. Naturw. Ges. 61: 144 pp.

Landolt, E. 1971. Oekologische Differenzierungsmuster bei Artengruppen im Gebiet der Schweizerflora. Boissiera 19: 129–148.

Landolt, E. & K. Urbanska-Worytkiewicz. 1971. Zytotaxonomische Untersuchungen an Cardamine pratensis L.s.l. im Bereich der Schweizer Alpen und des Jura. Ber. Dtsch. Bot. Ges. 84: 683–690.

Lövkvist, B. 1956. The Cardamine pratensis complex. Outlines of its cytogenetics and taxonomy. Symp. Bot. Ups. 14/2: 131 pp.

Marsden-Jones, E. M. & W. B. Turrill. 1954. British knapweeds a study in synthetic taxonomy. Quaritch Ltd. London: 201 pp.

Meyer, M. 1967. Pflanzensoziologische und ökologische Untersuchungen an insubrischen Trockenwiesen karbonathaltiger Standorte. Veröff. Geobot. Inst. ETH, Stiftung Rübel 57: 145 pp.

Pignatti, S. 1968. Die Verwertung der sogenannten Gesamtarten für die floristische Systematik. Ber. Symp. Int. Ver. Vegk. 1964: 71–73. Junk, Den Haag.

Ruggli-Walser, A. 1976. Vikariierende Arten auf Kalk und Silikat. Diplomarbeit Geobot. Inst. ETH, Stiftung Rübel, Zürich: 127 pp.

Saarisalo-Taubert, A. 1966. A study of hybridization in Centaurea, section Jacea in eastern Fennoscandia. Ann. Bot. Fenn. 3: 86–95.

Turesson, G. 1922. The species and variety as ecological units. Hereditas 3: 100–113.

Turesson, G. 1925. The plant species in relation to habitat and climate. Hereditas 6: 147–236.

Turesson, G. 1930. The selective effect of climate upon the plant species. Hereditas 14: 99–152.

Urbanska-Worytkiewicz, K. & E. Landolt. 1974. Biosystematic investigations in Cardamine pratensis L.s.l. I. Diploid taxa from Central Europe and their fertility relationships. Ber. Geobot. Inst. ETH, Stiftung Rübel 42: 42–139.

Westhoff, V. 1947. The vegetation of dunes and salt marshes on the Dutch Islands of Terschelling, Vlieland and Texel. Thesis, Utrecht: 131 pp.

Westhoff, V. & A. J. den Held. 1969. Plantengemeenschappen in Nederland. Thieme, Zutphen: 324 pp.

Westhoff, V. & E. van der Maarel. 1973. The Braun-Blanquet approach. In: Whittaker, R. H.: Handbook of Vegetation Science 5: 617–707. Junk, The Hague.

117

WEED SPECIES AND WEED COMMUNITIES*

Wolfgang HOLZNER**' ***

Botanisches Institut, Universität für Bodenkultur, Vienna, Austria

Keywords: Calcicole/Calcifuge plants, Classification, Compensation, Competition, Ecological amplitude, Herbicides, Weeds

Bowing to the mower,
Yet they know nothing about classification.
Happy little weeds.

Hō-Ru Tsu-na

Introduction

One of the most satisfying experiences of phytosociological work is the moment when out of the mass of amorphous data begins to crystallize a lattice of general principles and an understanding of vegetation and its ecology. But if one works still further and gathers more and more knowledge in one and the same field, one reaches a deeper level. The more one knows, the more the boundaries of classes are broken down, the crystals loose their defined shape again and one begins to understand the impossibility of classifying nature in a generally satisfying way. If one tries to retreat to the classes of individuals called species, one has soon to admit that, here too, with growing knowledge the boundaries become insecure, especially where the notoriously heterogeneous weed species are concerned. Here one approaches the boundary between science and philosophy. As this is to be a scientific paper we have to stop here and try to maintain classification as a mere practical means to get a general view and to summarize and pass on our knowledge to others.

Weeds are a class of plants that is difficult to define (if one knows too much about them) as it has no sharp boundaries. A vast literature deals with this problem

* Nomenclature follows Ehrendorfer (1973), Phytosociological units according to Westhoff & Den Held (1969).
** Contribution to the Symposium on Plant Species and Plant communities, held at Nijmegen, 11–12 November 1976, on the occasion of the 60th birthday of Professor Victor Westhoff.
*** Field studies were partly supported by a grant of the Austrian Federal Ministry of Agriculture and Forestry.

(summarized in Harlan & de Wet 1965, King 1966). For our purpose we shall define weeds as plants adapted to man-made habitats and interfering there with human activities. Weed species are, therefore, plant species that meet this definition in at least a part of their area. Many, but not all weeds are typical colonizing species, but not vice versa as often maintained, as there are many colonizing plant species that cannot be considered as weeds.

Weeds may occur in three general types of vegetation: (1) As agrestals (or segetals) in arable land; (2) As ruderals in one of the large range of possible ruderal sites; (3) In natural vegetation, from which they originate or into which they have been able to invade.

The results presented here are the general conclusions from the data of extensive phytosociological field research into weed vegetation on arable land and ruderal sites all over Austria (about 2.000 relevés), in Southern Europe and (together with Prof. Dr. E. Hübl) Southwest Iran. The phytosociological results have previously been published in part only (Holzner 1970–1974b, Hübl & Holzner 1974) but some of the general aspects have already been published briefly elsewhere (Holzner 1974a, 1977).

Ecological and sociological behaviour of two opposite types of agrestals in the different parts of their geographical range

Type A: It is generally accepted that many of the European agrestals, especially those of autumn-sown cereals, were brought from the Middle East, the "cradle of agriculture", and the Mediterranean to Central Europe. From there they were spread over the whole world. It is a group of annuals adjusted to a climate with an adverse season, which they endure as seeds. They occur in different types of natural and semi-natural communities from the Orient to Southern

Europe, some of them even in Central Europe, in places where conditions for perennials are so severe that they do not form dense stands, leaving enough space among them for shortlived pioneers. Their range has been enlarged by man as agriculturist, who created open habitats, distributed their seeds and brought them to each country in Europe. In many places they can only survive by the involuntary help of man, creating for them sites free from the competition of most species of the native flora. In these countries we can observe a great difference between the flora of cultivated land and the flora of the natural and seminatural vegetation, which is not the case in the Mediterranean and particularly the Middle East where many of the weeds occur in cultivated land and in the adjacent more or less natural vegetation (Kühn 1972, Zohary 1973). This seems to be an important argument, but we must not overlook the fact that these areas have been under intensive human influence for thousands of years and that overgrazing in particular opened the natural vegetation to colonizing annuals. Thus, an answer to the question of the origin of many weed species will always be a rather doubtful one and it is very likely that many weed species have evolved rather recently under cultivation pressure ("homeless weeds": Zohary, M. 1973). (As it is not the purpose of this paper to discuss the origin of weeds I can only refer to the literature (e.g. Baker 1965, 1974, Zohary 1973).)

From a Central European point of view the weeds of type A are thermophilous species of southern origin, occurring as weeds mainly in winter crops, where they are adapted to the germination conditions and to the seasonal rhythm of the cereal species that have come from the same area. In the classical Phytosociological System (cf. R. Tüxen 1950, Westhoff & den Held 1969), the species of type A are in general identical with those characteristic for the *Secalietalia*. If we consider the whole range of this order we can observe a gradual impoverishment mainly related to the declining summer temperatures, from the south (and east) to the north (and west) as one species after the other drops out but very few are added, acting, in a way, as indicators of cool and humid climates. The same phenomenon can be observed with increasing altitudes in the alps. In a somewhat exaggerated sense we could say with Kühn (1972) that the European weed communities of winter crops on basic soils are just depauperate forms of the oriental ones.

As the complex climatic gradient that is relevant to this group of weeds usually shows a smooth transition, the alteration of the species-composition in weed communities is a very gradual one leading to a "continuum" between the rich, southern and the poor, northern communities. Thus it

is difficult to classify distinct community types within this transition zone, if one is considering a large area.

Before the species reach the northern (western, upper) limit of their range they show a typical preference for calcareous soil, while in their optimal climatic region they are indifferent to this soil factor. Thus in Eastern Austria for instance the same *Caucalidion* communities can be found on acid as well as on calcareous substrates, because the floristic dissimilarities between the weed communities on the different substrates are much smaller than between the communities in cereals or row-crops (see below).

Thus, the general ecological description of the *Caucalidion* as an alliance of weed communities on calcareous substrates as is generally accepted in the literature (Tüxen, R. 1950) is valid only for areas near to its climatic boundary.

I have the impression that at least for weeds the number of "calciphilous" species is much smaller than usually accepted (if there are any at all), as most of the many weeds that are labelled as calci- or basiphilous in the European literature grow well on acid soils if the climate is optimal for them.

One of the effects on weeds of intensified agriculture is that many sensitive species diminish their range of distribution by retreating towards its centre because they are more sensitive and less vigorous at the limit of their range. *Secalietalia*, especially *Caucalidion* species, have become extinct in many parts of Northern, Western and Central Europe because climatic conditions there are not optimal for them.

Type B represents a similar gradient to A not from warm to cool but from oceanic to continental climate* as is shown by species of the *Aperetalia* and lower syntaxa belonging to this order, species which have an oceanic-suboceanic distribution centre and are acidophytes, a combination often found in European plants. Their origin is even more mysterious than that of the former group. It is probably not the Middle East (Kühn 1972) but Southwestern Europe, where they have their natural habitats in open communities on poor sand or shallow granitic soil. With human help they were able to invade as weeds the continental areas of the rest of Europe. Today they are retreating again because they cannot withstand the double stress of unfavourable climate and intensified agriculture. They are especially sensitive to intensified fertilization

* To maintain the theme and general approach of this paper, geographical matters and chorological peculiarities of the species are simplified in a rather crude manner. For details see Meusel, Jäger & Weinert (1965) and Weinert (1973).

because it enhances the strength of other weeds that were not able to compete with them on the poor soils.

While calciphily often seems to be a function of climate rather than a physiological feature of the plants, many acidophytic species really need acid soils, or better expressed, have to avoid calcareous ones because of their special physiological properties (cf. Kinzel 1968, 1969, 1971). If the climate is optimal and competition is low some of them are able to compete on rich, even neutral soils. But in continental and subcontinental areas they are able to compete only in very acid and poor soils.

These results of ecological field-surveys show us that the reaction of plant distribution to the complex of soil properties, indicated by the pH values, is a complicated one and a function of (complex) climatic factors and the competition of other species, and it can therefore vary from one site or part of the species area to another.

If one attempts to derive general principles from findings on such a special group of plants as annual weeds, it could be said that the number of calciphilous species will turn out to be much smaller and that of genuinely calcifuge species (for physiological reasons) larger than usually accepted, if the whole area of the species and the complex dependence on climate and competition are taken into account. Though alpine vegetation has always been the primary and the most impressive area for demonstrating the floristic differences between carbonaceous and siliceous substrates, the same results can be expected there, as has been shown recently for instance by our own observations (Holzner & Hübl 1977) in eastern Austria and more particularly by the experiments of Gigon (1971) in the Swiss alps.

Distribution of weed species in ruderal and agrestal vegetation types respectively depending on climatic factors

While the discussion has so far been mainly about agrestal species, weeds that may occur on ruderal sites as well as on cultivated land will now be dealt with. Many of them are typical pioneer species (characteristics cf. Baker 1965, Ehrendorfer 1965, Harper 1965). In arable land they occur primarily in so-called "row-crops" (maize, turnips, potatoes), crops that are sown late in spring and harvested in autumn, and, at least in former times, were hoed several times. The row-crop weeds are characterized mainly by high temperature requirements for germination (Lauer 1953), some of them also by short life cycles and high nutrient requirements. In the phytosociological system the row-crop weed communities are united in the *Polygono-Chenopodietalia*.

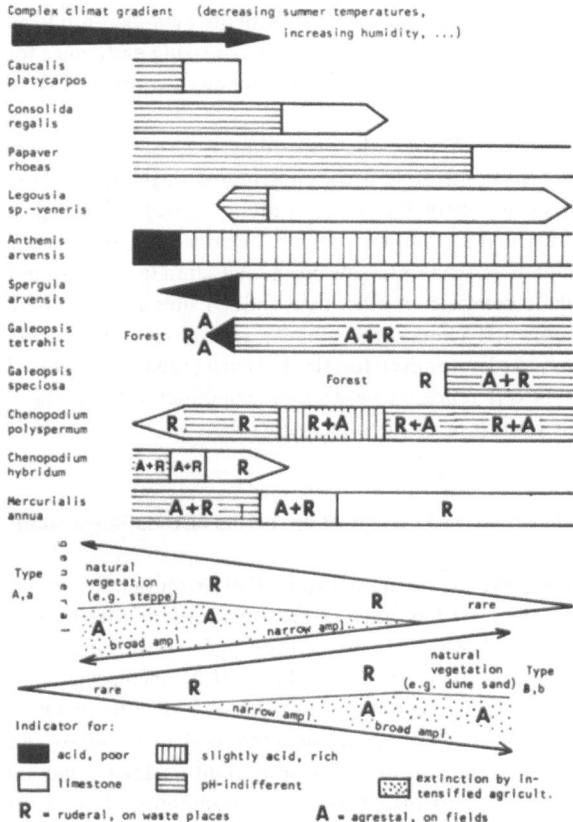

Fig. 1. Variability of ecological and sociological behaviour of some weed species in Austria with the climate.

Type a: Termophilous species requiring high summer temperatures, mainly species of *Eragrostidion* but also *Panico-Setarion*, occur in their optimal climatic areas in ruderal as well as agrestal vegetation. The cooler the climate, the more they are restricted to ruderal sites only. For this phenomenon there could be three reasons: Many ruderal sites have a microclimate warmer than the general climate of the area (Grosse-Brauckmann 1953), and many ruderal sites are richer in nutrients and often contain more lime (from walls, etc.) than the agrestal sites nearby. The most important reason in my opinion is that in young or often disturbed ruderal sites which offer the suitable environment for the colonizing plant species in question, there is much less competition than on cultivated land, where the weeds are not only subjected to the competition from other weeds but also from the cultivated plants, which are sown as densely as possible.

If we observe the sociological and ecological behaviour of species in this group, we see a combined effect of increasing restriction to communities with little competition from other species on ruderal sites and increasing preference for calcareous soils with decreasing summer temperatures (see Fig. 1).

Type b: Among the ruderal/agrestal species we are talking about, there is also a group with its distribution center in subatlantic (or even atlantic) areas. While they occur in their optimal climatic regions as agrestals and ruderals as well, in a subcontinental climate they are not able to compete on other than ruderal sites and in a more continental climate they finally retreat into very shady sites, gardens or even forests. J. Tüxen (1958) has reported on the differences in weed species combinations in gardens and row-crops of the same area.)

Influence of modern agriculture on the distribution of weeds

During the last few decades a kind of revolution in agricultural methods has taken place. Most tools and techniques that have been in use for centuries more or less unaltered have been abandoned or replaced. With increasing standards of living, high wages and shortage of workers, farmwork had to become rationalized and mechanized. As human actions are the most important ecological factors for weeds, their distribution and communities have been subject to strong alterations, a development that is still going on.

One of them has been the extinction of many species in Central Europe (R. Tüxen 1962, Westhoff & Zonderwijk 1960, Zonderwijk 1975, Kump 1970) that have accompanied the cropplants there for centuries. As I have pointed out in a previous section these plants are only locally eradicated and are retreating towards their climatic optimum, where most of them will be able to survive somewhere in the vegetation outside the fields or perhaps even within them.*

Characteristically those species vanish first that were already rare in the area followed later by those that have been indicators of extreme conditions (especially of the soil). There is a whole complex of agricultural measures that caused these alterations. Their effect can be summarized in general with "levelling of ecological conditions". What the farmer simply does is to try to get the conditions

---

* There are some extremely specialised species (crop-mimics) that are actually threatened by extinction in their whole area; weeds that have lost most of their colonizing abilities and are dependent on being sown with the crops.

122

in the field as near as possible to the physiological optimum of his crops which means for most factors trying to get away from extremes: acid soils are supplied with lime, wet ones drained, dry ones irrigated and fields in areas that cannot be meliorated by artificial means are abandoned or used for other purposes. This levelling means that weeds with requirements near to those of the crop species will thrive, while those disappear that are adapted to worse conditions but cannot compete. Thus the argument often cited that weed communities are no indicators any more, is clearly wrong; uniform weed communities indicate the uniformity of environmental conditions. The indicators of optimal conditions not only supress the other weeds, they are also strong competitors with the cultivated plants. Thus a necessity for intensified weed control arises.

Usually the drastic alterations in the composition of weed communities have been attributed to the use of herbicides. As I have already pointed out there is a complex of other factors that is one of the main causes for this development. The use of herbicides is not the only one but the most drastic factor. (For a more detailed survey of the influence of modern methods on weed distribution see Bachthaler 1968).

Influence of herbicides

Sensitive species are driven back to areas where they can find refugial sites, as described above. In many countries they are completely eradicated. The outcome is the decrease of competition of other weeds with the resistant species. Thus the use of herbicides has afforded competition experiments on a huge scale. How do species in a plant community behave if most of their competitors are removed?

Resistant species have now become able to grow in greatly increased densities and with single individuals much larger than before (as is described e.g. for the Netherlands by Zonderwijk 1975, for Austria by Neururer 1966, Szith 1977). This development is called compensation (Rola 1973), resulting in weed communities poor in species, but with high densities of individuals. The second phenomenon is that resistant species are also able to enlarge their range of distribution and to fill the niches of the eliminated species, conquering new areas where they were not able to compete before. This development is a further proof of the theory, especially elaborated by Ellenberg (1950, 1963, 1968), that the ecological behaviour of species is dependent on the competition of others and vice versa, the competitive power of a species depending on the environmental conditions.

From an ecological point of view the enlargement of areas of the resistant weeds takes place in two general directions:

(1) Towards ecologically adverse conditions

(1, 1) climatically adverse, e.g. migrating to the north, *Sorghum halepense, Avena fatua*, ...

(1, 2) Adverse soil conditions e.g. *Avena fatua* occurring on poor soils, where it never occurred before (Prante 1970, Holzner 1973), or ruderals becoming agrestals (*Datura, Descurainia, Galium aparine*, ...)

(2) Towards their physiological optimum: Poorly competing species that have been only able to survive at the limit of their physiological tolerance are able to invade better sites, because of the lack of competitors (e.g. *Digitaria ischaemum*).

The following scheme is used to illustrate the general development of weed vegetation after several years of regular usage of herbicides.

General example for a compensation series:

Weed species   A, B, C, D, E, F, G, H, I, J, K, L, M, N, O, P.

               A ... dominant, a ... sparse

               R = ruderal weeds

               F = agrestal weeds from adjacent region

               E = introduced exotic species

Herbicides   1, 2, 3, 4, ...

$$A, B, C, d, g, h, i, j, k, l, m, n, O, P \xrightarrow{1} B, c, g, h, k, m$$

$$\xrightarrow{2} G \xrightarrow{3} C, R, F, E \xrightarrow{4} \ldots \ldots$$

The evolution of weed species is still in progress and is furthered unintentionally by man as he is bringing together species and races that have been separated geographically or ecologically, giving them the opportunity for hybridization, introgression and polyploidization, and by increasing the mutation rates of weeds by the use of herbicides (Mohandas & Grant 1972, Harper 1956, Grant 1970). The current speed of weed evolution is best demonstrated by the increasing genesis of resistant ecotypes (Szith 1977) ("herbicidotypes") in hitherto sensitive species. There will be a continued shifting of weed distributions, resistant weeds invading communities where they did not occur before and the formation of quite different, new weed communities.

Problems of classification

Classification of weed communities into abstract community types according to the rules of the Zürich-Montpellier school (Braun-Blanquet 1964, Westhoff & van der Maarel 1973, Werger 1974) has always been difficult, as is proved by the many alterations that have been made in this part of the system and the new concepts specially developed to press the "obstinate" weed communities into the system (e.g. Brun-Hool 1966, Kopecky & Hejny 1974). The main reasons for these difficulties are:

(1) Weeds are not only dependent on the so-called "natural" environmental factors but also strongly on the complex of anthropogenic factors that are difficult to comprehend. "Chance" plays a big role influencing the occurrence of weed species in communities.

(2) Agrestal weeds, most of them being often short-lived annuals, react quickly to alterations of the environment. Thus, the composition of weed communities can vary strongly from year to year, depending on the weather and human measures, but also from season to season of one year, forming distinct spring, summer and autumn aspects.

(3) Weeds often are "ubiquitous" species with a large amplitude often caused by phenotypic plasticity and heterogeneity within the species, which is typical for colonizing species.

(4) The phytosociological system for weed communities was propounded in Central Europe, an area where many agrestal weed species occur at the very edge of their distribution. The rich weed vegetation of Southern Europe and the Middle East is practically unknown to science.

(5) The alterations in agrestal weed communities described above have made weed sociology even more difficult.

(6) Many agressively colonizing ruderals tend to form dense stands dominating large areas by the combined means of vigorous vegetative reproduction and allelopathic influence (Numata 1974, 1975, Rice 1974). These communities dominated by one species are difficult to integrate into the floristic system.

A typical example of the problems which weed communities offer in classification is the scientific controversy about the integration into the system of three different weed communities which may be developed within one year on one and the same field.

Especially in warmer areas of Europe, e.g. in Eastern Austria, in spring a rich community (1) may be observed of short-lived winter annuals (e.g. many *Veronica* spp.) that germinate mainly in autum, ripen their seeds in late spring and very soon perish. If relevés are made in June no "traces" of them are left whereas in the table the spring relevés show a clear, distinctive block of species. This community is followed (2) by one of weeds that also germinate in autumn or in spring but ripen their seeds with the grain crops. Groups (1) and (2) are best developed in autumn-sown cereals. Group (3) consists of summer annuals requiring a high soil temperature for germination (Lauer 1953). They germinate and ripen with the row crops but can also be found in the stubble of the cereals.

Different authors have had different concepts to describe these communities and to fit them into the system. Today two extreme versions have become established: (A) The three communities are to be regarded as aspects of one association and (B) the first community is ignored and (2) and (3) are separated at the highest level of the system, they are regarded as different classes (*Secalietea* and *Chenopodietea*). Both views have good arguments to offer and a lot of paper was used to discuss them. But the question whether (A) or (B) is true has no meaning at all as it is only a question of conventions. It is therefore useless to discuss the arguments presented by the supporters of these diametrically opposite points of view.(A discussion of the extensive literature on this topic has already been given by Hilbig (1967), Schubert & Mahn (1968), Kropác, Hadac & Hejný (1971).

The purpose of this example was to illustrate one of the severe problems of contemporary phytosociological weed community systematics. The original aim of the system was to give a practical and reasonable order and comprehension and this is even more obstructed by the proliferation of associations that are described from a very local point of view and are very similar in species combination (especially within the *Panico-Chenopodietalia*). Even worse is the habit of making alterations in the higher ranks of syntaxa without a complete review of the whole range of distribution of the syntaxon concerned.

Thus, if we are to have a useful system of weed communities we shall need more flexibility avoiding too rigid rules, and the inclusion of ecological knowledge as it is e.g. also proposed by Kojić (1976).

Conclusions

The area a weed species can occupy depends on ecological factors (including anthropogenic ones) and competition of other species. As most of them are colonizing species, it can be assumed that they are able to invade each suitable habitat within the distance they can reach with their diaspores in a rather short time. Transport by man over vast distances plays an important role by forming new centres of dissemination. The main factor determining areas of distribution is a climatic one (first level factor, Numata 1962, 1967). In their optimal climatic area, which is often also their presumable native one, weeds show the strongest competitive power and the widest ecological and sociological amplitude and are able to occur within a wide range of different soil conditions and within many types of plant communities.

It is well known (Ellenberg 1950, 1963, 1968, Werger & van Gils 1976) that many plant species have a narrower ecological amplitude towards the edge of their range than in the center. Weeds are plants spread by man far beyond their original range, thus occupying a large "border area" (e.g. for many weeds the whole of Central and Northern Europe), where climatic conditions are suboptimal to them and where they can only persist with the help of man as agriculturist. Here they are rather weak competitors and have a narrow ecological and sociological amplitude. Some occur on soils with extreme conditions because they are especially sensitive to the competition of other species and cannot compete in better stands, others are restricted to the habitats nearest to their physiological optimum or offering conditions able to replace the most important missing factor at least partly (Walter & Walter 1953).

In this geographical (and ecological) border area many agrestals are especially sensitive to the measure of rationalized agriculture (in general the ecological levelling of soil conditions and the use of herbicides), and retreat towards their optimal (and original) areas, becoming extinct in many parts of Central Europe. The behaviour of resistant species after the extinction of many of their competitors by herbicides proves Ellenberg's theory, that the distribution of plant species is dependent on environmental factors and the competition of other species. The competition of other species weakens the power to withstand adverse environmental conditions and vice versa adverse environmental conditions weaken the competitive power of species.

Thus, the observation of weed species and weed vegetation provides us with good insight into the mutual action of competition and environmental factors and its results, the distribution, ecological and sociological behaviour of plant species.

Summary

With weeds as with many plant species the main or first
level factor determining the area of distribution is a (complex) climatic one. As they have an artificially enlarged
area of distribution, they have a huge border area (in an
ecological sense), where the climate is not optimal for them,
and where they have a narrow ecological and sociological
amplitude and are especially sensitive to some measures of
modern intensified agriculture. In their northern border
areas species of southern origin are restricted to calcareous
substrates and to agrestal and finally ruderal communities,
while in their optimal climate they are indifferent to that
soil factor and able to compete with other species even in
natural vegetation types. Species presumably of origin in
atlantic areas are restricted with increasing continentality
to very poor and acid soils, as they cannot compete with
other species on better sites any more, because of their
physiological properties. Thus weed distribution demonstrates the complicated reaction of plant species to the
complexes of soil-climatic factors and to the competition of
other species. As far as weeds are concerned, species may
be only relatively calciphilous, but genuinely calcifuge
species, the control being climatic in the former case and
physiological in the second.

The measures of modern agriculture bring about a
gradual extinction of sensitive species from the limit of
their range towards their centre of distribution, where they
can find refuge habitats in the natural vegetation. The
sensitivity of such species (also against herbicides) seems to
increase towards their limits. Resistant species occur with
increasing densities after the removal of their competitors.
In addition, they are able to enlarge their area and to
invade sites, where they had not been able to compete
before, or sites where they could not previously bear the
environmental conditions together with the competition of
the rich weed flora.

As the complex climatic gradients responsible for the
ranges of weed species show smooth transitions, the
alteration of species composition in weed communities is
also a gradual one. This is one of the problems of weed
phytosociology briefly discussed.

References

Bachthaler, G. 1968. Die Entwicklung der Ackerunkrautflora in
Abhängigkeit von veränderten Feldbaumethoden. I. Der
Einfluss einer veränderten Feldbautechnik auf den Ackerunkrautbesatz. Z. Acker u. PflBau 127: 149–170. II. Untersuchungen über die Ausbreitung grasartiger Unkräuter und
ihre Bekämpfung. Ibid. 127: 326–358.

Baker, H.G. 1965. Characteristics and Modes of Origin of
Weeds. In: H.G. Baker & G.L. Stebbins. The Genetics of
Colonizing Species. pp. 147–168. Academic Press, New York-
London.

Baker, H.G. 1974. The evolution of weeds. Ann. Rev. Ecol. Syst.
5: 1–24.

Braun-Blanquet, J. 1964. Pflanzensoziologie. 3. Aufl. Springer,
Wien.

Brun-Hool, J. 1966. Ackerunkraut-Fragmentgesellschaften. In:
R. Tüxen, Anthropogene Vegetation. Ber. Int. Symp. Stolzenau, 1961. pp. 38–47. Junk, The Hague.

Ehrendorfer, F. 1965. Dispersal Mechanisms, Genetic Systems,
and Colonizing Abilities in Some Flowering Plant Families.
In: Baker, H.G. & G.L. Stebbins (see Baker, 1965), pp.
331–351.

Ehrendorfer, F. (ed.) 1973. Liste der Gefässpflanzen Mitteleuropas. Fischer, Stuttgart.

Ellenberg, H. 1950. Physiologisches und ökologisches Verhalten
derselben Pflanzenarten. Ber. Dtsch. Bot. Ges. 65: 350–361.

Ellenberg, H. 1963. Vegetation Mitteleuropas mit den Alpen.
Einführung in die Phytologie IV/2. Ulmer, Stuttgart.

Ellenberg, H. 1968. Wege der Geobotanik zum Verständnis der
Pflanzendecke. Naturwissensch. 55: 462–470.

Gigon, A. 1971. Vergleich alpiner Rasen auf Silikat- und auf
Karbonatboden. Veröff. Geobot. Inst. ETH, Stiftg. Rübel. 48:
1–159.

Grant, W.F. 1970. Pesticides and Heredity. Macdonald Journal
31: 211–214.

Grosse-Brauckmann, G. 1953. Untersuchungen über die Ökologie, besonders den Wasserhaushalt von Ruderalgesellschaften.
Vegetatio 4: 245–283.

Harlan, J.R. & J.M.J. de Wet. 1964. Some thoughts about weeds.
Econ. Bot. 18: 16–24.

Harper, J.L. 1956. The evolution of weeds in relation to resistance
to herbicides. Proc. 3rd. Brit. Weed Control Conf., Blackpool,
pp. 179–188.

Harper, J.L. 1965. Establishment, Agression, and Cohabitation
in Weedy Species. In Baker H.G. & G.L. Stebbins, 1965. pp.
243–265 (see Baker 1965).

Hilbig, W. 1967. Die Ackerunkraugesellschaften Thüringens.
Feddes Rep. 76: 83–191.

Holzner, W. 1970. Die Ackerunkrautvegetation des nördlichen
Burgenlandes. Wiss. Arb. Bgld. 44: 196–243.

Holzner, W. 1971. Niederösterreichs Ackervegetation als umweltzeiger. Die Bodenkultur 22: 397–414.

Holzner, W. 1972. Einige Ruderalgesellschaften des oberen
Murtales. Verh. Zool.-Bot. Ges. Wien. 112: 67–85.

Holzner, W. 1973. Die Ackerunkrautvegetation Niederösterreichs. Mitt. Flor. Arbgem. Linz. 5: 1–157.

Holzner, W. 1974a. Über die Verbreitung von Unkräutern auf
Ruderal und Segetalstandorten. Acta Inst. Bot. Acad. Sci.
Slovac. 1: 75–81.

Holzner, W. 1974b. Das Anthemo ruthenicae-Sperguletum,
eine eigenartige Ackerunkrautgesellschaft des mittleren Burgenlandes. Wiss. Arb. Bgld. 53: 21–30.

Holzner, W. 1977. Plant Ecological Considerations about the
use of Herbicides in Agriculture. In: A. Miyawaki & R. Tüxen
(eds.), Vegetation Science and Environmental Protection.
Proc. Symp. Tokyo. pp. 295–297. Maruzen, Tokyo.

Holzner, W. & E. Hübl. 1977. Zur Vegetation der Kalkalpengipfel des Westlichen Niederösterreich. Verein z. Schutze der Bergwelt, München, Jahrb. 42: 247–269.

Hübl, E. & W. Holzner. 1974. Vorläufiger Überblick über die Ruderalvegetation von Wien. Acta Inst. Bot. Acad. Sci. Slovac. 1: 233–238.

King, L.V. 1966. Weeds of the World. Interscience, New York.

Kinzel, H. 1968. Kalkliebende und kalkmeidende Pflanzen in stoffwechselphysiologischer Sicht. Naturw. Rdsch. 21: 12–16.

Kinzel, H. 1969. Ansätze zu einer vergleichenden Physiologie Mineralstoffwechsels und ihre ökologischen Konsequenzen. Ber. Dtsch. Bot. Ges. 82: 143–158.

Kinzel, H. 1971. Biochemische Ökologie - Ergebnisse und Aufgaben. Ber. Dtsch. Bot. Ges. 84: 381–403.

Kojić, M. 1976 Über die syntaxonomische Gliederung segetaler Pflanzengesellschaften vom ökologischen und ökophysiologischen Standpunkt. Ber. Dtsch. Bot. Ges. 89: 391–399.

Kopecký, K. & S. Hejný. 1974. A new approach to the classification of anthropogenic plant communities. Vegetatio 29: 17–21.

Kropač, E., E. Hadač, & S. Hejný. 1971. Some remarks on the synecological and syntaxonomic problems of weed plant communities. Preslia 43: 139–153.

Kühn, F. 1972. Wildgetreidebestände and ihre Bedeutung für die Gliederung der Ackerunkrautvegetation. In: E. van der Maarel & R. Tüxen (eds.), Grundfragen und Methoden in der Pflanzensoziologie. Ber. Int. Symp. Rinteln 1970. pp. 435–422. Junk, The Hague.

Kump, A. 1970. Verschollene und seltene Ackerunkräuter in Oberösterreich südlich der Donau. Mitt. Bot. Arbgem. Linz 2: 25–40.

Lauer, E. 1953. Über die Keimtemperatur von Ackerunkräutern und deren Einfluss auf die Zusammensetzung von Unkrautgesellschaften. Flora 140: 551–595.

Meusel, H., E. Jäger & E. Weinert. 1965. Vergleichende Chorologie der zentraleuropäischen Flora. I. Fischer, Jena.

Mohandas, T. & W.F. Grant. 1972. Cytogenetic effects of 2,4-D and amitrole in relation to nuclear volume and DNA content in some higher plants. Can. J. Genet. Cytol. 14: 773–783.

Neururer, H. 1966. Beobachtungen über Veränderungen in der Unkrautgesellschaft als Folge pflanzenbaulicher und pflanzenschutzlicher Massnahmen. Tätigkeitsber. d. Bundesanst. fur Pflanzenschutz 1961–1965. pp. 77–79.

Numata, M. 1962. Chikurin no seitaigaku. (Ecology of bamboo forest). Jap. J. Ecol. 12: 32–40.

Numata, M. 1967. Seitaigaku hohoron. (Methodology of Ecology). Kokihshokan, Tokyo.

Numata, M., A. Kobayashi & N. Ohga. 1974. Studies on allelopathic substances concerning the formation of the urban flora. In: M. Numata (ed.), Studies in Urban Ecosystems. pp. 22–25. Chiba.

Numata, M., A. Kobayashi & N. Ohga. 1975. Studies on the role of allelopathic substances. Ibid. 1975. pp. 38–41

Prante, G. 1970. Die Verbreitung des Flughafers (Avena fatua L.) in Schleswig-Holstein und Untersuchungen zur Varietätenbildung. Diss. Kiel.

Rice, E. 1974. Allelopathy. Academic Press, London.

Rola, J. 1973. Der Einfluss der Intensivierung der Landwirtschaft auf die Segetalgemeinschaften. In: R. Schubert (ed.), Probleme der Agrogeobotanik. M. Luther Univ., Halle/S. pp. 139–145.

Schubert, R. & E.-G. Mahn. 1968. Übersicht über die Ackerunkraugesellschaften Mitteldeutschlands. Feddes Repert. 80: 133–304.

Szith, R. 1977. Die Bekämpfung von Problemunkräutern bei Mais in der Steiermark. Der Pflanzenarzt (Wien) (in press).

Tüxen, J. 1958. Stufen, Standorte und Entwicklung von Hackfrucht- und Garten-Unkrautgesellschaften und deren Bedeutung für die Ur- und Siedlungsgeschichte. Angew. Pflanzensoziol. 16: 1–167.

Tüxen, R. 1950. Grundriss einer Systematik der nitrophilen Unkrautgesellschaften in der eurosibirischen Region Europas. Mitt. Flor-Soz. Arbgem. N.F. 2: 94–175.

Tüxen, R. 1962. Gedanken zur Zerstörung der mitteleuropäischen Ackerbiozönosen. Mitt. Flor.-Soz. Arbgem. N.F. 9: 60–61.

Walter, H. & E. Walter. 1953. Einige allgemeine Ergebnisse unserer Forschungsreise nach Südwestafrika 1952/53: Das Gesetz der relativen Standortskonstanz; das Wesen der Pflanzengemeinschaften. Ber. Dtsch. Bot. Ges. 56: 227–235.

Weinert, E. 1973. Herkunft und Areal einiger mitteleuropäischer Segetalpflanzen. Arch. Naturschutz u. Landschaftforsch. 13: 123–139.

Werger, M.J.A. 1974. The place of the Zürich-Montpellier method in vegetation science. Folia Geobot. Phytotax. 9: 99–109.

Werger, M.J.A. & H. van Gils. 1976. Phytosociological classification in chorological borderline areas. J. Biogeogr. 3: 49–54.

Westhoff, V. & A.J. den Held. 1969. Plantengemeenschappen in Nederland. Thieme, Zutphen.

Westhoff, V. & E. van der Maarel. 1973. The Braun-Blanquet approach. In: R.H. Whittaker (ed.), Handbook of Vegetation Science. Vol. 5: 616–726. Junk, Den Haag.

Westhoff, V. & P. Zonderwijk. 1960. The effect of herbicides on the wild flora and vegetation in the Netherlands. Proc. IUCN Symp., Warschau. pp. 69–78.

Zohary, M. 1973. Geobotanical foundations of the Middle East. Fischer, Stuttgart.

Zonderwijk, P. 1975. Ecologische aspecten van de selectieve bestrijding van onkruiden. (Ecological aspects of selective weedkilling). Gewasbescherming 1975 (6): 107–119.

LE CONCEPT DE SIGMASSOCIATION ET SON APPLICATION A L'ETUDE DU PAYSAGE VEGETAL DES FALAISES ATLANTIQUÈS FRANÇAISES*

Jean-Marie GEHU**

Laboratoire d'Ecologie végétale, Faculté de Pharmacie, 59045 Lille cedex, et
Station de Phytosociologie, 59270 Hendries, Bailleul, France

Mots-clef (Keywords): Complexe de groupements végétaux (Plant community complex), Falaise (Cliff), Paysage végétal (Vegetation landscape), Sigmassociation (Sigma-association), Syntaxonomie (Syntaxonomy)

Introduction

Le concept de sigmassociation apparaît bien à travers la combinaison répétitive d'associations végétales au sein de paysages homogènes et analogues. Ce sont toujours les mêmes listes de groupements végétaux que l'on peut établir, par exemple dans les dunes picardes, dans les estuaires de la Manche orientale ou sur les collines crétaciques de l'Artois.

La sigmassociation est la somme des groupements présents dans de telles unités paysagères homogènes. Elle est l'unité de base de la Symphytosociologie*** ou Phytosociologie globale.

* Contribution to the Symposium on Plant species and plant communities, held at Nijmegen, 11–12 November 1976, on the occasion of the 60th birthday of Professor Victor Westhoff.
** Parce que le concept de sigmassociation paraît spécialement fructueux pour le développement d'une science chère et tout particulièrement utile dans l'approche scientifique des problèmes de l'environnement végétal, j'ose dédier, très amicalement et à l'occasion de son 60ème anniversaire, cette modeste note, au Professeur V. Westhoff, éminent spécialiste de la végétation néerlandaise et inlassable protecteur de la Nature.
Cette note doit beaucoup aux fructueuses discussions qui eurent lieu lors de la Session en Valais (8/1976) de l'Association Amicale francophone de Phytosociologie avec nos amis, Cl. Beguin, O. Hegg, M. Costa, S. Rivas-Martinez et ne fut possible que par la connaissance du manuscrit de la note magistrale de notre maître, le Professeur R. Tüxen (1976).
*** The prefix sign is spelled 'sym' when a ph follows: symphenology, symphytosociology (edit. note).

Le concept de Sigmassociation

La filiation et la parenté de la "Sigmassociation" avec le "complexe de groupements végétaux" (Gesellschaftskomplex) sont évidentes.

Le complexe de groupements végétaux

Dès 1928, Braun-Blanquet précise clairement, à l'aide d'exemples, sa conception des "complexes d'associations": mosaïque de groupements de tourbières, de pozzines, de dunes, d'amas de blocs rocheux, de combes à neige.

L'auteur insiste sur la liaison des groupements du complexe entre eux et la répétitivité du phénomène dans un paysage géomorphologiquement déterminé et homogène. Quand le complexe de végétation possède quelque étendue il se prête parfaitement à l'expression cartographique. A la même époque la définition apportée par le "vocabulaire de sociologie" (Braun-Blanquet & Pavillard 1928) est on ne peut plus explicite:

"Le complexe de groupements (Gesellschaftskomplex) est une mosaïque de groupements (alliances, associations ou fragments d'associations), déterminée surtout par la diversité locale des facteurs géomorphologiques et se répétant plus ou moins identiquement en des localités diverses". On retrouve donc dans ces définitions, le souci de combinaison d'unités, de répétitivité, d'échelle déterminée de surface, et d'homogénéité, qui sont à la base des définitions syntaxonomiques.

Après guerre, Braun-Blanquet (1951) introduit dans le cadre des complexes de groupements d'autres notions, telle que celles des groupements de contact (Kontakgesellschaften) ou des groupements de ceintures (Gürtelung, Zonationskomplex) et c'est sous l'appellation de mosaïque

(Gesellschaftsmosaik) qu'il traite plus particulièrement des complexes de groupements tels qu'il les avait définis en 1928. Il cite cependant bien d'autres exemples que ceux des tourbières, des pozzines, des amas de bloc rocheux, des dunes, des combes à neige. Ces nouveaux exemples marquent un changement et un élargissement de la conception des complexes – mosaïques de végétation.

Il s'agit notamment des "groupements disposés en mosaïque, sous l'évidente influence anthropogène, du complexe *Querceto-Lithospermetum-Bromion* des pentes sud de l'Europe moyenne, du complexe *Juniperion nanae – Rhodoreto-Vaccinion* des Alpes centrales, du complexe *Thero-Brachypodion* des garrigues méditerranéennes, du complexe *Quercion roboris – Ulicion* des landes de l'ouest de l'Europe". Il y a donc ici prise en compte de l'influence anthropogène dans le paysage et une singulière extension spatiale du phénomène (Braun-Blanquet 1928). Les premiers exemples de complexe étaient explicitement situés en dehors des séries locales d'évolution, les derniers exemples cités montrent bien leur appartenance à un ensemble climacique précis.

Braun-Blanquet (1964) nous apporte de nouvelles données. D'abord il synonymise pratiquement Gesellschaftsmosaik et Gesellschaftskomplex en priorisant ce dernier terme. Tout en regrettant qu'ils soient trop peu étudiés en raison des difficultés à les cerner et à les définir, il précise que les complexes de groupements, "reflet de la mosaïque de la végétation, correspondent à des mélanges et à des coexistences spatiales dans des régions à climat et à histoire floristique homogène". Et de citer, comme exemple de telles unités phytosociologiques complexes bien définies "en Europe moyenne, l'Insubrie océanique, la plupart des vallées alpines, la Suisse moyenne, le Jura, les plaines rhénanes, les régions sèches de l'Allemagne centrale.... et dans le sud-ouest de l'Europe, le Nord du Portugal avec la Galice, les Iles Majorques, les steppes de l'Ebre, la région méditerranéenne française, les plateaux des Causses, mais aussi les petites coulées basaltiques d'Agde et Béziers".... "Le prodrome des groupements végétaux de la France méditerranéenne donne un aperçu des divers complexes de groupements de ce type dans leur multiplicité et leur enlacement".

Braun-Blanquet ajoute que "les complexes spatiaux de groupements correspondent à des synoécosystèmes ou à des groupes de synoécosystème", et qu'ils peuvent "comme l'a montré Schmithüsen (1959), caractériser des divisions territoriales géobotaniques comme les secteurs de végétation". D'autre part une distinction est faite entre ces grands complexes de végétation et les complexes locaux

de végétation pour lesquels les exemples de 1928 sont repris. Le concept de complexe de groupements connait donc en 1964 un nouvel élargissement puisqu'il peut s'appliquer à des régions entières. Le terme de synoécosystème introduit, par ailleurs, dans le cadre biocoénotique, l'idée d'un mécanisme, d'un fonctionnement. Enfin, il est évident que les grands complexes régionaux de groupements, peuvent couvrir plusieurs polyséries climaciques.

En résumé, les "complexes de végétation" sont donc pour Braun-Blanquet, l'expression des mosaïques du tapis végétal dans des cadres géographiques homogènes, mais de dimensions variables (paysage local, régional ...), s'inscrivant ou non dans des séries dynamiques plus ou moins complexes et présentant un caractère significatif de répétitivité. Leur définition apparaît déductive. Ce concept génial manque pourtant, jusqu'alors, d'une méthodologie. Seule celle-ci en permettra l'application réelle à l'étude de la végétation et rendra possible une définition plus stricte des complexes de végétation.

La sigmassociation
C'est au biogéographe sarrois, Schmithüsen (1968), parfait connaisseur de la méthodologie phytosociologique traditionnelle et passionné par l'étude des paysages que revient le mérite de l'idée de l'analyse inductive (i.e. phytosociologique) des complexes paysagers de végétation. Mais c'est le chef de file de la phytosociologie présente, le Professeur R. Tüxen, qui le premier a su élaborer une méthodologie précise d'étude des complexes de groupements (Tüxen 1973). Cette méthode est calquée sur la technique phytosociologique usuelle. Elle comprend donc une phase analytique au terrain (prise de relevés selon des transects dans le paysage) et une phase synthétique (élaboration de tableaux). Trois exemples de relevés de complexes d'associations sont donnés dans cette note initiale. Les associations présentes dans les territoires étudiés remplacent dans la liste du relevé les espèces végétales. Elles sont affectées d'un chiffre d'abondance selon une échelle d'évaluation quantitative inspirée de celle de Braun-Blanquet.

Par ailleurs, le niveau d'homogénéité du paysage étudié est situé à l'intérieur d'un territoire occupé par un type donné de végétation potentielle naturelle (notion qui correspond, selon l'auteur, au climax complexe de Braun-Blanquet & Pavillard 1928). Ainsi, des exemples sont pris dans le "*Betulo-Quercetum roboris*-Gebiet" ou dans le "*Querco-Carpinetum luzuletosum*-Gebiet".

Peu après paraissent les premiers tableaux détaillés et synthétiques élaborés selon cette nouvelle méthodologie (Géhu

1974a, 1974b, 1976, Beguin & Hegg 1975, 1976). L'auteur du présent travail préféra dès 1974 utiliser le terme d'association d'associations, à celui de complexe de groupements. Le développement méthodologique avait, en effet, introduit une précision taxonomique jusqu'alors inexistante qui nécessitait une amélioration nomenclaturale parallèle. Il y avait en outre risque de confusion avec listes, relevés, associations, "complexes" dues à des erreurs méthodologiques en Phytosociologie traditionnelle.

C'est au cours du Colloque de Rinteln, consacré en 1974 à la définition des paysages à l'aide de la végétation, que le Professeur W. Haber proposa pour ces associations d'associations, le terme de "sigmassociation" dont l'avantage lui paraissait double puisque sigma (lettre grecque Σ) symbolise la somme (= somme des groupements du paysage) et correspond au sigle de la Station de recherches du fondateur de la phytosociologie (S.I.G.M.A. = Station Internationale de Géobotanique méditerranéenne et alpine du Dr. J. Braun-Blanquet). Beguin & Hegg (1975, 1976), puis Géhu (1976) et Tüxen (1977) vont bientôt reprendre cette expression pour nommer les complexes de groupements définis par la méthode des tableaux phytosociologiques.

La sigmassociation apparaît alors comme l'unité fondamentale de la Phytosociologie paysagère globale encore appelée symphytosociologie (Beguin & Hegg 1976, Beguin, Géhu & Hegg, à paraître). Elle est à la symphytosociologie ce qu'est l'association à la phytosociologie ou l'espèce à la taxonomie. Pourtant, il n'est pas certain que le terme de sigmassociation, qui rencontre quelques oppositions, survive longtemps. Le terme de synassociation lui sera peut-être préféré dans l'avenir.

Quoiqu'il en soit et entre temps Tüxen (1976) développe sa conception des sigmassociations. Il démontre qu'à l'instar des associations, les sigmassociations possèdent des caractères propres, tels que la surface moyenne de l'aire minimale, le degré d'homogénéité de leur tableau, le nombre spécifique moyen d'associations constitutives. Ces caractères viennent renforcer la combinaison caractéristique des associations dans la définition de l'ensemble. Il montre aussi que le degré de diversité du paysage sera fonction du nombre d'associations présentes dans la sigmassociation ou plutôt de la valeur du rapport du nombre d'associations possibles de la sigmassociation au nombre d'associations réellement observées. Il propose surtout un double système hiérarchisé pour la typologie des sigmassociations:
– un système syntaxonomique, analogue au système phytosociologique, dans lequel les sigmassociations (ou synassociations) sont groupées en sigmalliances (ou synalliances), en sigmordres (ou synordres) et en sigmaclasses (ou synclasses).
– un système chorologique où les unités symphytosociologiques précédentes servent à définir, selon les conceptions de Schmithüsen (1968), des divisions territoriales de grandeur variable.

Tableau 1 résume, tout en rappelant leur parenté et leur filiation, les différences principales entre les complexes de groupements et les sigmassociations. On voit donc:
– la sigmassociation est une coupure intégrante des caractères du tapis végétal beaucoup plus précise et signifiante que le complexe de groupements.
– la sigmassociation est un complexe de groupements de rang élémentaire et déterminé appartenant à un (ou plusieurs) système typologique.

Tableau 1. Différences entre complexe de groupements et sigmassociation

	Complexe de groupements (sensu Braun-Blanquet 1928-1964)	Sigmassociation (sensu Tüxen 1973-1976)
Principe	Combinaison répétitive de groupements végétaux divers dans un paysage homogène	Combinaison répétitive de groupements végétaux divers dans un paysage homogène
Cadre spatial	Non déterminé, local à régional	Local (mais pouvant devenir régional pour les Unités supérieures)
Cadre dynamique	Extra sérial à pluri-sérial polyclimacique	Végétation potentielle déterminée (climax complexe)
Définition	Déductive	Inductive
Typologie et hiérarchisation	Inexistantes	Doubles selon: système syntaxonomique / système chorologique
Méthodologie	Inexistante	Méthode des relevés et des tableaux; analytique puis synthétique

– elle est au complexe de groupements ce qu'est l'association vis-à-vis du groupement végétal.

L'utilité et la supériorité de la symphytosociologie et des sigmassociations pour la cartographie à petite échelle et les comparaisons entre paysages végétaux vicariants paraissent assez évidentes pour qu'il ne soit point besoin d'insister ici. Les problèmes de la symphytosociologie feront d'ailleurs l'objet de développements ultérieurs (Beguin, Géhu & Hegg, à paraître).

Un exemple de sigmassociations: les paysages végétaux des falaises atlantiques françaises

Introduction

Les groupements végétaux du littoral atlantique français qui a fait l'objet de nombreuses publications phytosociologiques depuis une douzaine d'années (notamment Géhu 1962, 1964a, 1966b, Géhu & Géhu-Franck 1973) sont suffisamment connus pour que la définition de sigmassociations valables soit possible. Il y a deux ans (Géhu 1974a), dans une première approche, j'ai d'ailleurs pu montrer que les sigmassociations des côtes d'érosion (falaises) différaient totalement de celles des côtes de sédimentation (levées de galets, dunes, vases salées). L'absence de tous groupements de liaison les situait dans des classes paysagères (ou symphytosociologiques) différentes. Une étude plus précise des sigmassociations des prairies salées des côtes du Nord-Quest de la France fut, d'autre part, réalisée au début de cette année (Géhu 1976). Définissant l'existence de deux grandes sigmassociations de part et d'autre du Cotentin, elle confirmait le rôle essentiel de limites phytogéographique et géobotanique joué par cette région.

Le but de la présente note est de réaliser, au niveau des falaises atlantiques de France, un travail comparable.

Cadre des recherches

Les relevés de sigmassociations ont été réalisés en larges transects, sur des falaises géomorphologiquement assez homogènes et spatialement assez vastes. La technique en est aussi simple que celle des relevés phytosociologiques. Chaque association présente est notée et quantifiée dans l'espace considéré selon l'échelle d'abondance-dominance de Braun-Blanquet. Sa forme dans le paysage est schématisée selon trois symboles (cf. légende du tableau 2). Pour déceler d'éventuelles variations des paysages végétaux

de falaise et définir leurs unités régionales, les relevés ont été répartis le long des côtes du Nord-Ouest de la France (cf. fig. 1).

Les emplacements examinés appartiennent à deux systèmes géographiques très différents:
– côte du bassin sédimentaire de Paris (Manche orientale), au climat de type nord atlantique plus rude,
– rivages du massif armoricain cristallin (Manche occidentale et atlantique), au climat océanique, beaucoup plus tempéré.

Les neuf premiers relevés appartiennent au deuxième système et correspondent donc à des falaises de roches cristallines. Les deux premiers relevés ont été réalisés à Belle Ile en mer, sur des falaises de micaschistes, sous le climat tempéré et lumineux du Morbihan. Les sept suivants appartiennent aux côtes nord et ouest de la Bretagne, sous un climat plus frais mais plus uniforme. Les relevés de Crozon (3, 4) concernent des falaises de grès quartziteux, ceux de la région d'Erquy-Fréhel (5, 6, 8, 9) des falaises de grès rose ancien (permien), celui de Cancale (7), une falaise de gneiss granitique. Les cinq derniers relevés appartiennent au littoral du bassin parisien. Trois d'entre eux (10, 11, 12) proviennent des falaises de craie sénonienne (avec lit de silex) de Haute Normandie. Un relevé (13) a été effectué sur les craies turoniennes (plus marneuses et plus friables) du Cap Blanc Nez et le dernier (14) provient des falaises jurassiques du Gris Nez (sable et grès portlandien surmontant les marnes Kimmeridgiennes).

Le paysage directement perçu sur ces falaises, diffère considérablement d'un type à l'autre, même pour un non spécialiste. Il était donc intéressant à l'aide des méthodes de la symphytosociologie d'essayer d'objectiver scientifiquement ces différences d'impression paysagère et de voir dans quelle mesure la combinaison des groupements végétaux reflétait les variations, géographique et physionomique, de ces falaises.

Résultats obtenus

L'étude des ensembles de groupements végétaux pose, de façon plus aiguë pour les falaises que pour les prairies salées, ou même que pour les dunes, le problème de l'échelle d'homogénéité à laquelle le symphytosociologue doit s'efforcer de travailler. A titre expérimental et démonstratif, le Tableau 2 a été construit en deux parties transversales distinctes, correspondant chacune à une homogénéité écologique plus stricte. La partie supérieure concerne les groupements des échelons halophiles, tandis que l'inférieure regroupe les associations non (ou peu) halophiles généralement développées sur le plateau ou ses rebords.

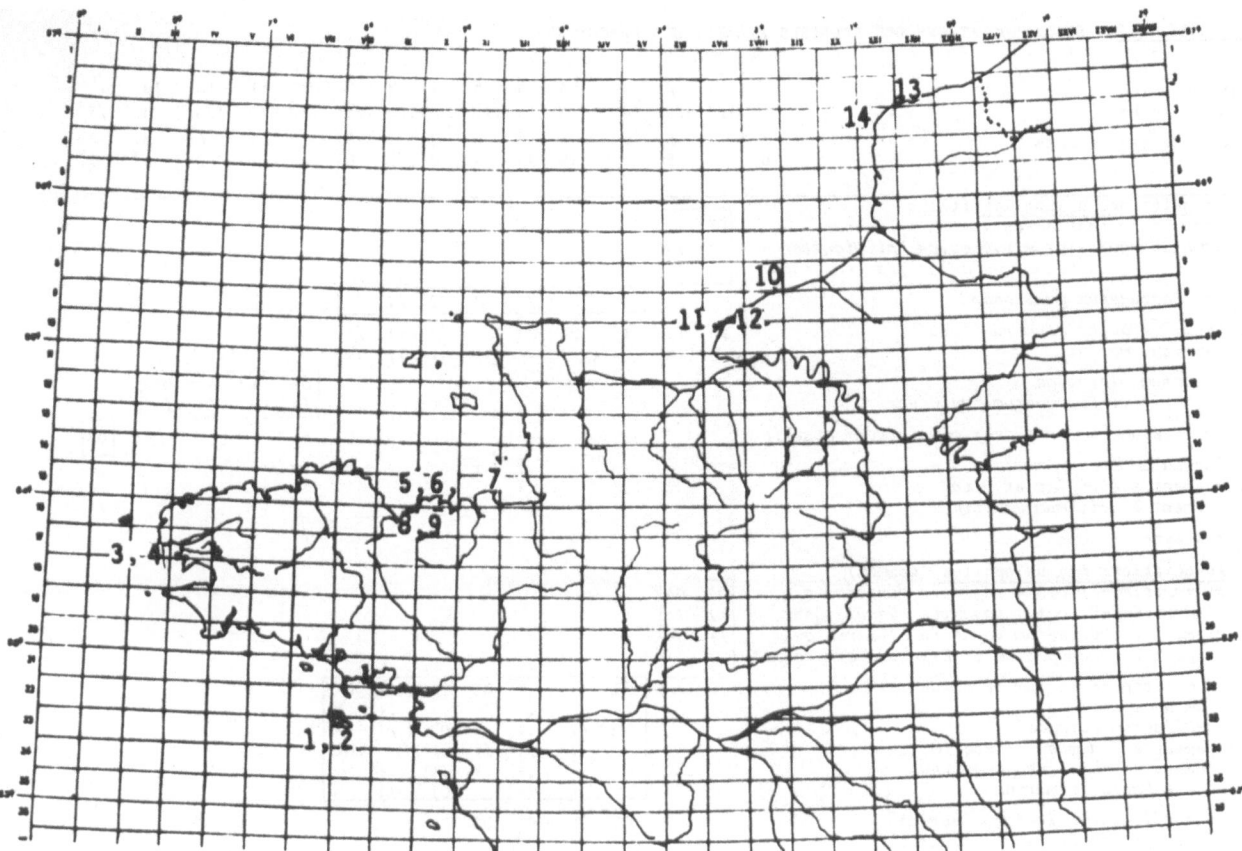

Fig. 1. Localisation des falaises étudiées.

La première observation qui s'impose, à la lecture de ce tableau, c'est que les coupures symphytosociologiques de ces 2 échelons d'homogénéité écologique, talus et plateau, sont rigoureusement parallèles et peuvent servir à définir, ensemble de larges sigmassociations, et séparément de petites sigmassociations plus homogènes (que l'on pourrait nommer synassociations). Dans une telle perspective une sigmassociation serait formée d'éléments plus petits et plus homogènes: les synassociations.

La deuxième observation c'est que les sigmassociations (larges et restreintes) des falaises atlantiques qui ont été étudiées ici se répartissent en quatre ensembles bien distincts que l'on peut considérer comme quatre sigmassociations (ou quatre groupes de synassociations).

Le paysage des falaises de Belle-Ile (Rel 1–2 du Tableau 2) (= sigmassociation: Σ *Plantagini carinatae-Festucetum pruinosae/Ulici maritimi-Ericetum vagantis*).

Cette unité paysagère correspond aux falaises abruptes des micaschistes de la côte occidentale de Belle-Ile. Elle est très caractérisée dans son échelon halophile par la pelouse aérohaline endémique du *Plantagini carinatae-Festucetum pruinosae*, vicariante du *Sileno-Festucetum pruinosae* et riche en taxons spéciaux comme *Daucus gadeceaui*. La sous-association *halimionetosum* du *Crithmo-Spergularietum rupicolae*, formant de véritables petits "schorres suspendus" dans les zones de retombées de paquets de mer projetés en geysers sur la falaise, possède également une intéressante valeur différentielle de ce paysage. Par contre la sous- association *frankenietosum* de cette même association transgresse plus que ne l'indique le tableau dans les formes thermophiles de la sigmassociation suivante.

Sur le plateau, en retrait des pelouses, les landes maritimes à *Erica vagans* (*Ulici maritimi-Ericetum vagantis*) qui sont également des groupements endémiques locaux, sont très caractéristiques. Cette sigmassociation n'existe,

131

TABLEAU 2 : SIGMASSOCIATIONS DES FALAISES ATLANTIQUES FRANCAISES

		1	2	3	4	5	6	7	8	9	10	11	12	13	14
Numéros		1	2	3	4	5	6	7	8	9	10	11	12	13	14
Surface en km2		1/2	1/2	1/4	1	1	1/2	1/10	1/4	1/5	2	1/3	1/5	1	1/5
Nombre de groupements	a)	7	6	6	3	5	4	3	3	3	4	5	6	3	4
	b)	4	5	5	10	11	7	9	7	6	2	3	3	5	7

a) Associations halophiles (talus)

	1	2	3	4	5	6	7	8	9	10	11	12	13	14
Plantagini carinatae-Festucetum pruinosae	02	02												
Crithmo-Spergularietum rupicolae halimionetosum	/1	/1												
Crithmo-Spergularietum rupicolae frankenietosum	/1	/1	/+											
Sileno-Festucetum pruinosae				/2	/1	/1	/1	02	/1	/+				
Crithmo-Spergularietum rupicolae	/1	/1	/1	.5	/1	/1	/+	/1	/+					
Asplenietum marini	.+	(.r)	.r	(.r)	.+	.r	.r	.r	.r			(.r)		
Brassicetum oleraceae										02	/+	.+	02	
Dauco maritimi-Festucetum pruinosae										/1	03	03	/+	
Sileno-Festucetum pruinosae cirsietosum acaulis														03
Sagino-Catapodietum marinae	.+		.+		.r	.r				.r	/+	.r		.+
Groupement à *Atriplex* et *Beta*	.r	.r	.r		.+						.+	.r	.r	.r
Groupement à *Crithmum maritimum*											.r	.r		/+
Autres associations										1				

b) Associations non halophiles (sommet)

	1	2	3	4	5	6	7	8	9	10	11	12	13	14
Ulici maritimi-Ericetum vagantis daucetosum	04	04												
Ulici maritimi-Ericetum vagantis cuscutetosum	/2	/1												
Ulici maritimi-Ericetum vagantis ciliarietosum	/+	.+												
Groupement à *Brachypodium* et *Eryngium campestre*	.+	.+												
Ulici maritimi-Ericetum cinereae			/1	/1	.+	03	02	01	04					
Sedetum anglici			.+	.+	.1	/+	.+	.+	.r					
Rubio-Ulicetum europaei			(.r)	/+	.+		03	/1	.+					
Groupement à *Endymion* et *Pteris*				.r	/1	/+	.+		02					
Groupement à *Umbilicus pendulinus*					.+	.+	.+	.+						
Pelouse à *Dactylis marina*					/+		01	/1						
Ulici humilioris-Ericetum cinereae			04	05	04	04	.+							
Ulici humilioris-Ericetum ciliaris			.r	.+	03	/+								
Festuco tenuifoliae-Galietum littoralis										04	03			
Festuco-Brachypodietum pinnati											03	03	02	03
Ulici-Prunetum spinosissimae											/1	.r	/1	(.r)
Groupement à *Tussilago farfara*														02
Draperies de *Hedera helix*					.+	.+	.+	.+	.+	.r		.r	(.r)	
Groupement à *Teucrium scorodonia* et *Centaurea*					.r	.r		.r						
Osmundo-Salicetum atrocinereae fragmentaire					.r	.r								
Autres associations	1		1										3	6

LEGENDE : 1 : Belle Ile, Port Coton (56) ; 2 : Belle Ile, Apothicairerie (56) ; en outre : groupement à *Agrostis stolonifera* .+ ; 3 : Crozon, Pointe de Dinan (29) ; 4 : Crozon, Cap de la Chèvre (29), en outre : *Dactylo-Sarothamnetum maritimum* /1 ; 5 : Cap Fréhel (22) ; 6 : Cap d'Erquy (22) ; 7 : Cancale, Pointe du Grouin (35) ; 8 : Erquy, Pointe de la Laheussaye (22) ; 9 : Erquy, le Portuais (22) ; 10 : Fécamp, falaise d'Aval (76), en outre : *Suaedetum verae* /1 ; 11 : Etretat, falaise d'Aval (76) ; 12 : Etretat, falaise d'Amont (76) ; 13 : Cap Blanc Nez (62), en outre : groupement à *Rumex obtusifolium* et *Agrostis* /+, groupement à *Nasturtium officinale* .+, groupement du *Cratoneurion* /+ ; 14 : Cap Gris Nez, le Rident (62), en outre : groupement à *Apium graveolens* et *Agrostis* /+, *Helosciadietum nodiflori* .+, groupement du *Cratoneurion* /+, groupement à *Epilobium hirsutum* .r, fourré d'*Hippophae* .+, groupement de *Sedum acre* .+.

Les symboles indiquent la "forme" de l'association dans le paysage : / = linéaire ; 0 = spatial ; . = ponctuel.

Les chiffres donnent la "quantité" des groupements dans le paysage selon l'échelle d'abondance dominance de BRAUN-BLANQUET : + = présent ; r = rare.

en dehors de Belle-Ile, qu'à la pointe nord-occidentale de l'Ile de Groix. Elle possède donc une haute signification de paysages côtiers endémiques des deux grandes iles morbihannaises. Elle s'inscrit dans un contexte climacique général où les irradiations du *Quercion ilicis* pénètrent le *Fraxinion* et le *Quercion robori-petraeae*.

Le paysage des falaises bretonnes (Rel 3 à 9 du Tableau 2)

132

(= sigmassociation: *Sileno-Festucetum pruinosae/Ulici maritimi-Ericetum cinereae*).

C'est l'unité paysagère des falaises cristallines (granite, grés, quartzite . . .) du pourtour armoricain. Elle est caractérisée dès l'échelon halophile par la pelouse salée du *Sileno-Festucetum pruinosae* et sur les plateaux littoraux ou le haut de la falaise, par un ensemble de groupements de lande, de vires rocheuses et de fourrés comme l'*Ulici maritimi-Ericetum cinereae*, les ptéridaies à *Endymion*, le *Sedetum anglici*, le *Rubio-Ulicetum europaei*. Elle est présente tout autour des côtes bretonnes sur les roches précédemment citées. Souvent réduite à un liseré plus ou moins large (rel. 7) sur les côtes peu escarpées, elle prend tout son développement sur les grandes falaises des côtes nord et ouest de la Bretagne.

Elle peut présenter d'autres variations affectant sa composition et sa physionomie. Deux de ces variations apparaissent bien dans le tableau.

– sur les grands promontoires finistériens (Crozon) ou nord bretons (Fréhel) – Relevés 3 à 6 du tableau – le plateau sommital est occupé par des landes littorales à *Ulex gallii* (*Ulici humilioris – Ericetum cinereae* et parfois *Ulici humilioris – Ericetum ciliaris*) qui marquent fortement le paysage estival (floraison tardive de cet ajonc).

– sur les falaises proches des anses sableuses – Relevés 8 et 9 du tableau – l'ascension éolienne des sables peut induire l'apparition de groupements liés à l'arrière-dune (*Koelerion albescentis* et notamment *Festuco-Galietum litoralis*) au milieu des landes maritimes. On note généralement une exclusion entre ces groupements liés aux plaçages d'arènes, et les landes à *Ulex galli*.

Il en résulte une différence d'aspect notable qui correspond à une autre combinaison des associations. Ces deux variations peuvent être considérées comme relevant de deux sous sigmassociations (ou deux sous synassociations) différentes.

En dehors des côtes bretonnes, ces paysages peuvent aussi être observés en quelques points des côtes du nord-ouest du Cotentin dont l'appartenance au district phytogéographique de basse Bretagne a été démontrée depuis longtemps par Des Abbayes sur des bases floristiques. Il sera intéressant d'étudier les variations de ces paysages sur les côtes anglaises occidentales.

Cette sigmassociation s'incrit dans le contexte climacique général des forêts du *Quercion robori-petraeae* à fortes nuances occidentales. Elle est absente des rivages et falaises à couverture limoneuse sur lesquels la forêts littorale appartient à l'Ormaie.

Le paysage des falaises crétaciques normandes (Rel. 10 à 13 du Tableau 2)
(= sigmassociation: Σ *Brassicetum oleraceae/Festuco-Brachypodietum*).

C'est l'unité paysagère des falaises crétaciques du Bassin Parisien. Elle est parfaitement caractérisée par le *Brassicetum oleraceae* des parois crayeuses et la pelouse à *Daucus × maritimus* et *Festuca pruinosa* dans l'étage halin, ainsi que par les pelouses du *Centaureo-Brachypodion* (*Festuco-Brachypodietum*) piquetées de fourrés atlantiques (*Ulici-Prunetum*) sur le plateau. C'est en Haute Normandie que ce paysage littoral est le mieux développé. Il se retrouve légèrement modifié (influence des marnes) en Artois, au Cap Blanc Nez. Ses éventuelles variations sur les côtes du Sud-Est anglais, où il est fréquent, seront intéressantes à étudier. Il s'inscrit dans la potentialité générale des forêts du *Fagion*.

Le paysage des falaises jurassiques gréso-marneuses (Rel. 14 du Tableau 2)
Cette sigmassociation probable, notée au Cap Gris Nez, devra être confirmée par d'autres relevés à effectuer sur les falaises analogues du Boulonnais, du Sud d l'Angleterre et de Basse Normandie. La pelouse halophile y est pénétrée d'espèces des pelouses calcaires (*Sileno-Festucetum pruinosae circietosum acaulis*). Les groupements traduisant l'instabilité du milieu (*Tussilago*) et les suintements d'eau douce liés aux marnes, sont différentiels de l'ensemble. Ce type de paysage littoral s'inscrit dans le contexte climacique général du *Fraxinion*.

Conclusions

Sur le tableau, trois ensembles de sigmassociations ne présentent entre eux aucunes liaisons (ou très faible par quelques groupements non significatifs; deux groupements thérophytiques présents aussi dans d'autres milieux que les falaises et le faciès de lierre non spécifiquement littoral): ceux des falaises cristallines, des falaises crayeuses, des falaises marneuses. On pourrait donc les considérer, dans un sigmasystème, comme relevant de classes différentes; ce qui correspond bien en fait à l'impression, si fortement ressentie intuitivement, d'originalité de ces trois paysages.

Par contre, au sein des falaises cristallines armoricaines, une liaison existe, au moins à l'échelon halophile, entre la sigmassociation de Belle-Ile et celle de la Bretagne continentale. Ces deux sigmassociations appartiendraient alors à la même sigmalliance. Elles ont d'ailleurs le même aspect de falaises couvertes de landes à leur sommet.

Cependant aucune liaison n'apparaît au niveau des

associations du plateau. On peut dès lors se demander si l'échelle d'analyse des petites sigmassociations (ou synassociations) n'est pas un peu trop fine. A moins d'être considérées comme éléments constitutifs des sigmassociations, ces synassociations pourraient apparaître comme un obstacle aux étapes synthétiques ultérieures dans la construction des sigmasystèmes taxonomiques, sinon chorologiques.

C'est tout l'art du symphytosociologue, comme du phytosociologue, de savoir travailler à l'échelle d'homogénéité la plus opportune et la plus appropriée aux problèmes à résoudre.

Résumé

Le concept de sigmassociation (sensu Tüxen 1973–1976) est discuté dans une première partie. Il dérive de la notion des complexes de groupements, progressivement développée par Braun-Blanquet dans les éditions successives de sa "Pflanzensoziologie" (1928, 1951, 1964). Le Tableau 1 résume les différences essentielles entre sigmassociation et complexe de groupements.

Dans la deuxième partie de l'exposé, des exemples de sigmassociations sont étudiés sur les côtes atlantiques françaises: ceux, notamment, des falaises cristallines armoricaines et des falaises crétaciques normandes. Le tableau 2 explicite bien les variations dans la combinaison des groupements pour chaque cas. A une combinaison différente des groupements constitutifs des sigmassociations correspond une physionomie totalement autre du paysage. Quelques problèmes concernant les sigmasystèmes sont présentés en conclusion.

Summary

The concept of sigma-association (Géhu 1974a)* or association of associations is discussed; it is derived from Braun-Blanquet's concept of plant community complex (cf. Westhoff & van der Maarel 1973), as it was developed by Braun in the subsequent editions of his textbook. Sigma-associations are delineated in the field according to the general features of the landscape, notably land form, land use and estimated potential natural vegetation type). Table 1 summarizes the differences between sigma-association and community complex:

* In English the spelling sigma-association should be preferred (edit. note).

	community complex	sigma-association
– spatial framework	not determined local-regional	local, regional for higher syntaxonomical units.
– dynamical framework	extra-serial to pluri-serial poly-climax	determined potential vegetation (climax complex)
– type of definition	deductive	inductive
– typology, hierarchy	none	syntaxonomical and chorological system
– analytical approach	none	method of relevés and tables

Examples of sigma-associations from cliffs along the West and North-West French Atlantic coast are given. Table 2 shows the results. Obviously, a sigma-association table has a similar structure as a normal association table. Each local vegetational unit is characterised syntaxonomically as far as possible (usually on the association and subassociation level).

Its occurrence in the entire landscape stand (order of size one sq. km) is noted with two symbols: the first one indicates the pattern it forms: l = linear, 0 = large, polygon, . = small, pointlike pattern; the second symbol gives the abundance or dominance within the stand according to the usual Braun-Blanquet combined estimation scale.

Bibliographie

Beguin, Cl. & O. Hegg. 1975. Quelques associations d'associations (sigmassociations) sur les anticlinaux jurassiens recouverts d'une végétation naturelle potentielle (essai d'analyse scientifique du paysage). Doc. Phytosoc. 9–14: 9–8.

Beguin, Cl. & O. Hegg. 1976. Une sigmassociation remarquable au pied du premier anticlinal jurassien. Doc. Phytosoc. 15–18: 15–24.

Beguin, Cl., J.-M. Géhu & O. Hegg. 1977. La symphytosociologie (à paraître).

Braun-Blanquet, J. 1928. Pflanzensoziologie. 1ère éd., Berlin.

Braun-Blanquet, J. 1951. Pflanzensoziologie. 2ème éd., Wien.

Braun-Blanquet, J. 1964. Pflanzensoziologie. 3ème éd., Wien.

Braun-Blanquet, J. & J. Pavillard. 1928. Vocabulaire de sociologie végétale. 3ème ed., Montpellier, 23 p.

Géhu, J.-M. 1962a. Quelques observations sur la falaise crétacée du Cap Blanc Nez et étude de la végétation de la paroi abrupte. Brassicetum oleraceae. Bull. Soc. Roy. Bot. Belg. 95: 109–129.

Géhu, J.-M. 1964a. Sur la végétation halophile des falaises bretonnes. Rev. gén. Bot. 71: 73–77.

Géhu, J.-M. 1964b. L'excursion dans le Nord et l'Ouest de la France de la Société Internationale de Phytosociologie. Vegetatio 12: 1–95.

Géhu, J.-M. 1974a. Sur l'emploi de la méthode phytosociologique sigmatiste dans l'analyse, la définition et la cartographie des paysages. C. R. Acad. Sc. 279: 1167–1170.

Géhu, J.-M. 1974b. Essai de définition de quelques associations d'associations sur les côtes de la Manche. Coll. Int. Rinteln (à paraître).

Géhu, J.-M. 1976. Sur les paysages végétaux ou sigmassociations des prairies salées du Nord-Ouest de la France. Doc. Phytosoc. 15–18: 57–62.

Géhu, J.-M. & J. Géhu-Franck. 1973. Apport à la connaissance phytosociologique des Landes littorales de Bretagne. Coll. Phytosoc. II. La Végétation des Landes, Lille, p. 183–200.

Schmithüsen, J. 1968. Allgemeine Vegetationsgeographie. Lehrbuch der Allgemeinen Geographie. 4. Berlin.

Tüxen, R. 1973. Vorschlag zur Aufnahme von Gesellschaftskomplexen in potentiell natürlichen Vegetationsgebieten. Acta Bot. Acad. Sc. Hung. 19: 379–384.

Tüxen, R. 1977. Zur Homogenität von Sigmassociationen ihrer syntaxonomischen Ordnung und ihrer Verwendung in der Vegetationskartierung. Doc. Phytosoc. N.S. 1: 321–328.

Westhoff, V. & E. van der Maarel. 1973. The Braun-Blanquet approach. In: R. H. Whittaker (ed.), Handbook of Vegetation Science, Part 5. Classification and Ordination, pp. 617–726. Junk, The Hague (2nd ed. in press).

A VEGETATION MAP OF THE NETHERLANDS, BASED ON THE RELATIONSHIP BETWEEN ECOTOPES AND TYPES OF POTENTIAL NATURAL VEGETATION* **

A.H.P. STUMPEL & J.T.R. KALKHOVEN

Research Institute for Nature Management (RIN), Leersum, The Netherlands
with the co-operation of S.E. Stumpel-Rienks, Leersum, and E. van der Maarel, Division of Geobotany, University of Nijmegen

Keywords: Ecotope, Landscape ecology, Potential natural vegetation, Vegetation complex, Vegetation map, Vegetation series

Introduction

To adjust the physical planning in The Netherlands to new views of this time, including ecological views, the National Physical Planning Agency has started a study in process planning with the natural environment and human society as most important components. Within this framework two additional ecological studies were initiated:
a. The draft of a so-called general ecological model to be used in physical planning (cf. Van der Maarel & Vellema 1975, Van der Maarel 1977, Anon. 1977).
b. The set up of an 'Environmental Survey of The Netherlands' with the aim to typify and evaluate the Dutch natural environment.
This survey that has been finished now (Kalkhoven, Stumpel & Stumpel-Rienks 1976), is an example of the so-called landscape ecological studies, which have been initiated in many parts of our country (Van der Maarel & Stumpel 1975).

As it was impossible to obtain a complete survey of all occurring terrestrial biotic communities and the main species forming part of them in the two years of field work available, we had to confine ourselves almost entirely to a broad survey of the vegetation. The most important considerations to this restriction were on the one hand that vegetation is supposed to reflect the combined action of

climate, relief, soil, water and the influence of animals and men, on the other hand that biotic communities are mainly determined by their vegetation, particularly its structure (cf. Westhoff 1955).

The vegetation was characterized in such a way that it was possible to construct a small-scale vegetation map. A scale 1:200,000 has been chosen for this map because Dutch physical planning on a national level works on the same scale and because this scale leaves sufficient room to express the different characters of the Dutch landscape. With a predominantly agricultural landscape, so strongly scattered as the Dutch one, the actual vegetation is difficult to map on this small scale (cf. Küchler 1973). To solve this problem we have chosen for a method, in which the actual vegetation is described in relation to the potential natural vegetation (Tüxen 1956, Kalkhoven 1974, Stumpel 1974, cf. Westhoff & Van der Maarel 1973).

The vegetation map with explanatory text can be found in the final report of the survey project (Kalkhoven, Stumpel & Stumpel-Rienks 1976). In the present paper some starting points and the method of mapping are discussed.

Potential natural vegetation and vegetation series

For a good understanding of what a potential natural vegetation actually means, it is necessary to go a little further into the term 'natural vegetation'. This is the vegetation which directly relies on the environment of the site,

* Contribution to the Symposium on Plant Species and Plant Communities, held at Nijmegen, 11–12 November 1976, on the occasion of the 60th birthday of Professor Victor Westhoff.
** Nomenclature follows Heukels-van Ooststroom, Flora van Nederland, 18e druk, 1975, Wolters-Noordhoff, Groningen; nomenclature of syntaxa follows Westhoff & den Held (1969)

assuming that it has not been influenced by man (cf. Tüxen 1956). It is the kind of vegetation which develops entirely spontaneously where both the composition of the species and the arrangement of the growing plants are concerned; it is determined by climate, relief, soil character, water regime and the autochtonous fauna, but not by man and his (in)direct influences. When man's influence becomes noticeable, 'natural vegetation' cannot be spoken of anymore. In the Dutch cultural landscape human influence has continuously been exerted nearly everywhere for many centuries. The result is, that natural vegetation is hardly found in this country nowadays.

In relation to the degree of human influence, a number of substitute and companion communities exist in the landscape which, according to the amount of human impact, can be classified into various categories of naturalness. Usually we distinguish between natural, near-natural, semi-natural and hardly natural vegetation or landscape (Westhoff 1968, 1971). Van der Maarel (1975) suggested a six-category system based on Sukopp's (1972) hemerobiotic degrees: natural, near-natural, semi-natural, agricultural, near-cultural, cultural. These categories can be characterized by the spontaneousness of structure and floristic composition as well as the amount of alien species. If human influence would have gone, succession gets started towards the climax stages. This future climax is designated as potential natural vegetation. In view of the planological background of this study some restrictions have been added to the concept of potential natural vegetation. A first restriction concerns the factor time. In his definition of potential natural vegetation, Tüxen (1956) excluded any changes in time by indicating that it should be imagined that the final stage of succession is reached at once. If, however, final stages or interim-stages of succession are to be realized, this will take time. We therefore have coupled the attainment of a (provisional) final stage to a development period of 50 to 150 years because there will likely occur no climatic changes within that period. An additional advantage of taking only a relatively short time into consideration is the avoidance of all uncertenties about the long-term changes in the environment caused by vegetation itself, which we cannot (anymore) deduce from present-day climax vegetation (simply because we do not have such climax vegetation left for most of the forest regions). With this limitation, our concept of potential natural vegetation comes close to Gaussen's (1955) term 'plesioclimax', a developmental stage of succession reached within 100 years after man's disappearance. This needs not always to be the final climax. From a practical point of

view it is mainly the direction in which a vegetation will develop, that matters. Our description implies, for instance, that the *Querco roboris-Betuletum* can be considered as a type of potential natural vegetation, though it is still uncertain whether this type of forest is a final stage of succession in our country or whether it will develop in the long run into a *Fago-Quercetum*. Something similar can be said of the *Stellario-Carpinetum* in the South of the province of Limburg, of which Van der Werf (in: Trautmann 1972) supposes that this will become ultimately a *Luzulo-Fagetum* whilst the present form is the result of human exploitation. We see this also as a rather long-term development. A second restriction concerns the extent to which human influence is excluded (Westhoff & Van der Maarel 1973). In the definition of potential natural vegetation only human impact is taken into account, resulting from activities directly affecting a certain site or its immediate surroundings. Such activities like manuring, sprinkling, cutting, mowing, burning and grazing can be stopped immediately and the effects can be turned slowly. However, during the last few centuries, man has also interfered with the landscape as a whole, which has altered the character of many sites irrevocably. For instance in the north-east and south-east of the country extensive bog areas occurred. Owing to reclamation their soil and water regime have completely and irreversibly changed. In these reclamation areas, bog can therefore no longer be considered the potential natural vegetation. Also in the many heath reclamation areas man has, by disturbing the upper soil layer and by manuring, changed the character of the site so much that this influence will continue to be reflected in the vegetation for at least the first 150 years.

Another influence which can hardly be repelled, is the establishment of self-rejuvenating exotic species, such as *Prunus serotina* and *Quercus rubra* in our deciduous woodlands (cf. Sissingh 1976, 1977).

Another form of influence, which may be considered of a permanent character and which will therefore also determine the potential natural vegetation, is the indirect large-scale impact following from 'the making of the Dutch landscape' (Lambert 1971). A large part of the west of The Netherlands lies below sea-level and is protected against the influence of the sea by dunes and dikes, combined with large-scale polder drainage. While estimating the climax vegetation in these areas, we must realize that drainage will not be stopped and that dikes will not be pierced.
In conclusion, the potential natural vegetation of a certain site is defined as the vegetation which would come into existence there within a period of 50 to 150 years when all

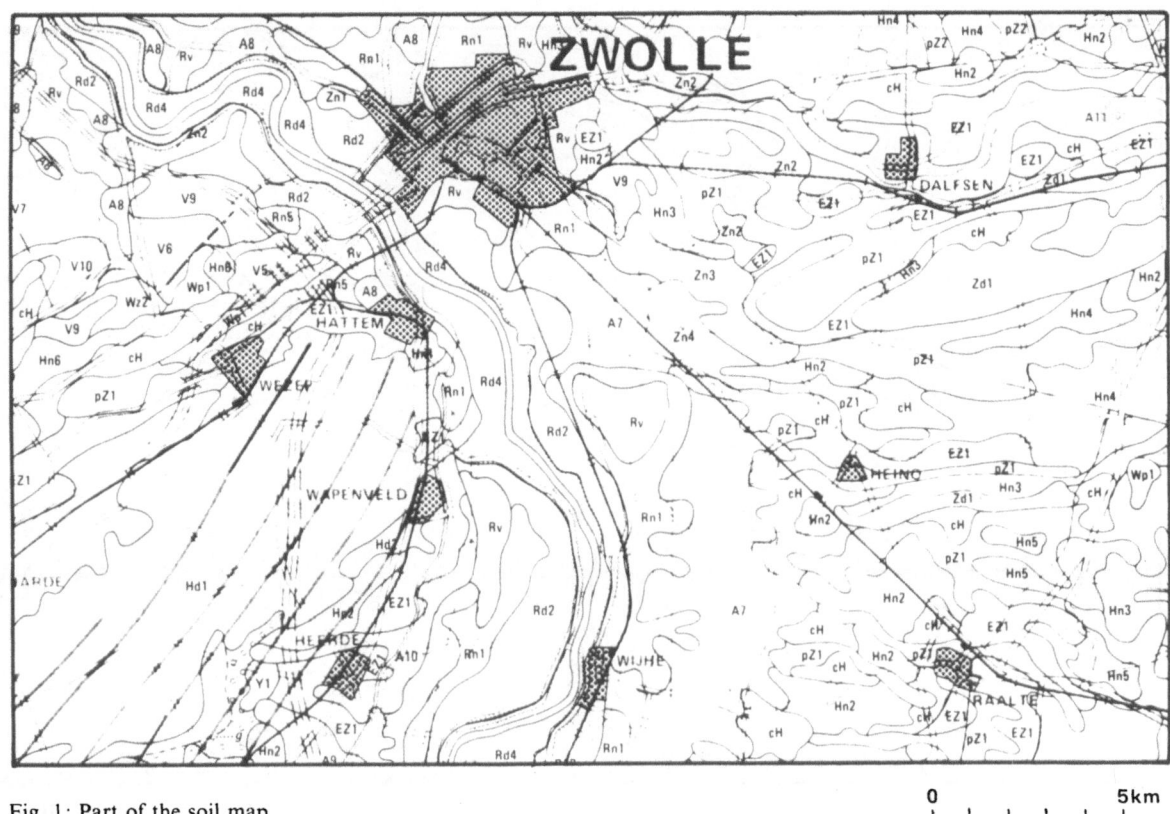

Fig. 1: Part of the soil map.

Legend, shortened, in classic terms:

V	peat soils
EZ	plaggen soils
cH	intergrade podzol soils – plaggen soils
Y	brown podzolic soils
Hd	podzol soils
Hn	hydromorphic podzol soils
Wp	peaty podzol soils
Wz	peaty gley soils
pZ	humic gley soils

Zd	inland dune sand
Zn	gley soils
Rv	alluvial gley soils overlying peat
Rn	alluvial gley soils
Rd	brown alluvial gley soils
A7	Rn + Zn
A8	sandy 'wash' from scours
A9	EZ + cH
A10	Zn + Wz

0   5km

direct human influence upon the place itself or upon its immediate surroundings would stop.

The type of potential natural vegetation depends mainly on climate (temperature, light), relief, water (ground water, rain water) and nutrients. Data on water and nutrients, and, to a limited extent, on relief, can be taken from soil maps of The Netherlands, scale 1 : 50,000, which are based on the classification system by De Bakker & Schelling (1966). These maps were scaled down and generalized into scale 1 : 200,000. Kloosterhuis & Pape (1976) describe the method and its pedologic consequences. The map, obtained in this way, a fraction of which is shown in Fig. 1, has been used as the basis for mapping the poten-

tial natural vegetation.

The actual vegetation was recorded on 600 sites, using Tansley's (1965) 'codom' method, which were distributed representatively over the various soil types distinguished on the 1 : 200,000 soil maps. Relevés were comparatively large, covering at least 300 sqm and representing mainly woodlands, scrubs and semi-natural grasslands. The relevés were interpreted as syntaxonomical units of the Braun-Blanquet system according to Westhoff & Den Held (1969), using the association level as far as possible. The association Luzulo-Quercetum, distinguished in South-Limburg, has been adapted from Noirfalise & Sougnez (1956), who described this type from the Belgian Ardennes. At the site

139

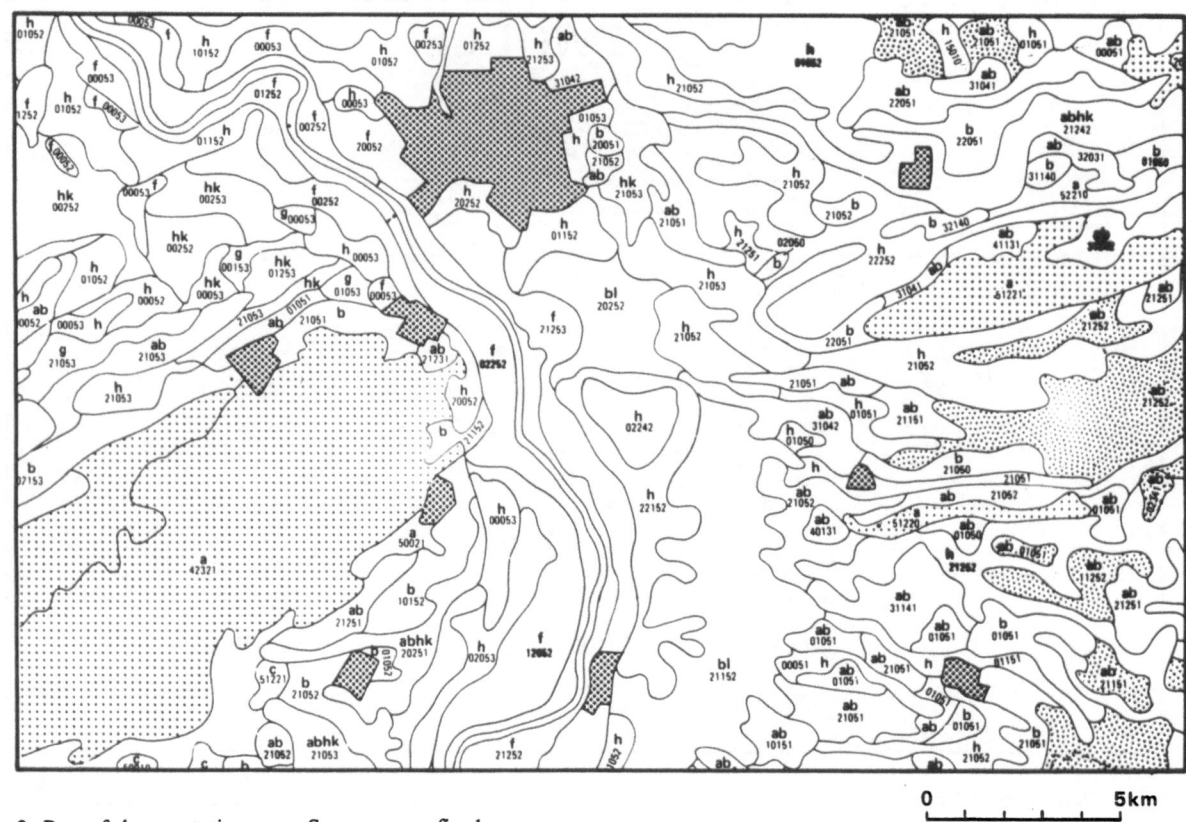

Fig. 2: Part of the vegetation map. Same area as fig. 1.

Legend:

a	*Querco roboris-Betuletum*
ab	Complex of *Querco roboris-Betuletum* and *Fago-Quercetum*
b	*Fago-Quercetum*
bl	Complex of *Fago-Quercetum* and *Alno-Padion*
c	*Fago-Quercetum*, type poor in species
f	*Fraxino-Ulmetum*
g	*Anthrisco-Fraxinetum*
h	*Circaeo-Alnion*
hk	Complex of *Circaeo-Alnion* and *Alnion glutinosae*
abhk	Complex of *Querco roboris-Betuletum*, *Fago-Quercetum*, *Circaeo-Alnion* and *Alnion glutinosae*

predominantly moist and wet vegetation types

predominantly dry vegetation types

urban area

for figures, see text and Fig. 6.

of each relevé a boring was carried out to a depth of 1.20 m, to determine the corresponding soil type and the ground water regime.

Generally a clear correlation between a certain type of vegetation and certain soil units was found. Soil water conditions appeared to play a major differentiating role here. On places with the same soil type but with differences in ground water regime (particularly fluctuations), different vegetation types may occur. On humic gley soil the *Circaeo-Alnion*, the *Fago-Quercetum*, or a complex of both types

was found, dependent on the occurring ground water regime. Various soil types show regional variation, which has often been neglected because of the predominant use of the soil classification for agricultural purposes. This variation could be syntaxonomically interpreted, together with regional climatic differences, however only on levels below that of the association. They can not be recognized in a vegetation typology adapted to a mapping scale of 1 : 200,000.

140

Fig. 2 shows the potential natural vegetation of the area, the soil of which is shown in Fig. 1.

Each potential natural vegetation type is linked with a number of vegetation types which can be considered as substitute and companion communities. All types of vegetation which may develop on the same soil type and therefore belong to the same potential natural vegetation, can be classified into a series the elements of which are arranged according to the kind of human influence. Van Gils & Van der Maarel (in: Werkgroep G.R.I.M. 1974) made a start with the use of vegetation series. This method has been worked out and refined by the authors. Such a vegetation series is similar to, though still distinct from the 'Gesellschaftsring' (community ring) described by Schwickerath (1954, cf. Schmithüsen 1968, Seibert 1968), because 1. within the series not only the climax and substitute communities are involved, but also certain companion communities (forming 'vegetation complexes' with the former ones) that do not form a direct part of that succession series. 2. the idea of ring is not appropriate because it is uncertain that original climax from which the series started and potential natural vegetation to which it develops are identical (Seibert 1968, Werkgroep G.R.I.M. 1974).

The syntaxonomical units arrived in the way described above were first arranged into 38 vegetation complexes, each of them characterized by one or several mosaic-like occurring potential natural vegetation types. These complexes are the mapping units on the 1 : 200,000 vegetation map. The vegetation complexes were again arranged into seven main vegetation series. Within these main series a number of sub-series are distinguished which have the greater part of the units in common with the main series, but differ in their presumed final stage. The main series as well as the sub-series have been called after their final stage of development (= potential natural vegetation). The existing units have been classified both on the alliance and association level (see Fig. 3).

The following main series are distinguished:

I. *Quercion robori-petraeae* series
   4 sub-series: a. *Querco roboris-Betuletum*
               b. *Fago-Quercetum*
               c. *Fago-Quercetum*, type poor in species
               d. *Luzulo-Quercetum*

II. *Carpinion betuli* series
   This alliance has only 1 association in The Netherlands:
               a. *Stellario-Carpinetum*

III. *Alno-Padion* series
   5 sub-series: a. *Ulmion carpinifoliae*
               b. *Fraxino-Ulmetum*
               c. *Anthrisco-Fraxinetum*
               d. *Circaeo-Alnion*
               e. *Macrophorbio-Alnetum*
               a and d have been used in areas where the vegetation could not further be specified.

IV. *Alnion glutinosae* series
   no sub-series

V. *Salicion albae* series
   Only 1 association in The Netherlands:
               a. *Salicetum albo-fragilis*

Fig. 3: Main vegetation series of the Quercion robori-petraeae (simplified from Kalkhoven, Stumpel & Stumpel-Rienks 1976).

A. Forest vegetation	B. Scrub vegetation	C. Semi-natural land vegetation	D. Hardly natural or non-natural land vegetation	E. Water and marsh vegetation
Quercion robori-petraeae Querco roboris-Betuletum Fago-Quercetum Fago-Quercetum, type poor in species Luzulo-Quercetum Dicrano-Pinion	Lonicero-Rubion sylvatici Dicrano-Juniperetum	Nanocyperion flavescentis Epilobion angustifolii Spergulo-Corynephorion Thero-Airion Ericion tetralicis Violion caninae Calluno-Genistion pilosae Empetrion nigri	Polygono-Chenopodion Sisymbrion Polygono-Coronopion Arnoseridion Aphanion Lolio-Plantaginion Agropyro-Rumicion crispi Arction Aegopodion podagrariae Arrhenatherion elatioris	Potamion graminei Littorellion uniflorae Bidention Cardaminion

VI. Bog vegetation series
Bog is syntaxonomically characterized as a complex of *Erico-Sphagnion*. *Sphagnion fusci* and *Rhyncho-sporion albae*, sometimes in combination with *Betulion pubescentis*.

VII. Coastal dune vegetation series
Within the coastal dunes, two separate vegetation complexes are distinguished, namely those of the dunes rich in lime with the sub-series
      a. *Convallario-Quercetum dunense*
      b. *Crataego-Betuletum*
and those of the dunes poor in lime with a complex of *Empetrion nigri*, *Violo-Corynephoretum* and *Quercion robori-petraeae* (*Violo-Corynephoretum* is the only association of the *Galio-Koelerion* existing within this complex).

The plant communities of salt marshes have been added to this survey as an appendix. Environmental dynamics are so strong here that no higher stages of succession can develop and the proper idea of vegetation series cannot be applied here.

Within every vegetation series, the plant communities have subsequently been divided into five structural groups. This division is similar to that of Rey & Izard (1972). It also resembles the division into degrees of naturalness by Westhoff (1973). For the vegetation series I-V these groups include:

A. Forest vegetation
This group includes all the potential natural forest types. For practical reasons the planted woods have been also included in this group. Especially in the older plantations, the layers of shrubs and herbs correspond with those of the more natural forest.

B. Scrub vegetation
This group includes communities of scrubs, forest margins and hedges. Also single trees and shrubs along the roads and parcels are placed into this group.

C. Semi-natural land vegetation
Vegetation of this type shows a largely spontaneous species composition, but a structure determined by man through a constant or regularly returning human influence. Among these are heath lands, extensively used grasslands (particularly hay meadows) and the margin-communities of forests.

D. Hardly natural or non-natural land vegetation
This group includes the communities strongly influenced by man, such as fields, intensively used grasslands, ruderal

places, disturbed margins of oligotrophic pools, etc.

E. Water and marsh vegetation.
These are the communities of open water, trembling-bog and the communities of the margins of oligotrophic to eutrophic waters. The swamp vegetations of the peat-moor and peat-bog landscape also belong to this group.

For the vegetation series VI and VII some deviations were necessary. The two first groups of the main series VI have been composed in a different way, namely: A. Bog vegetation, B. Forest and scrub vegetation. Group C of the main series VII deviates from the other series and is called Natural to semi-natural dwarf scrub and herb vegetations.

Fig. 3 gives an example of a simplified description of main series I. In our final report details are included on the correlated soil types and the distribution in The Netherlands of all vegetation complexes occurring in the series. Also guidelines for management are added, which are needed to have a type developed to a further stage of succession or to maintain a substitute community. The sequence from group A to D runs roughly parallel to the sequence in degrees of naturalness mentioned above.

With increasing human impact on vegetation the corresponding plant community types are less typical for particular series. A number of communities from group D are found in nearly all series. An example of this is the *Poo-Lolietum*, the cultivated grassland poor in species which is now found in large parts of The Netherlands on all kinds of soil. Within every group we find a variation in composition of the vegetation complex due to differences in microclimate and mineral richness of the soil. A subsequent classification can thus be made within every group e.g. according to the degree of oligotrophy or the degree of environmental dynamics. This has not yet been worked out further because, in view of the scale of this map, such a classification is not relevant.

Ecotope and ecotope complex

As mentioned above, plant communities in the landscape are found in spatially and functionally coherent vegetation complexes. Such a complex stands for a group of plant communities which can be deduced from one type of potential natural vegetation and which are connected with each other ecologically and physiognomically (cf. Hofmann 1965). Thus a vegetation complex is a characteristic and discernable part of the vegetation with an internal structure and an ecologically interpretable position in the landscape.

142

This conception of vegetation complex comes close to the 'ecological community complex' in Seibert's (1974) system. Since the term is used in different meanings (e.g. Oberdorfer 1964, Schmithüsen 1968, Tüxen 1973, see Seibert 1974 for a review) it may be useful to avoid the term and to use the term ecotope instead, as is frequently done in landscape ecology. The framework of landscape ecological concepts, mainly developed in physical geography, is appropriate indeed (cf. Van der Maarel & Stumpel 1975, Tjallingii 1976, Anon. 1977). Landscape ecology deals with 'geographic areas' which can be considered as the smallest landscape ecological units. These units are defined with help of environmental factors, such as climate, soil, water and relief, as well as vegetation and fauna, and they form a functional system: an ecosystem (cf. Neef 1970). In natural and near-natural situations, it is relatively difficult to distinguish one ecosystem from another; in the semi-natural and agricultural landscape, this is often less difficult because of the many sharp boundaries induced by man. For such easily recognizable sites with a particular ecosystem the term ecotope was suggested, probably independently by Tansley and Troll in 1938 (cf. Tansley 1965, Troll 1970, see also Schmithüsen 1948, Leser 1976). We consider an ecotope as an ecologically uniform part of the landscape, characterized by one or more interrelated ecosystems, i.e. the entire set of interrelated abiotic and biotic factors.

A vegetation complex, as described above, can be considered as the vegetational component of the local ecotope. It should be noted that the meaning of the term ecotope in the present scope deviates from the meaning attached to it by Whittaker, Levin & Root (1973) who used the term ecotope for the set of environmental factors determining the occurrence of an organism or a biotic community as a whole.

Stumpel-Rienks (1974) published a list of ecotopes on behalf of the environmental survey on which we report here. Most ecotopes of this list comprise a complex of plant communities (on the levels of association and sub- association).

For mapping landscape ecological units on small scales it will be necessary to distinguish complexes of landscape ecological units: ecotope complexes. Heathland and heath-pool for instance, are separate ecotopes, but, on a map from scale 1 : 100,000, a heathland with a few scattered pools in it, will have to be represented as an ecotope complex. In the survey below, no further distinction is made between ecotopes and ecotope complexes.

In this stage of research we consider this approach appropriate on scale 1 : 200,000. Further work will have to be directed towards both analytical and synthetic research of landscape ecological units on larger scales.

The list of existing ecotopes and ecotope complexes in the landscape, has been obtained first of all by interpreting topographical maps on scale 1 : 25,000. This interpretation has been supplemented by national inventories of ecotopes which cannot be read off from these maps. They mostly concern rare, small remainders of semi-natural to near-natural ecosystems, such as wells, limestone grasslands, peat-moors etc. Fig. 4 gives an example of ecotopes and ecotope complexes in the mainly agricultural landscape of The Netherlands, arranged according to the groups classified above.

In order to obtain an idea about the distribution of the actual vegetation types, an estimate was made of the size of the areas (or length for linear elements) taken in by the five structural groups distinguished within every vegetation series. This was done for every area on the map and expressed in a code of five figures. A distinction was made between plane-, line- and point-shaped elements. Of course this distinction is entirely determined by the scale of the map. The presence figures are given according to the standards in Fig. 5. An example of the use of this code is given in Fig. 6.

Discussion

The vegetation map resulting from this study gives information about both the actual vegetation and the potential natural vegetation and it constitutes the first vegetation map covering the total surface of The Netherlands. The map is very global and when applying this method of mapping on a larger scale, one is faced with a number of problems.

One of the main problems is the great discrepancy between the number of legend units of the soil map (134) and those of the vegetation map (38). The soil classification used here is particularly useful for agricultural purposes, but less not relevant for correlation with vegetation data. A more detailed map of the potential natural vegetation requires further research on the relationship between soil and vegetation on the basis of an adapted soil classification.

As a result of the poverty in more or less natural woodland of the Dutch landscape, the potential natural vegetation of some areas cannot be classified below the level of alliance. The present woodland has been planted nearly always by man in recent times and is therefore not natural.

143

Fig. 4: Examples of ecotopes and ecotope complexes, characterized by their vegetation (as far as possible).

A. Forest vegetation	B. Scrub vegetation	C. Semi-natural land vegetation	D. Hardly natural or non-natural land vegetation	E. Water and marsh vegetation
dune forest	wooded bank	dry heath	field	spring
springwood	hedge	wet heath	orchard	heath-pool
fenwood	dune-scrub complex	shifting dunes	recent dike	dike-burst pool
acidophilous decidu-ous forest	plantations along roads and yards	river-dune	cultivated grassland	duck-decoy pool
basiphilous decidu-ous forest	recent plantation of deciduous trees	old river-, sea- and polder-dike	peat-cutting area	former river-bed
coniferous forest		peat-ridge	horticultural area	brook
willow-coppice		limestone grassland	factory area	canal
duck-decoy forest		old unmanured hayfield	dry sand-excavation	creek
		river foreland grass-land	ruderal vegetation	pond
		coastal ridge		lake
		dune-slack		dune-lake
		dune grassland		fen complex
		complex of dune-heath, -grassland and -valley		marsh
				reed-swamp

Only in a few cases the older planted forests have been developed into a more natural vegetation. In large parts of The Netherlands (especially in the polders, except in the peat-polders) there has never been any woodland owing to the constant use as culture land. Nevertheless the spontaneous growth of trees and scrubs indicates the direction of development into a natural vegetation. For application on a larger scale, more research on this subject is necessary.

In this connection there is another problem concerning the relatively young age of a part of the soils in The Nether-lands. A further development of the soil will change the soil characteristics and, as a result, the prediction on the potential natural vegetation. We confine ourselves to a period of 50–150 years and therefore exclude this problem, which is expected to be a longer term development.

The influence of ground water fluctuations on vegetation is still not known in detail. A global classification of ground water regimes has been applied for the legend of the soil maps, but its relevance for vegetation is still uncertain. Recent ground water lowering will influence the site and hence the vegetation so much, that vegetation types from another series can develop, and another type of potential natural vegetation will arise. The developing communities from that vegetation series can likely be recognized within a period of 50 years.

As mentioned earlier, there are regional differences within the vegetation types that manifest themselves on a level lower than that of associations. Well-known examples are the plant communities of fields, moist oligotrophic grasslands and acidiphilous deciduous forests. They are worth being mapped on a larger scale.

The ecotopes are listed as a survey of the main types of landscape ecological units and can be used as a general basis for an evaluation of the natural environment, as is done in our project by estimating the national significance of each ecotope. Still further research is necessary and will be carried out by the authors in characterizing ecotopes in a more inductive way. It should be noted that a lot of our ecotopes are typical for the Dutch landscape and hence difficult to compare with situations elsewhere, even in Northwestern Europe. For use on local maps, these ecotopes can be described more specifically. Especially the ecotopes of the various aquatic systems, so typical for The Netherlands, are to be characterized in a more ecological way.

The various ecotopes can be as distinguished in this study considered as basic landscape ecological units. A more complete ecological interpretation of such units may be obtained on the basis of phytosociological relevés, and subsequent analyses of the respective animal communities. These data have to be integrated into units which represent ecosystems or their complexes. However, much basic

Fig. 5.: Code for the occurrence of the ecotope groups per map area.

plane-shaped elements	>75% of the total surface	5
	50-75%	4
	25-50%	3
	<25%	2
	< 5%: see point-shaped elements	

line-shaped elements	very many	3
	rather many	2
	few	1

| point-shaped elements | many | 2 |
| | few | 1 |

The categories very many, many, rather many and few are specified as followes:

for hedges and brushwood	>1500 m per km^2	3
	800-1500 m per km^2	2
	< 800 m per km^2	1

for ditches and little waters	>8 km per km^2	3
	<8 km per km^2 + minimal 3 smaller pools	3
	2-8 km per km^2	2
	<2 km per km^2	1

| for groves, heaths and small pools | >5 per km^2 | 2 |
| | <5 per km^2 | 1 |

When a group is not represented, the figure 0 is given.

research will be necessary on the relationship between vegetation and fauna before interpretations on the level of cenoses are possible.

As a derivation from our vegetation map, an ecological evaluation is made, first of all meant as a general basis for the judgement of the various impacts on the natural environment, resulting from developments in physical planning (cf. Van der Maarel & Vellema 1975, Van der Maarel 1978). Details of this evaluation are also presented in our final report (Kalkhoven, Stumpel & Stumpel-Rienks 1976).

A further application of the vegetation map is an outline of the perspectives for the development of vegetation complexes within certain ecotopes and ecotope complexes, in order to increase their ecological quality. These data may become important in certain areas where nature manage-ment techniques are applied, in order to start a development aiming at the promotion of natural elements above agricultural and civil interests.

Summary

The method of mapping the vegetation on scale 1 : 200,000 and the starting points in relation to the potential natural vegetation and ecotopes, are discussed.

In view of the planological background of this study, some restrictions have been added to the concept of potential natural vegetation, concerning the period of development and the human influence.

The relationship between soil, ground water and vegetation was studied, which resulted in the map of the potential natural vegetation.

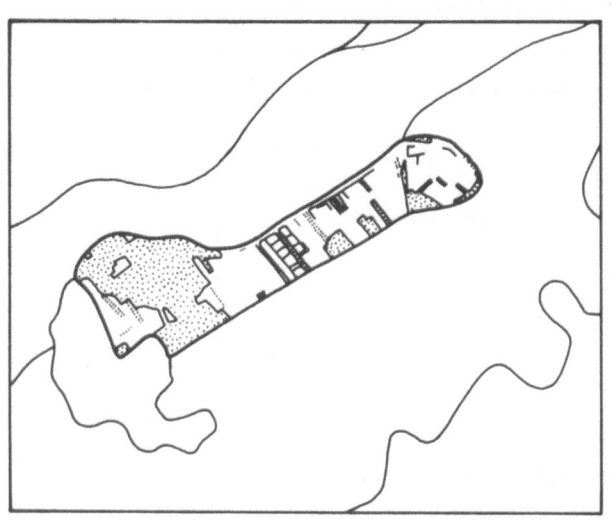

1 km

Fig. 6.: Example of the use of the ecotope occurrence code in one area on the vegetation map, in which the ecotopes are drawn in.

Legend for example area:

forest

wooded bank and single trees

ditches

fields and cultivated grassland

code: 31042
soil: hydromorphic podzol soil
potential natural vegetation: Complex of *Querco roboris-Betuletum* and *Fago-Quercetum*

In this case the code means:

3....	25–50 % oak and coniferous forest
.1...	few banks of oak coppice
..0..	no heath
...4.	50–75 % cultivated land
....2	a number of ditches

N.B. Another area may have the same code, but a different potential natural vegetation. In that case the structural groups are represented by other vegetation types!

Each type of potential natural vegetation stands for a series of vegetation types on the same site. Seven main series, with a number of sub-series are distinguished. Within each vegetation series the plant communities have

146

been spread over five groups, according to their structure and naturalness.

Ecotopes and ecotope complexes are considered as landscape ecological units. A list of ecotopes was obtained by interpreting topographical maps and by inventory data.

The actual vegetation was mapped by estimating the size of the ecotopes within the separate areas. It was expressed in a five figure code for the five groups from the vegetation series. The information on potential natural vegetation and ecotopes is combined into the vegetation map of The Netherlands.

Interpretation problems, some of them specific for The Netherlands, are discussed and some remarks are made on the necessity of further research.

References

Anonymus, 1977. Samenvatting Globaal Ecologisch Model. Summary General Ecological Model. Part 3 of the Series General Physical Planning Outline, Ministry of Housing and Physical Planning, The Hague, 42 pp.

Bakker, H. de & Schelling, J. 1966. Systeem van bodemclassificatie voor Nederland. Soil Survey Institute, Wageningen, 217 pp + bijl.

Gaussen, H. 1955. Rapport général sur la cartographie écologique. Ann. Biol. 31: 221–231.

Hofmann, G. 1965. Über Vegetationskomplexe unter besonderer Berücksichtigung der Trockenwaldkomplexe. Feddes Repert. Beih. 142: 216–222.

Kalkhoven, J.T.R. 1974, Eine vorläufige Übersicht der potentiellnatürlichen Pflanzengesellschaften der Niederlande und der wichtigsten von denen abgeleiteten Vegetationskomplexe und Vegetationsreihen. Paper International Symposium 'Landschaftsgliederung mit Hilfe der Vegetation', Rinteln. (in press).

Kalkhoven, J.T.R., Stumpel, J.T.R., & Stumpel-Rienks, S.E. 1976. Landelijke Milieukartering. Environmental Survey of The Netherlands, a landscape ecological survey of the natural environment in The Netherlands for physical planning on national level. Staatsuitgeverij, The Hague, 141 pp + bijl.

Kloosterhuis, J.L. & Pape, J.C. 1976. Toelichting bij de bodemkaart van Nederland, schaal 1 : 200 000, ten behoeve van de Landelijke Milieukartering. Report Nr. 1285 Soil Survey Institute, Wageningen, 85 pp.

Küchler, A.W. 1973. Problems in classifying and mapping vegetation for ecological regionalization. Ecology 54 (3): 512–523.

Lambert, A.M. 1971. The Making of the Dutch Landscape. Seminar Press Ltd, London, XII + 412 pp.

Leser, H. 1976. Landschaftsökologie. Ulmer, Stuttgart, 432 pp.

Maarel, E. van der 1975. Man-made natural ecosystems in environmental management and planning. In: W.H. van Dobben & R.H. Lowe-McConnell (eds): Unifying concepts in ecology, p. 263–274. Junk, The Hague/Pudoc, Wageningen.

Maarel, E. van der 1978. Ecological principles for physical planning. In: M.W. Holdgate & M.J. Woodman (eds): The breakdown and restoration of ecosystems. Plenum, New-York, p. 413–450.

Maarel, E. van der & Stumpel, A.H.P. 1975. Landschaftsökologische Kartierung und Bewertung in den Niederlanden. Verh. Ges. Ökologie, Erlangen 1974 p. 231–240 Junk, The Hague.

Maarel, E. van der & Vellema, K. 1975. Towards an ecological model for physical planning in the Netherlands. In: Ecological aspects of economic development planning. Report Seminar U.N. Economic Commission for Europe, Rotterdam, p. 128–143. E.C.E., Geneva.

Neef, E. 1970. Zu einigen Begriffen der Ökologie. Arch. Naturschutz und Landschaftsforschung 10: 233–240.

Noirfalise, A. & Sougnez, N. 1956. Les Chênaies de l'Ardenne verviétoise. Pedologie 6: 119–143.

Oberdorfer, E. 1964. Der insubrische Vegetationskomplex, seine Struktur und Abgrenzung gegen die submediterrane Vegetation in Oberitalien und in der Südschweiz. Beitr. Naturk. Forsch. SW. Deutschl. 23: 141–187.

Rey, P. & Izard, M. 1972. Notions générales d'utilisation des cartes de la végétation. 2nd edition. Centre national de la recherchescientifique, Paris 26 pp.

Schmithüsen, J. 1948. 'Fliesengefüge der Landschaft' und 'Okotop'. Berichte Deutsch. Landesk. 5: 74–83.

Schmithüsen, J. 1968. Allgemeine Vegetationsgeographie. 3. Auflage. Walter de Gruyter, Berlin, XXIII + 463 pp.

Schwickerath, M. 1954. & Die Landschaft und ihre Wandlung auf geobotanischer und geographischer Grundlage entwickelt und erläutert im Bereich des Messtischblattes Stolberg. R. Georgi, Aachen, 118 pp.

Seibert, P. 1968. Gesellschaftsring und Gesellschaftskomplex in der Landschaftsgliederung. In: R. Tüxen (ed): Pflanzensoziologie und Landschaftsökologie, Stolzenau 1963, p. 48–60. Junk, The Hague.

Seibert, P. 1974. Die Rolle des Massstabes bei der Abgrenzung von Vegetations-einheiten. In: R. Tüxen (ed): Tatsachen und Probleme der Grenzen in der Vegetation, Rinteln, p. 103–118. J. Cramer Verlag, Lehre.

Sissingh, G. 1976. Betekenis en gevolgen van menselijke ingrepen voor de samenstelling en instandhouding van bossen, speciaal onder Nederlandse omstandigheden. Ned. Bosbouw Tijdschr. 48 (3): 86–96.

Sissingh, G. 1977. Optimal woodland development on sandy soils in the Netherlands. Vegetatio 35: 187–191.

Stumpel, A.H.P. 1974. Kartierung landschaftsökologischer Einheiten mit Hilfe von Vegetationskomplexen, Vegetationsreihen und Ergebnisse von Bodenkartierungen. Paper International Symposium 'Landschaftsgliederung mit Hilfe der Vegetation', Rinteln. (in press).

Stumpel-Rienks, S.E. 1974. De botanische waardering van ecotopen als bijdrage tot een globale waardering van het natuurlijk milieu. Gorteria 7: 91–98.

Sukopp, H. 1972. Wandel von Flora und Vegetation in Mitteleuropa unter dem Einfluss des Menschen. Ber. Landwirtschaft 50: 112–139.

Tansley, A.G. 1965. The British Islands and their Vegetation. 4. Imp. University Press, Cambridge, 930 pp.

Tjallingii, S.P. 1976. Enkele opmerkingen over het begrip ecotoop. Gorteria 8: 31–35.

Trautmann, W. (1972): Erläuterungen zur Karte 'Vegetation' (potentielle natürliche Vegetation). Deutscher Planungsatlas Band I: Nordrhein-Westfalen. Lieferung 3. Gebrüder Jänecke Verlag, Hannover.

Troll, C. 1970. Landschaftsökologie (Geoecology) und Biocoenologie. Eine terminologische Studie. Rev. Roum. Géol., Géophys. et Géogr., Sér. Géogr. 14 (1): 9–18.

Tüxen, R. 1956. Die heutige potentielle natürliche Vegetation als Gegenstand der Vegetationskartierung. Angewandte Pflanzensoziologie 13: 5–42.

Tüxen, R. 1973. Vorschlag zur Aufnahme von Gesellschaftskomplexen in potentiell natürlichen Vegetationsgebieten. Acta Bot. Acad. Scient. Hungar. 19: 379–384.

Werkgroep G.R.I.M. 1974. Landschapsecologische basisstudie voor het streekplangebied Midden-Gelderland. Provinciale Planologische Dienst, Arnhem, 156 pp.

Westhoff, V. 1955. Samenhang tussen vegetatie en milieu, in het bijzonder de bodem. TNO-nieuws 106: 9–16.

Westhoff, V. 1968. Die 'ausgeräumte' Landschaft, Biologische Verarmung und Bereicherung der Kulturlandschaften. In: K. Buchwald & W. Engelhardt (eds): Handbuch für Landschaftspflege und Naturschutz, Band 2, p. 1–10. Bayerischer Landwirtschaftsverlag München.

Westhoff, V. 1971. The dynamic structure of plant communities in relation to the objectives of conservation. In: E. Duffey & A.S. Watt (eds): The scientific management of animal and plant communities for conservation, p. 3–14. Blackwell, Oxford.

Westhoff, V. 1973. Vegetatie-ontwikkeling. In: A.D. Voûte & J. F. de Vries Broekman (eds): Natuurbeheer in Nederland, p. 46–54, 180–189. Samsom, Alphen aan de Rijn.

Westhoff, V. & Held, A.J. den 1969. Plantengemeenschappen in Nederland. Thieme, Zutphen, 324 pp.

Westhoff, V. & Maarel, E. van der (1973): The Braun-Blanquet Approach. In: R.H. Whittaker (ed): Handbook of Vegetation Science, Vol. 5, p. 617–726. Junk, The Hague.

Whittaker, R.H., Levin, S.A. & Root, R.B. (1973): Niche, habitat and ecotope. The American Naturalist 107: 321–338.

APPLICATIONS PHYTOSOCIOLOGIQUES A L'AMENAGEMENT DU TERRITOIRE*

J. LEBRUN

Laboratoire d'Ecologie Végétale, Université Catholique de Louvain, 4 Place Croix du Sud, 1348 Louvain-la-Neuve, Belgique

Mots-clèf (Keywords):
Aménagement du territoire (Land management), Carte de végétation (Vegetation map), Phytosociologie appliqué (Applied phytosociology)

Introduction

L'aménagement du territoire est le bon usage de ses ressources naturelles en vue de créer et de conserver un cadre propice à la satisfaction et, le cas échéant, au développement des besoins et des aspirations socio-économiques, moraux, intellectuels et esthétiques de tous ses habitants. Définition très ample et fort vague utilisée à dessein pour suggérer la complexité de cette discipline. Ses règles sont loin, en effet, d'être définitivement établies. Son caractère synthétique, éminemment écologique en fait, apparaît bien dès qu'elle se fonde avant tout sur le comportement de l'homme dans son milieu. Les objectifs eux-mêmes de l'aménagement et les moyens de les atteindre sont fort divers et, par là, nécessairement multidisciplinaires. Les sciences biologiques y sont profondément impliquées par de multiples facettes.

Le couvert végétal, qu'il soit naturel, semi-naturel ou purement domestiqué, est l'un des principaux jalons de l'aménagement. Ce n'est pas seulement par sa production qu'il intervient – qu'il s'agisse de sylviculture, de praticulture ou d'agriculture – mais encore par ses caractéristiques affectant la salubrité de l'air, l'assainissement du sol, tout comme par ses attributs culturels, esthétiques ou récréatifs.

Le manteau d'Arlequin des végétations, encore, est l'un des déterminants des paysages; son analyse est à la base même de la connaissance paysagère.

Mais davantage toujours, les communautés végétales sont les révélateurs des biocénoses; ce sont aussi les porte-drapeaux des écosystèmes. Ce sont elles qui produisent, –

* Contribution to the Symposium on Plant species and plant communities, held at Nijmegen, 11–12 November 1976, on the occasion of the 60th birthday of Professor Victor Westhoff.

et de loin – la majeure partie de la biomasse, contribuent le plus aux cycles biogéochimiques, alimentent au maximum les flux d'énergie et de matière entre biosphère d'une part, lithosphère et atmosphère d'autre part. Tout le monde sait l'importance de la fonction chlorophyllienne, notamment dans l'équilibre des constituants de l'air, comme le rôle de l'absorption et de la transpiration des plantes dans le cycle naturel de l'eau.

La composition précise des unités de végétation, la connaissance de leurs particularités synécologiques et phénologiques, la détection de leur dynamisme comme de leurs interférences sur le terrain circonscrivent le corps et les attributs de la phytosociologie. Cette branche vigoureuse de la biologie est donc au fondement des concepts et des réalisations de l'aménagement du territoire. Elle est le truchement le plus sûr de l'exploitation de toutes les données écologiques qui en constituent le piédestal.

Ces généralités sembleraient conférer à l'étude de la vie en commun des plantes, dans le cadre envisagé, un impact réservé aux seules zones campagnardes, rurales ou forestières, là où la verdure des frondaisons ou des feuillages impose, durant la plus grande partie de l'année, le coloris dominant des paysages, en excluant par le fait même les terroirs urbanisés ou industriels, et a fortiori, les cités et les usines. Qu'on ne s'y trompe point! Même quelque peu méconnues et délaissées jusqu'il n'y a guère, les applications phytosociologiques sont aussi nombreuses dans ces derniers cas. La fonction sociale et hygiénique des espaces verts, des plantations d'alignement, des parcs et jardins, des terrains vagues même, des pièces ou cours d'eau enherbés au coeur et dans la périphérie des agglomérations urbaines ou manufacturières n'est maintenant plus sous-évaluée.

Ce sont là des éléments non négligeables qui étayent les 'plans de secteur' voire la rénovation ou réhabilitation d'anciens quartiers dans les villes. La nature souvent très mouvante des végétations rudérales révèle des appartenances sériales et des conditions mésologiques très significatives quant à la connaissance des sites, à l'appropriation des espaces disponibles, à l'introduction des essences qui s'indiquent le mieux... C'est encore faire oeuvre de phytosociologue averti que de composer les haies vives et rideaux protecteurs contre le bruit, les poussières, les pollutions, les intempéries ou de préconiser les implantations végétales dissimulant au mieux les emprises techniques et conférant un aspect agréable au tissu urbain ou industriel.

Il est une caractéristique encore de l'aménagement du territoire – qui n'est pas toujours mise en évidence comme il le faudrait – et qui n'est pas sans retentir sur les interventions de l'écologiste. C'est que les plans élaborés sont sujets à révisions successives, au moins dans le cadre des options fondamentales retenues. L'organisation du secteur, de la contrée, du pays doit, en effet, se plier aux changements démographiques, s'accomoder des transformations des usages, satisfaire de nouveaux besoins, suivre la croissance et le développement économiques. Il s'agit donc, en définitive, d'une oeuvre permanente, à remettre sans cesse sur le métier pour en polir les détails, en nuancer les teintes. Comme les végétations qu'ils portent, les terroirs aménagés sont sujets à transformations dans le temps et dans l'espace.

Participation de la phytosociologie

Les participations de la phytosociologie à l'aménagement du territoire au sens large sont déjà nombreuses et diverses. Une d'entre elles est naturellement la cartographie des communautés végétales (C.N.R.S. 1973). Les cartes de végétation actuelle ou potentielle ne sont pas seulement des éléments importants du diagnostic écologique mais elles servent encore de tremplin à nombre d'autres préoccupations de l'aménagiste.

Parmi les thèmes fort abondants que rapporte la littérature à ce sujet, on se bornera à citer ceux qui ont été entretenus ces dernières années par le Laboratoire d'Écologie Végétale de l'Université Catholique de Louvain (Lebrun 1968).

L'aménagement forestier classique fondé sur la connaissance des types forestiers réels ou potentiels est, par excellence, un objectif souvent poursuivi. La définition des associations épiphytiques et de leur sensibilité particulière aux altérations atmosphériques permet, selon une voie qui a été largement défrichée par notre collègue J. J. Barkman,

de dresser des cartes synthétiques de la salubrité de l'air. Un peu selon les mêmes principes, c'est l'étude aussi de la pureté des eaux courantes en considérant aussi bien les cryptogames que les hydrophytes supérieurs (De Sloover 1964, De Sloover et al. 1977). En relation avec l'inventaire des ressources en eau, il était important que des mesures soient faites durant un temps suffisant, au sein d'associations végétales bien définies, en vue de déterminer leurs paramètres propres, à insérer dans des modèles aussi précis que possible (Renard 1971). Tracé et aménagement des grandes voies de communication, tellement à l'honneur maintenant, gagneraient beaucoup à s'inspirer des règles dégagées du message de la végétation spontanée : quelques recherches à ce propos tendent précisément à le comprendre et à l'interpréter (Anon 1975, Vandiest-Wallon & De Sloover 1976). Notre Service s'est aussi intéressé à une question abordée en de nombreuses contrées et qui ne peut d'ailleurs recevoir que des solutions locales. Il s'agit de l'effet attractif ou répulsif des diverses formes du couvert végétal entourant les aéroports sur l'avifaune. On sait le danger que peut représenter pour les aéronefs, au décollage comme à l'atterrissage, la pullulation sur ces sites de certains types d'oiseaux (Stenbock-Fermor 1973).

Dans le cadre des aménagements de l'espace rural, et quels que soient les objectifs du projet, un aspect important est l'appréciation des paysages tels qu'ils existent (Noirfalise 1970, Nef 1975, Froment & Nef 1976). Or, la végétation n'est-elle pas l'élément paysager le plus apparent ? On reviendra plus loin sur cet aspect de l'application de la science des communautés végétales. On citera encore divers thèmes relatifs à la conservation des peuplements biotiques, et, par là, à la gestion des réserves naturelles (Brasseur et al. 1977). Il s'agit là précisément d'un sujet que notre éminent collègue V. Westhoff a traité avec autorité et brio et dont il a démontré d'une manière convaincante l'opportunité, les voies analytiques et les solutions tout entières fondées sur la discipline phytosociologique (Westhoff 1971).

Trois exemples de l'application

Notre propos aujourd'hui, pour étoffer cet exposé à caractère fort général, est de développer et d'illustrer trois exemples de ces applications, choisies parmi les plus originales que nous ayons abordées.

Plan pour une ville nouvelle

Le premier porte sur un objectif d'aménagement tout à fait exceptionnel dans notre pays : celui de la fondation d'une ville nouvelle. En 1968, à la suite de décisions politiques très pressantes, l'Université de Louvain fut scindée en deux unités, en fait autonomes, et sa Section francophone fut appelée à quitter la ville qui était son siège depuis plus de cinq siècles. Le corps principal de l'Université était dès lors amené à s'installer en pays roman et la région d'Ottignies dans le Brabant wallon fut retenue. Un site essentiellement campagnard fut choisi: celui du Plateau de Lauzelle occupant 800 ha. Les autorités universitaires, confrontées à un problème de transfert intégral et rapide, choisirent finalement, avec l'approbation des instances administratives, la solution d'installer leur vieille Alma Mater au sein d'une ville nouvelle, polyvalente, entièrement à créer: Louvain-la-Neuve. Fort opportunément, les bureaux d'étude chargés des approches d'implantation estimèrent qu'il leur convenait de disposer d'un maximum d'informations écologiques sur un large périmètre où serait fait le choix définitif du site urbain et de ses annexes. C'est le Laboratoire d'Écologie végétale qui fut chargé de réaliser cette étude préalable portant sur une surface de quelque 5000 ha. Une contrainte nous était imposée: la célérité! En effet, la mission nous était attribuée en juin 1968 et nous devions fournir nos rapports et documents finals dès le mois d'octobre de la même année (Anon, 1968). Toute une équipe du Laboratoire, y compris d'ailleurs des étudiants de Licence, fut appelée à participer à cette activité. Outre les documents topographiques habituels et les photographies aériennes, nous disposions encore d'une série de cartes socio-économiques résultant déjà des diligences des organismes de planification et, – circonstance heureuse – de la planchette pédologique déjà levée mais encore inédite, couvrant la région (Van Wambeke 1970). Une carte phytosociologique d'un secteur pas tellement éloigné était aussi disponible (Dethioux 1959).

L'aire étudiée s'étend sur une partie de la vallée de la Dyle et sur une portion de son bassin méridional surtout. C'est une zone de sables et argiles tertiaires, souvent calcaires mais promptement décalcifiés dès qu'ils affleurent, recouverts d'un manteau rarement fort épais de limon quaternaire nivéo-éolien. De profonds vallons latéraux découpent ce socle et entaillent le plateau dont l'altitude moyenne est nettement supérieure à 100 m.

La première tâche consista dans le levé d'une carte phytosociologique au 1/20.000ᵉ. Malgré qu'il s'agisse d'une région essentiellement agricole, à l'exception de la vallée elle-même fort urbanisée et industrialisée, la diversité de la végétation est appréciable. Quelques bois intéressants couvrent encore les vallonements latéraux et divers bosquets plus ou moins rudéralisés subsistent sur le plateau.

Sans tenir compte des groupements représentés à l'état purement linéaire ou ponctuel, 27 communautés végétales furent cartographiées. Parmi celles-ci, 15 associations ou sous-associations différentes se rapportent à des types forestiers ou subforestiers. Certains de ceux-ci présentent même un intérêt exceptionnel pour le territoire géobotanique concerné.

Très classiquement, et sur la base des séries dynamiques reconnues sur le terrain et interprétées à la lumière des observations originales et des données de la bibliographie, une carte des végétations potentielles fut esquissée; elle prend naturellement appui sur le plan phytosociologique lui-même. Six types fondamentaux sont ainsi représentés.

Un troisième genre de carte fut encore élaborée. Sur la base des observations malheureusement sporadiques réalisées au cours de la prospection, en fonction aussi des degrés de pente du terrain et de son exposition, en estimant les écarts des flux d'ensoleillement, en tenant compte également de l'allure du relief et des abris naturels, on a tenté de représenter la répartition des mésoclimats. S'agissant d'une large vallée, relativement humide à brouillards fréquents, d'un plateau découvert et très venteux et de vallons latéraux encaissés dont les flancs manifestent un violent contraste d'exposition, un tel document montrant des différences fort appréciables, nous a paru s'imposer pour être mis entre les mains des aménagistes et des urbanistes.

Il importait aussi que soit fourni un état de la salubrité de l'air. Dans l'ensemble, la région est relativement saine sous ce rapport encore que dans les vallées de la Dyle et de certains de ses tributaires existent des sources de pollution industrielle. Par ailleurs, les zones fort dégagées peuvent être influencées par des émanations éloignées. Une carte de la pureté de l'air a donc été dressée sur la base de la richesse et de la composition spécifique des communautés d'épiphytes corticoles, Lichens et Bryophytes surtout, relevées sur les arbres fruitiers ou d'alignement. Un indice de pureté atmosphérique est calculé en chaque point de sondage. Les valeurs obtenues sont finalement regroupées en classes et représentées cartographiquement (Iserentant & Margot 1974). Les résultats ainsi dégagés ont notamment bien mis en évidence l'effet protecteur des rideaux d'arbres et des grosses haies par rapport aux vents dominants transporteurs de fumées ou émanations polluantes. On s'en est largement inspiré pour fonder certaines recommandations d'aménagement écologique.

Au cours des investigations sur le terrain, tout fait notable est consigné et cartographié dès lors qu'il peut être mis à profit pour l'organisation du terroir. Toutes les suggestions d'ordre écologique ou biologique complètent cet inventaire qui fait aussi l'objet d'un carton supplémentaire.

Toutes ces informations ont-elles finalement été retenues; ont-elles servi au choix et à l'utilisation de l'espace? Architectes, urbanistes, ingénieurs, paysagistes, forestiers et horticulteurs, auteurs et responsables des réalisations en cours ont-ils tenu compte du message que nous leur avions délivré? Dans une large mesure, certainement! Un seul trait à cet égard, mais significatif en lui-même. Nous avions proposé la constitution d'une réserve naturelle sur un espace limité mais suffisamment protégé, où les types de biocénoses les plus notables étaient bien représentés. Cette suggestion a été intégralement suivie par les autorités et les aménagistes, à notre entière satisfaction.

Plan d'aménagement socio-touristique

La petite ville de La Roche-en-Ardenne, blottie dans une large courbe de la vallée très encaissée de l'Ourthe, est un centre touristique très fréquenté. Sa population de quelques 2000 âmes décuple durant les mois d'été. La municipalité possède un patrimoine forestier de plus de 800 ha, bien géré, comme le veut la Loi, par l'Administration des eaux et forêts. Très préoccupés naturellement par tout ce qui touche au tourisme, le Syndicat d'initiative et l'échevinat compétents de la ville, ont mis sur pied diverses réalisations destinées à intéresser les amis de la Nature, toujours plus nombreux parmi les villégiateurs. Un 'Club d'Écologie' aussi, récemment fondé, souhaitait que le domaine communal serve de cadre propice à l'initiation de ses membres comme de centre d'accueil et d'information pour tous les vacanciers.

Notre Laboratoire d'Écologie végétale poursuit précisément – et depuis plusieurs années – un programme de recherches au Plateau des Tailles, vaste sommet qui surplombe la vallée de l'Ourthe notamment. C'est pourquoi le Président du Syndicat d'initiatives de La Roche pria nos collaborateurs d'élaborer un plan d'aménagement socio-touristique et culturel de son domaine communal (Dumont & Stein 1973).

Une fois encore, c'est la carte phytosociologique qui a servi de point de départ à l'étude. Plus de la moitié de la surface boisée est encore couverte de feuillus. La hêtraie domine largement sous divers faciès propres à la Moyenne-Ardenne. Par places, se rencontrent encore d'anciens taillis de chêne à écorce, lesquels étaient autrefois soumis à l'essartage. On recourut encore à cette pratique ancestrale, très localement et en quelques points de la région, durant la dernière guerre. Erablières de ravin, frênaies et aulnaies alluviales se rencontrent également. En nombre d'endroits,

dans la vallée surtout, la hêtraie est remplacée par sa forme secondaire de substitution: la chênaie à charmes.

L'étude aboutit au choix d'un 'canton privilégié' qui, à proximité de la ville est aisément accessible, réunit un maximum de diversité, d'intérêt biologique, de valeurs esthétiques ou didactiques et de ressources récréatives. Plusieurs éléments y étaient déjà rassemblés: parcs d'animaux sauvages indigènes, arboretum . . . De nombreuses recommandations ont été faites pour corriger ou améliorer ce qui peut l'être, pour créer du neuf et mettre en valeur tout ce qui concourt aux objectifs indiqués. Il serait fastidieux de les détailler ici. Mentionnons cependant qu'un souci particulier a porté sur la conservation ou la restauration de toutes les emprises sur la nature qu'impliquaient les usages du passé: anciennes terrasses de culture ou d'irrigation disparaissant sous l'envahissement des fourrés ou plantation d'essences de rapport, traces au sol de l'écobuage et de l'essartage, ravinements dûs aux passages répétés des charriots des charbonniers d'antan, modes divers de traitements forestiers comme le taillis simple en voie totale de disparition . . . Toutes les formes intermédiaires de végétation devraient ainsi être maintenues. C'est un conservatoire archéologique de la nature, en quelque sorte, que l'on a visé à créer, en même temps qu'une zone d'intérêt scientifique incontesté.

Les projets d'aménagement à base écologique portant sur l'ensemble du domaine ont surtout concerné certains aspects parfois négligés. Parmi ceux-ci, on épinglera les suivants: Le mode de chasse tel qu'actuellement pratiquée doit être entièrement transformé en une méthode rationnelle de maintien d'équilibre entre phyto- et zoocénoses. Là où les enrésinements ont été faits à mauvais escient: zone de sources, fonds de vallons notamment, ils seront supprimés. Des dégagements seront effectués partout où ces plantations masquent les points de vue. Les lisières ou manteaux forestiers naturels, si accueillants aux populations biotiques les plus diverses, seront maintenus ou favorisés, assurant ainsi une transition harmonieuse entre types différents d'unités contiguës du paysage. Cette règle vaut surtout pour les périmètres si géométriquement tranchés des plantations résineuses. Enfin, de nombreuses autres dispositions ont pour but d'améliorer la propreté du domaine et d'éviter toute source de pollution.

Plan de remembrement agricole

Le troisième et dernier exemple présenté a trait à une 'évaluation des sites' dans le cadre d'un remembrement agricole. Des dispositions réglementaires récentes imposent, en effet, que toute réaffectation des terres dans une telle entreprise soit précédée d'une appréciation des sites naturels ou historiques qui risqueraient d'être touchés. Il revient alors aux aménagistes de prendre leurs décisions en toute connaissance de cause. Le Laboratoire d'Écologie végétale a tout récemment entrepris une telle évaluation conformément au cahier des charges préparé par le Comité de remembrement (Dumont 1976).

On pourrait s'étonner qu'une Institution universitaire, à vocation de recherche, entreprenne une telle réalisation. En fait, l'évaluation des sites et des paysages est l'un des fondements de l'aménagement du territoire, mais sa pertinence et sa portée ne sont clairement apparues que dans des temps récents. Il est certain aussi qu'aucune méthode ne s'est définitivement imposée jusqu'ici. Il est douteux d'ailleurs qu'une façon unique puisse s'imposer car les critères utilisés sont nécessairement différents selon les circonstances et les objectifs. Il s'agit donc encore d'un domaine incomplètement défriché où la recherche appliquée doit demeurer vivace et active.

Le périmètre considéré (Haneffe-Jeneffe) s'étale sur un peu plus de 2000 ha englobant diverses sections d'une douzaine de Communes (avant les fusions de 1977). Situé au coeur de la Hesbaye, le pays, au relief peu accusé, est néanmoins formé d'une succession de mamelons.

Le substrat géologique est constitué par le Crétacé supérieur recouvert d'un placage souvent très épais de loess quaternaire. Quelques vallées peu profondes drainent la contrée; plusieurs d'entre elles sont maintenant asséchées; leur thalweg est sédimenté par des alluvions et colluvions récentes.

Une première carte fut dressée en cabinet, d'après les dossiers disponibles, géologiques et pédologiques entre autres, indiquant la nature des matériaux pédogénétiques. Nous avions supposé, au départ, que des écarts de fertilité native auraient pu différencier des terroirs, notamment là où le manteau limoneux moins épais n'aurait pas oblitéré l'influence du socle sous-jacent. Des investigations sur le terrain ont montré qu'il n'apparaissait point de divergences notables sous ce rapport.

Toute la région est dévolue à la culture des céréales et de la betterave à sucre principalement. Elle est de longue date le théâtre d'une agriculture active et la distribution des terres – hormis leur morcellement – comme l'organisation de l'habitat rural n'ont pratiquement point changé depuis plusieurs siècles ainsi que le montrent les plus anciennes cartes topographiques.

C'est dire qu'il s'agit d'un territoire bien peu favorisé, à première vue, sous le rapport de la diversité des types de végétation. Dresser dans ces conditions une carte phytosociologique paraissait une gageure. Cette prospection et ce levé ont néanmoins apporté une récolte inattendue d'informations dignes d'intérêt.

Une première étape de synthèse des renseignements ainsi accumulés a permis de dégager les unités fondamentales du paysage végétal, les sigmassociations en quelque sorte.

L'*habitat rural*, d'abord, où l'on peut distinguer le coeur du village typiquement bien concentré, avec jardins et petits vergers, entouré d'une ceinture de prés piquetés d'arbres fruitiers de haute tige prolongeant les nombreuses fermes villageoises. Ces vergers sont quasi délaissés et servent de parcours pour le petit bétail à proximité des porcheries et étables. Un plus large anneau extérieur est formé de pâtures clôturées de haies anciennes souvent, réservé au gros bétail, peu nombreux d'ailleurs car la tenure rurale typique de la contrée est à dominante culturale.

Vient ensuite la *campagne labourée* qui s'adjuge la part largement prépondérante de la superficie étudiée. La végétation commensale des moissons (froment, orge et avoine) et des cultures sarclées (betterave, maïs, plantes légumières) y est très uniforme du fait de l'emploi massif de fumure minérale et très pauvre par suite du recours généralisé et répété aux herbicides sélectifs. Mais un certain nombre de facteurs ou de sites particuliers jettent une note de diversification dans ce peuplement si profondément domestiqué. Ce sont d'abord les chemins creux entaillés dans le limon et qui traversent les mamelons. Apanage géographique très typique des terroirs hesbignons, ces anciennes voies d'accès et de circulation abritent, selon les circonstances, des végétations semi-naturelles ou rudérales et de grosses haies très caractéristiques. On y reconnaît, sans difficulté, les communautés propres aux formations préforestières et à leurs ourlets. Parmi d'autres traits encore, il en est un qui joue un rôle marquant sur la diversité floristique de la contrée. Ce sont des alignements de buttes de craie résultant du creusement de galeries souterraines dans la roche sous-jacente pour des captages d'eau alimentant la ville de Liège. Ces monticules, véritables tumuli, tranchent sur l'uniformité du couvert cultural; ils portent une végétation particulière allant de pelouses initiales et de groupements colonisateurs d'éboulis crayeux à des fourrés forestiers. Ce sont des gagnages naturels, véritables refuges pour nombre d'animaux sauvages, gîtes favorables à une foule d'oiseaux, lesquels interviennent d'ailleurs largement dans leur colonisation. Finalement – chose étonnante – la région est favorisée au point de vue ornithologique notamment, du fait de ces reposoirs comme aussi des grosses haies qui tapissent les talus des chemins creux.

Enfin, dernier groupe d'écosystèmes, les *dépressions des vallées* qui ne représentent que peu de surface dans le périmètre étudié. On y reconnaît des prairies pâturées, fraîches à humides, généralement cernées par des alignements de peupliers dont on retrouve aussi quelques massifs en plein. Des franges discontinues de saules et d'hélophytes le long des ruisselets et fossés actifs, des fragments de communautés d'hydrophytes aussi font également partie de cet ensemble.

Un premier élément d'évaluation a porté sur la richesse spécifique et biotique des divers cantons concernés. A cet effet, la carte de la zone de remembrement a été partagée par un réseau de carrés correspondant à un km de côté. Le nombre d'espèces (prospection et dépouillement de l'Atlas floristique (Van Rompaey & Delvosalle 1972)), et des communautés végétales reconnues dans chacun d'eux a fondé un premier indice. Un deuxième est basé sur la présence de plantes rares ou dignes de mention. Un troisième repose sur la diversité des écotopes et types de paysage. La combinaison de ces divers indices fournit finalement une valeur globale attribuée à chaque polygone du réseau. Condensées selon une échelle à six degrés et reportées sur carte, ces expressions chiffrées se groupent logiquement; elles représentent adéquatement la valeur biologique des sites. On notera qu'une telle méthode reposant entièrement sur l'analyse de la flore et de la végétation, sans négliger toutefois la variété des biotopes, suppose que la plante – ce que l'on a déjà admis – est bien le reflet de l'ensemble biotique. On admettra volontiers néanmoins qu'une enquête faunistique menée simultanément à nos propres recherches et portant sur certains groupes zoologiques au moins, enrichirait notablement l'information.

Il convenait ensuite d'apprécier cet élément primordial, agissant autant sur l'habitant que sur le visiteur, que nous désignons comme 'vues paysagères'. Modelé géomorphologique, couvert végétal, habitat humain, implantations techniques ou industrielles se combinent pour constituer le 'paysage'. Nous avons tenté d'en apprécier objectivement les qualités esthétiques. Il est clair que la méthode que nous allons décrire a été délibérément imaginée en fonction des conditions locales. Elle ne pourrait être généralisée sans nuances ni modifications appropriées. Les points culminants de la contrée ont été repérés et choisis de manière à couvrir régulièrement l'ensemble du périmètre. Du sommet de chacun de ces observatoires, et selon les quatres secteurs cardinaux une série d'éléments paysagers sont dénombrés: ligne de crête, couleur nettement contrastées, villages et agglomérations, types de grands écosystèmes tels que définis plus haut, tous autres termes positifs comme clochers,

moulins, chapelles ... Une échelle de points a été adoptée dont l'addition fournit un nombre global. Mais en même temps, et dans les mêmes conditions, des éléments négatifs sont également recensés: bâtisses insolites, usines, châteaux d'eau, lignes de force sur pylônes métalliques, alignements de conducteurs électriques ou téléphoniques, etc ... Le nombre ainsi obtenu est soustrait du précédent. Il peut d'ailleurs arriver que la différence fournisse un résultat négatif.

L'observation comparative, on l'aura compris, doit être faite par temps clair, en conditions à peu près identiques et à la même époque. Si l'hypothèse est respectée, les résultats obtenus par des observateurs différents, – on l'a vérifié – sont pratiquement identiques. L'objectivité de la méthode est donc garantie. Certes, on peut critiquer la portée esthétique des critères utilisés, mais le débat, dès lors, ne devient-il pas rapidement subjectif? Les valeurs obtenues pour chacun des secteurs sont reportés sur le plan et supposées semblables jusqu'à la rencontre avec un autre angle de vue, les limites étant tracées en suivant le modelé orographique. La technique visuelle à laquelle nous avons recouru répond en quelque sorte à un principe que nous empruntons à un Homme d'État bien connu (Giscard d'Estaing 1976) et que nous adaptons à notre propos: '... le point de départ (est) celui à partir duquel l'oeil cherche et aperçoit la perspective. Car tout tient à la fois dans ce que l'on regarde et dans le point d'où l'on regarde.'

Les cotations finales reproduites sur la carte des vues paysagères sont ramenées à une échelle à quatre niveaux (on pourrait naturellement en choisir davantage). Elles se regroupent aussi d'une façon cohérente.

Une dernière carte, enfin, constitue un 'plan de situation' requis d'ailleurs par les termes du contrat. Sur un parcellaire cadastral au 1/5.000ᵉ, les renseignements essentiels déjà obtenus et expliqués ci-avant sont reportés en même temps qu'une série de données précises sur les implantations de toute nature, les accidents du relief et du couvert, toute structure notable même isolée. Une légende détaillée justifie le cas échéant l'intérêt de chaque chose, qu'il s'agisse du secteur historique, culturel ou scientifique, du domaine esthétique ou de l'intérêt pratique. Il en résulte aussi des propositions précises quant à l'ajustement des écrans de verdure, au maintien à tout prix de certains sites de grand intérêt à quelque point de vue que ce soit, y compris celui de la tradition et des coutumes locales, à défaut de quoi le nouvel habit serait mal taillé.

Conclusion

A la base de chacun des trois aménagements bien particuliers et bien différents qui viennent d'être décrits se situe un facteur commun. Chaque fois, en effet, c'est l'étude phytosociologique de la contrée qui a servi de point de départ aux autres spéculations. Que celles-ci aient pu être étalées et poussées dans des directions fort différentes démontre la pertinence et l'opportunité du choix.

L'exposé avait aussi un autre objectif. Celui de montrer que les applications de la phytosociologie, notamment dans l'étude des paysages et plus généralement dans l'aménagement du territoire, sont nombreuses et en plein développement. La connaissance de la vie en commun des plantes demeure une discipline jeune et en pleine croissance. Elle n'a pas cessé de produire tous ses fruits!

Summary

Phytosociology is considered a basic biological science to be applied in concepts and realisation of land management. The significance of vegetation maps at various scales is explained. Three examples of phytosociological studies applied to land management and physical planning are discussed:

1. Plan for a new town, i.e., the city of Louvain-le-Neuve, where the new French-language University of Louvain is situated. Here, a vegetation map comprising of 27 communities was established for an area of 5,000 ha.
2. Plan for a recreation area surrounding a small village, La Roche-en-Ardenne.
3. Plan for agricultural improvement in an area of 2,000 ha in the Haneffe-Jeneffe region. Here, a biological evaluation was taken as the basis, with criteria such a floristic richness, rarity of species and ecotope diversity.

All vegetation studies were carried out by the Laboratory for Plant Ecology of the Catholic University of Louvain.

Bibliographie

Anon. 1968. Rapport sur le site et la région d'Ottignies. Laboratoire d'Ecologie Végétale de l'Université Catholique de Louvain. Louvain, 64 p., 6 cartes.

Anon. 1975. Autoroute et environnement. Laboratoires d'Ecologie Végétale et d'Ecologie Animale de l'Université Catholoque de Louvain. Louvain-la-Neuve, 166 p.

Brasseur, F., J. R. De Sloover, F. Devillez, M. Goossens, R. Iserentant, M.-F. Jouret & J. Lebrun. 1977. La végétation de la Réserve naturelle domaniale des étangs de Luchy. Bruxelles Service de la Conservation de la Nature 8, 63 p.

C.N.R.S. 1973. Colloques internationaux, 97: Méthodes de la cartographie de la végétation, Toulouse, 1961. UNESCO, Classification internationale et cartographie de la végétation. Paris, 93 p.

De Sloover, J. R. 1964. Végétaux épiphytes et pollution de l'air. Rev. Questions scientifiques (Bruxelles) 25: 531–561.

De Sloover, J. R., R. Iserentant & J. Lebrun. La renoncule à feuilles de lierre (Ranunculus hederaceus L.) au Plateau des Tailles. Bull. Soc. R. Bot. Belgique, à paraître.

Dethioux, M. 1959. Texte explicatif de la planchette Hamme-Mille. Carte de la végétation de la Belgique. Bruxelles, IRSIA, 56 p.

Dumont, J.-M. 1976. Remembrement de Haneffe-Jeneffe. Evaluation des sites. Louvain-la-Neuve, 65 p. + annexes.

Dumont, J.-M. & J. Stein. 1973. La forêt communale de La Roche-en-Ardenne. Etude écologique d'un aménagement scientifique, social et esthétique des bois et des paysages d'un territoire ardennais. Louvain, 23 p. + annexes.

Froment, A. & L. Nef. 1976. Méthodes d'évaluation écologiques des zones vertes commes base pour la gestion de l'environnement et la conservation de la Nature. Les Naturalistes Belges 57: 2–26.

Giscard d'Estaing, V. 1976. Démocratie française. Paris, Fayard.

Iserentant, R. & J. Margot. 1974. Carte de la pollution atmosphérique dans la région d'Ottignies en 1968. In: M. Goossens & J. R. de Sloover, Korstmossen en kartering van luchtverontreiniging. Extern (Hasselt) 3: 537–553.

Lebrun, J. 1968. Le rôle de l'écologiste dans l'aménagement du territoire. Bull. Cl. Sc. Acad. R. Belgique 54: 1131–1138.

Nef, L. 1975. Expression quantitative de la valeur biologique des zones vertes. Coll. Autoroutes et environnement. Louvain-la-Neuve, p. 127–134.

Noirfalise, A. 1970. Sauvegarde et traitement du paysage. C. R. Congrès sur la protection de la Nature, Féd. R. Soc. Hortic., Anvers, p. 5–18.

Renard, Ch. 1971. Les fluctuations saisonnières de la teneur en eau des diverses formations végétales en Haute Ardenne. Mém. Soc. R. Bot. Belgique 5: 139 p.

Rompaey, E. van & L. Delvosalle. 1972. Atlas de la flore belge et luxembourgeoise. Bruxelles, Jardin Botanique Nationale, 1530 cartes.

Stenbock-Fermor, K. 1973. Un problème écologique: Les oiseaux en tant que facteur dangereux des aéroports. Application à l'aéroport de Zaventhem. Louvain, 116 p., ronéogr.

Vandiest-Wallon, A. & J. R. De Sloover. 1976. Autoroutes et aménagements paysagers. Les Naturalistes belges 57, 45–60.

Wambeke, A. van. 1970. Texte explicatif de la planchette Wavre. Carte des sols de la Belgique. IRSIA, Bruxelles, 73 p.

Westhoff, V. 1971. Quelques aspects de la conservation de la Nature aux Pays-Bas. Natura Mosana 24: 33–35.

Westhoff, V. 1971. Choice and management of nature reserves in the Netherlands. Bull. Jard. Bot. Nat. Belgique 41: 231–245.

EVOLUTIONARY TRENDS IN MEDITERRANEAN FLORA AND VEGETATION *, **

Sandro PIGNATTI

Institute ed Orto Botanico, Università, Trieste, Italy I 34100

Keywords:

Ecosystems, Evolution, Mediterranean flora, Mediterranean vegetation, Paleoecology, Technological impact, Vegetational systems

Introduction

The Mediterranean Sea orginated at the beginning of the Tertiary as a result of a connection between the Tethys and the Atlantic Ocean. The Tethys was a comparatively shallow inland sea, which occupied the Central Asiatic area during most of the Mesozoic. At the beginning of the Tertiary the Tethys became gradually smaller because of the intensive evaporation under an arid or semi-arid climate and it shifted in western direction because of movements of the Earth's surface. Thus the Tethys strongly decreased in its Asiatic part (where the Aral Sea and the Caspian Sea still remain as its major relicts) and expanded westwards occupying the present site of the Mediterranean Sea.

Our knowledge of the flora which originated around the Tethys is based principally on the recent chorology. It was a continental flora with Indo-Malayan affinities and its relicts are preserved in the arid parts of Western Asia. The Atlantic Ocean originated at the end of the Paleozoic with the break-up of Gondwanaland. In the Tertiary the definitive separation between North America and Eurasia occurred and the Atlantic Ocean occupied its present site. The flora which originated around the Atlantic had an oceanic character and showed Patagonian affinities. Its relicts are chiefly preserved in the Canary Islands (Macaronesia).

When Tethysian and Atlantic waters reached each other also their floras began to mix. The Tertiary flora of the Mediterranean Basin (Braun-Blanquet 1923) shows subtropical characters in a predominance of evergreen woody plants. It was rich in *Palmae*, *Smilax*, *Lauraceae*, *Magnoliaceae* and similar groups, both of western and eastern origin. The Mediterranean Basin was characterized by a warm-temperate climate with summer drought, to which the Central Asiatic flora was ecologically adapted. On the other hand such districts with heavier rainfall as Liguria, Corsica or Colchis, were suitable also for the oceanic flora. Therefore both floras could be sources of primary importance for the present Mediterranean flora.

Evolution of the Mediterranean flora since the Tertiary

It is possible to distinguish several mechanisms of evolution in the Mediterranean flora since the Tertiary.

Speciation as a consequence of the appearance of new land masses

Geological and geomorphological (partly also climatic) mechanisms, which cause the formation of new niches, are particularly favourable for the fixation of new genotypes. Major geographical changes were followed by the development of large endemic floras as in the Appennines, in the Baleares, in Corsica, in the Aegean area and even on the European mainland (e.g. in Eastern Pyrenées, in the South-West Alps, in the Apuanic Alps, in Mount Pyndus and Mount Olympus).

Geographical segregation

Many taxonomical groups show geographical vicariance, which could be explained by a monophyletic origin followed by geographical fragmentation. For very old groups with a relict character, like *Pinus nigra* sensu latissimo or *Abies*, monophyletism combined with fragmentation seems to provide an efficient explanation for the appearance of new

* Nomenclature follows Pignatti (1978).

** Contribution to the Symposium on Plant species and plant communities, held at Nijmegen, 11–12 November 1976, on the occasion of the 60th birthday of Professor Victor Westhoff.

157

species. In other polymorphic groups such as *Centaurea cineraria*, *Armeria sardoa* or *Limonium multiforme*, the geographical differentiation remains below the species level (gamodemes).

Reduction of the vegetative apparatus

Series of reduction from a woody to a herbaceous habitus, from evergreen to summergreen and from perennial to annual can be observed in numerous groups in the Mediterranean flora, e.g. *Euphorbia, Alyssum, Biscutella, Iberis, Dianthus, Bupleurum, Potentilla, Primula, Centaurea, Androsace, Hieracium, Ononis, Medicago, Anthyllis, Viola, Stachys*, etc. (Meusel 1970, Pignatti in press, etc.). In most cases the original types (perennial, woody, evergreen) have more or less local, the reduced types broader distribution areas. Such cases of reduction are generally explained as an adaptation to more severe ecological conditions, but in this case it is very unlikely. During most of the Tertiary, the Mediterranean Basin had an arid climate, which was more favourable for therophytes than the present warm-temperate climate. The reduction of vegetative structures seems nevertheless to be a very general tendency in the Mediterranean flora.

Polyploidy, hybridisation, apomyxis

Speciation by polyploidy (partly as hybridogenous allopolyploids) has been described in many groups of the Mediterranean flora, e.g. *Galium* (Ehrendorfer), *Limonium* (Dolcher and Pignatti), *Campanula* (Damboldt, Podlech), *Biscutella* (Manton), ferns (Reichstein et al.). In many groups speciation through hybridisation has been the consequences of human activity during the last millennia, e.g. *Poa annua* (Tutin 1957), *Dactylis glomerata* (Stebbins & Zohary 1959), *Diplotaxis* (Harberd & McArthur 1972). In many groups a higher degree of allopolyploidy leads to apomyxis which exhausts the evolutionary chances.

The four mechanisms mentioned above and some others presumably of less importance (climatic variations, variations in coast-line and substrate as a result of transgressions and disappearance of old seas like the Transaegean trench or the former gulf which has become the present Po plain, etc.) are supposed to have acted synergetically in the evolution of a Mediterranean flora. (recent reviews in Davis et al. 1971 and in Raven 1973). The result is a very rich and diverse flora (at present more than 20.000 species) with a high degree of endemism.

Mediterranean vegetation at the end of the Pliocene (phase I)

Although the evolutionary patterns of the Mediterranean flora are identified, comparatively little is known about the vegetation history. Palynology has given fragmentary and partly contradictory results. The actual vegetation zones in the Mediterranean Basin (from the warmest to the coolest: evergreen forest, deciduous forest, alpine grasslands and scrub) are lacking or incompletely developed in most refugial areas (Sardinia, North Sicily, Crete) and seem to be of comparatively recent origin. Correlation between palaeontological and chorological evidence leads to the hypothesis that during the Tertiary the Mediterranean Basin was mostly covered with evergreen vegetation types. Traces of Pleistocene glacial activity in the Mediterranenan Basin are only found in the highest mountains, e.g. Rillo Planina and Gran Sasso. The deciduous forest, chiefly composed of *Quercus* spp. and *Fagus* became important in the cooler areas and mountain habitats during the Pleistocene as a result of climatic changes. At the same time a steppe-like vegetation developed along the coasts. With the gradual warming of the climate during the postglacial time, deciduous vegetation diffused gradually. The last considerable changes in vegetation have been caused by human activities: the forests were destroyed and replaced by scrub, grasslands and weed communities.

For a reconstruction of the plant cover of the Mediterranean Basin at the end of the Tertiary (Pliocene) some assumptions should be made:

1) The geographical situation was more or less similar to that at present. The Mediterranean at that time was a large inland sea connected with the Atlantic Ocean and containing many islands. The Iberian peninsula had already completely formed. The Balcanic peninsula, on the contrary, was less extensive than at present. The Appennines already existed in the Pliocene, but they were lower than at present and initially had an insular character. Land bridges connecting Africa with Europe did exist from time to time (chiefly between Sicily and Tunisia). Similarly the Balkans and Western Asia were connected (Termier 1960, Pasa 1953).

2) The climate was somewhat warmer than at present. A fossil Tertiary flora of subtropical character with predominance of evergreen Angiosperms has been found in Central Europe and fossils of deciduous woods are frequent even in Greenland. Therefore, the Pleistocene Mediterranean Basin belonged to the tropical zone. Since the climate was not constant, in that period (Emiliani 1955) an alternation of arid and humid phases must be inferred. During the

humid phases the climate was favourable for evergreen forests in most parts of the Mediterranean, while during the arid phases the evergreen forest was reduced to the advantage of xerophilous grasslands.

3) The flora was richer in tropical elements. The following tropical families and genera are frequent as fossils in the Mediterranean Basin (Mai 1965):

Anacardiaceae	*Icacinaceae*	*Menispermaceae*
Apocynaceae	*Malpighiaceae*	*Nyssaceae*
Burseraceae	*Mastixiaceae*	*Symplocaceae*
Calamus	*Eurya*	*Mallotus*
Cinnamomum	*Ficus*	*Musa*
Diospyros	*Lagerstroemia*	*Pandanus*
		Sterculia

Most of them contain only one or a few species sometimes with a restricted distribution in refuge areas, or they are entirely extinct in the present Mediterranean Basin. Under these tropical conditions it may be expected that a laurophyllous forest covered the Mediterranean coasts and the lower parts of the mainland even at the end of the Tertiary. The present vegetation (*Quercetea ilicis*) of these territories shows a similar composition. In fact the highest concentration of tropical relict species is found in the associations of *Oleo-Ceratonion* and *Quercion ilicis*, which presently form the climax vegetation at lower altitudes along the Mediterranean. Thus the *Quercetea ilicis* associations can be considered as laurophyllous vegetation types which survived since the Tertiary.

The tracing of mountain vegetation seems to be more difficult. In the largest mountain areas surrounding the Mediterranean (Pyrenées, Appennines, Balkans, Atlas, etc.) a large complex of deciduous forests, chiefly composed of *Quercus* and *Fagus*, has developed above the evergreen forest. This corresponds to the vegetation features of East Asia or North-East America under similar ecological conditions. Moreover these vegetation types seem to be relatively recent, judged by the few Tertiary relicts and species with tropical affinities that are found in there.

In view of the evidence summarized it is possible to postulate the existence of an evergreen vegetation complex with temperate character in the mountain areas of the Mediterranean at the end of the Tertiary. This „*Ilex-Taxus*" belt occupied the ecological space of the present deciduous forests. Its widest distributed surviving elements are *Ilex aquifolium, Taxus baccata, Daphne laureola, Ruscus, Buxus*, etc. Some facts may support this hypothesis:

a) The components of the *Ilex-Taxus* belt are evergreen and have tropical affinities although they are adapted to a temperate climate.

b) They are widespread all over the Mediterranean area, from Marocco to the Caspian Sea, be it mostly as secondary components of the deciduous forests.

c) Only in refuge areas, like the Baleares (Bolós & Molinier 1958), North Sicily on the Madonie (Di Martino, Marcenò & Raimondo in press), and Corse (Gamisans 1975), the elements of the *Ilex-Taxus* belt are locally dominant.

d) Floristical disjunctions as in the case of *Rhododendron ponticum* (presently in Portugal and Caucasus, but fossil also in intermediate areas) and *Rhododendron luteum*, or limited distribution areas as in *Prunus laurocerasus*, make it likely that the *Ilex-Taxus* belt has been more widespread in earlier times than at present.

Other vegetations types of the mountains represent an open community of conifers (*Pinus, Abies, Cedrus*) on the higher mountain slopes or formations of spiny shrubs (*Astragalus, Genista*) on the summits. Such vegetation types are well conserved on the southern mountains of the Mediterranean Basin, like Atlas (with *Cedrus atlantica* and *Erinacetalia*), Sierra Nevada (with *Abies pinsapo* and *Astragalus nevadensis*), Mount Olympus in Thessalia (with *Abies cephalonica* and *Astragalus angustifolius*) and Anatolia (with *Abies, Cedrus* and many spiny species of *Astragalus* and *Acantholimon*).

It is suggested that the climax zones in the Mediterranean Basin during the Pliocene were as shown in Fig. 1.

The vegetation and landscape of the Mediterranean Basin during the Pliocene can be regarded as an ecosystem, which has remained in a steady state during a very long period. This means that the climax associations occupied most of the surface available and specialized associations were limited to marginal habitats such as rocks, crevices, gravel plains, coastal cliffs and dunes, salt marshes, etc. The energy input in the system is high and the production

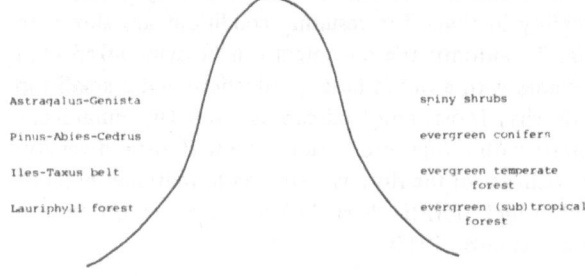

Fig. 1 Vegetation zones in the Mediterranean Basin at the end of the Pliocene (Hypothetical)

159

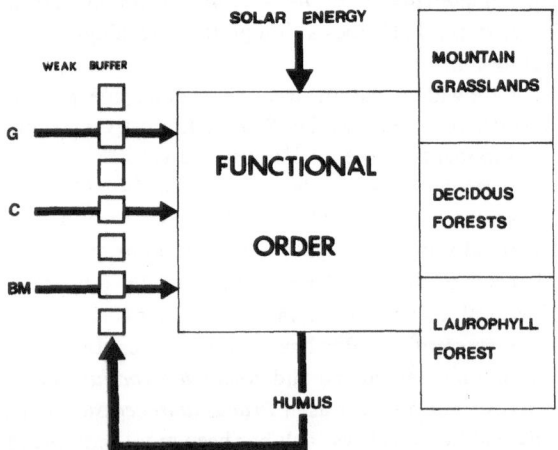

Fig. 2 The Mediterranean vegetational system during the Pliocene.

of organic matter under climax conditions is correspondingly high; most of the organic matter is accumulated as humus and thus conserved by the ecosystem (Fig. 2). The speciation mechanisms mentioned above, are supposed to have worked only on a low level in this climax ecosystem, because the major geological-geomorphological and climatological differentiation were counteracted by the humus in the soil and the vegetation respectively. One could say: the information available for the evolution of plant species is filtered by the buffering sub-systems and the speciation is repressed. It is a common belief that climatogene vegetation types are not favourable for the appearance of new species. The now existing complexes of ,,virgin forests'' (climatogene associations plus dependent associations) under temperate or subtropical climates include no more than 150–300 species of vascular plants; evolutionary processes are not effective beyond the subspecific or apomyctic level e.g. *Ranunculus auricomus, Hieracium sylvaticum*. This repressing effect was particularly great in the Pliocene ecosystems, characterized by a very high stability in time. The resulting conditions are shown in Fig. 7 (bottom): the ecosystem can be symbolized by a pyramid with a broad base (production) and a small top (diversity). Most of the landscape is covered by climatogene forests with a high production by low floristic diversity; the richness of the flora is restricted to marginal habitats with a high floristic diversity but a very low production (see Dickinson 1934).

The thermal 'crisis' of the Pleistocene (phase II)

The ice age is represented in the Mediterranean by cold periods, synchronized with Mindel, Riss and Würm; the evidence for a fourth period (Günz) is not clear. During the cold periods there was no formation of vaste ice sheets in the Mediterranean but only reduced circus activity on major mountain systems in the Spanish Sierras, in the Central Appennines and Balkans.

The climate of the ice ages was cool, probably 5° to 10 colder than at the present time, but not particularly humid. In the Mediterranean Basin there are no signs of pluvial periods. Secondary effects were the lowering of the sea level (up to about 300 m during Riss) and the emersion of new lands chiefly in the North Adriatic, around Sicily and in the Aegeis, as well as the erosion of stream valleys. Interglacial phases were somewhat warmer than nowadays and during those periods the sea level rose up to 10 m higher than at present. It is evident from pollen spectres that there is a maximum of *Artemisia* and *Chenopodiaceae* in the coolest phase, which probably corresponds to an extension of the (salt?) steppe. The glacial refuges of the South European forest did not exist as a zonal vegetation belt but only as isolated vegetation islands.

It seems that the vegetation zones were similar to those at present, but evidently with limits at lower altitudes (Fig. 3). The difference with regard to the Pliocene consists chiefly of the development of a belt of deciduous forests between the laurophyll evergreen forest and the evergreen belt of conifers. This deciduous forest expanded mainly in the ecological space of the *Ilex-Taxus* belt which lost importance and became a diffused vegetation type. Some elements of the evergreen temperate forest (like *Rhododendron*) disappeared nearly entirely from the Mediterranean Basin, others like *Ilex* of *Taxus* were absorbed as subordinate elements in the deciduous forest. Even some species of conifers (possibly newly segregated at this time) like *Abies alba* or *Pinus* gr. *nigra* were absorbed

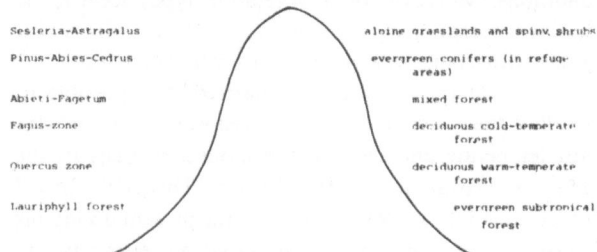

Fig. 3 Vegetation zones in the Mediterranean Basin during the glacial era (partly hypothetical).

in deciduous vegetation types. This phenomenon is particularly evident in Sicily where it is possible to compare the montane vegetation of the Madonie and that of the Etna.

In the Madonie the *Fagus*-zone is comparatively little developed and there are clear relics of the *Ilex-Taxus* belt, even a population of the endemic *Abies nebrodensis*. On the Etna the vegetation has a more Middle European character, with *Fagetum* and even a *Betula* community and without vestiges of the *Ilex-Taxus* belt. The difference is caused by the vegetation history: the Madonie (a very old mountain system) has a well-developed Tertiary vegetation, which partly has survived the ice age; the Etna with a maximum age of 1 million years, on the contrary, shows a completely post-glacial vegetation.

From an ecological point of view (Fig. 4) the Pleistocene was a period of rapid and radical changes. Each cool period brought two reciprocal climatic variations in ca. $1-2.10^5$ years, each of them larger than the complete climatic variation during the Tertiary (i.e. in $5-8.10^7$ years). Thus, the speed of these climatic changes was ca. 500–1000 times higher than in the Tertiary. The energy input of the system was strongly reduced and the production of organic matter became much lower. The strongly buffered vegetational system disappeared and its place was taken by new types of more open forest or often even by scrubs (e.g. the *Corylus*-phase at the end of Würm) or grasslands. New ecological niches appeared as a consequence of geographical, climatic and vegetational changes. The feedback effect by the accumulation of humus or stratification of the canopy was lowered. Consequently the evolutionary processes were stimulated. The general feature of the system remains the pyramid shown in Fig. 7

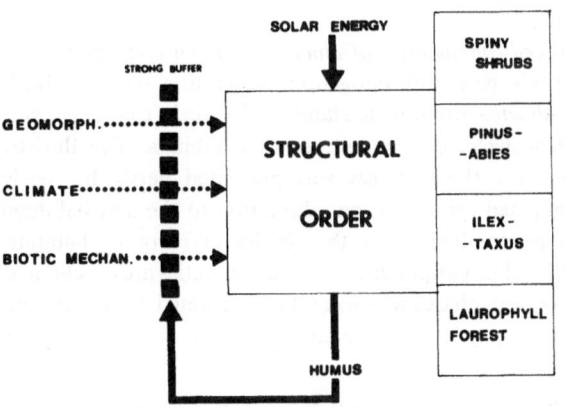

Fig. 4 The Mediterranean vegetational system during the glacial era.

(bottom), but in this case the ecosystem accumulates structure chiefly as temporal organization i. e. as coordination in time of the primary productivity (summergreen vegetation) and of the reproduction (synusial waves of flowering).

Man as a factor in forming vegetation and landscape (phase III)

At the end of the ice age Neandertal man appeared in the Mediterranean Basin and soon afterwards also Grimaldi man, a *sapiens* race with negroid characters. They formed loose settlements near the sea and lived as gatherers and hunters: their influence on the ecosystem was negligeable and also the vegetation of that time can be regarded as completely natural. The period between the end of the last cool phase of the Würm and the present (Postglacial) consisted of almost 20,000 years, in which the climate became progressively milder and (between 5000 and 2800 B.C.) even warmer than at present. This period was characterized by strong changes in vegetation and landscape, which occurred with ever-increasing speed because of the action of a new ecological factor: man. From this time onwards the description of the Mediterranean ecosystem on the basis of mere natural factors (climate, vegetation, etc.) will be incomplete: the features of the vegetation and landscape can be understood only if man too is included within the major factors of the system.

The vegetation zones reached their actual features: the *Quercetum ilicis*, an evergreen forest composed of the remnants of the Pliocene laurophyll forest, spread along the coasts and in the neighbouring mountains. On the mountains there was an extensive belt of forests dominated by deciduous oaks (*Quercus valentina* and *Q. faginea* in the Iberian peninsula, *Q. toza* and *Q. mirbeckii* in North Africa, *Q. farnetto* in Southern Italy and the Balkans, *Q. infectoria* in Anatolia etc.) and on higher levels by *Fagus sylvatica*. The forest limits were represented by *Fagus* alone or in mixture with conifers (chiefly *Abies alba*). In the southern part of the Mediterranean Basin the *Quercetum ilicis* was partly replaced by the *Oleo-Ceratonion* vegetation which covered large areas in North Africa. In the northern part of the present Sahara up to Fezzan, Hoggar and Tibesti, there existed a broad belt of grasslands and in the mountains even xeric forests. It is likely that the Sahara had a temperate climate and a landscape with forests, lakes and streams during the ice age as a consequence of the cooling. Paleontological and geomorphological evidence in this respect is abundant. On the other

161

hand, it is quite unexpected that grasslands persisted in the Northern Sahara during most of the Postglacial, after Frobenius (1933) up to 10,000–5,000 B.C., i.e. up to the beginning of historical times and even at a time when the climate in Southern Europe was warmer than nowadays. The inhabitants of the grassland belt were of the Grimaldi race, which lived as hunters and pastoralists of semi-domesticated ungulates: sheep, goats, bovids, antelopes and even buffalos (the latter require a comparatively moist environment). The history of these peoples is preserved in some graffito designs at many sites in this area. Before the beginning of the Egyptian civilisation (4000 B.C.) the Sahara became too dry to sustain any form of human life and negroid Grimaldi man disappeared. The later human settlements (by different peoples of the white race) occurred in islands and along the coasts, first in the eastern part of the Mediterranean. Agricultural societes with cultivation of cereals, extensive grazing and urbane development have existed in Egypt from the fifth millennium B.C., in Syria and Palestine from the end of the third millennium, in Anatolia and Crete from the second millennium; in Greece, Mykenes already existed in 1500 B.C. and can be considered the most ancient town in Europe. Italy and North Africa became densely populated at the end of the second millennium, Spain somewhat later.

The strategy of man in occupying the different territories in the Mediterranean area was always the same: the original forest was destroyed because wood was needed as fuel and as building material for houses and ships. The deforested lands were exploited for grazing and fire was used to prevent regeneration of the forest. The best soils were used for agriculture, which led to the accumulation of food and richness, development of trade and development of urbane systems. The largest towns in Egypt and

Mesopotamia had some hundreds of thousands of inhabitants in their period of maximum expansion, Athens, Carthago and Alexandria a million, Rome even 3 to 5 millions inhabitants. It is not difficult to imagine how large an area of land must have been exploited (with the primitive methods of the ancients) to sustain a town with a hundred thousand or a million inhabitants.

The ecosystem (Fig. 5) was greatly different from the preceding types. The energy input was reassessed on a high level, but the organic matter produced could not be used for the feedback of the system, because it was used by man and in many ways dissipated (e.g. typical uneconomical food chain herb-sheep-meat-food for man). The forest vegetation disappeared over large surface and was substituted by other vegetation types:

*Macchia (Maquis)** – Evergreen shrubs, from 2 to 6 m high, dense and impenetrable; by cutting every 10–20 years man obtained the raw material for producing charcoal. The vegetation was not very different from the natural one (*Quercetum ilicis, Oleo-Ceratonion*) but steadily conserved in a juvenile condition.

Garigue (Gariga, Garriga) – Evergreen and mostly aromatic shrubs, from 0,5 to 2 m high, alternated with small or large grasslands; shrubs were regarded as valueless; the grasslands were exploited as pastures for sheep and goats. Better pastures had a dense cover of perennial grasses and showed the tendency to regenerate to woody vegetation; consequently they were often burnt and the vegetation changed to unstable communities of annuals or geophytes. Entirely new semi-natural vegetation types appeared on this habitat: the shrub vegetation of *Ononido-Rosmarinetea* (on limestone) as well as *Cisto-Lavanduletea* (on acid soils) and grasslands of *Thero-Brachypodietea* (on limestone) as well as *Helianthemetea guttati* (on acid soils).

Weed-communities of annual plants on cultivated areas. Entirely new anthropogenous vegetation types evolved: *Secalinetea* on cultivated lands with cereals, *Chenopodietea*, *Artemisietea*, etc. on nitrate rich habitats. The floristic stock for these classes was presented partly by newly segregated species (as an adaptation to the new habitats) but predominantly by the species of marginal habitats.

Macchia, garigue and weed-communities are proclimacic vegetation phases which can be interpreted as disclimaxes. In a very general sense the evolution of the vegetation

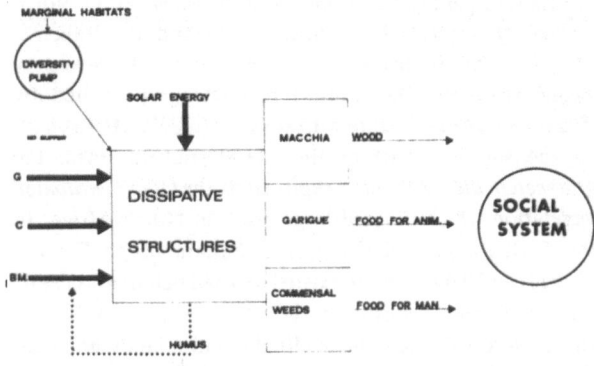

Fig. 5 The Mediterranean vegetational system during the ancient age and middle age (polis phase).

* Classical treatments of these vegetation types are presented in Braun-Blanquet (1933, 1936), Bharucha (1933) and Rikli (1942–1948); for a recent survey cf. Di Castri & Mooney (1973).

follows a logistic curve (Fig. 4 in Feoli, Lausi & Pignatti 1975); the production rate, expressed by the slope of the curve is low in A–B, which corresponds to the climax and high (very steep curve) in Al.Bl, which corresponds to proclimacic phases. Therefore, in this type of ecosystem the production is high, and also the uptake (or destruction) of organic matter by man is high. The ecosystem is charac- for vegetation disappeared over large surfaces and was

As man destroyed the forests, an ecological 'vacuum' was created which 'absorbed' new species from the neigh- bouring habitats. A severe selection took place during this colonization: only well adapted species or sufficiently variable complexes could occupy the new ecological niches, whereas others remained in the marginal habitats and became relicts. In this way, the floristic diversity which was accumulated during the Pliocene and the ice age expanded over most of the Mediterranean Basin and became a main feature of its landscape. It is a matter of fact that the appearance of new vegetation classes was a necessary consequence of the speedy turnover in the ecosystem. We can recognize in this case the formation of *dissipative structures in the ecosystem* in analogy to dis- sipative structures described by Prigogine (1967) for thermodynamic systems of the irreversible type.

The general feature of the ecosystem can be symbolized by an inverse pyramid: the highly ordered climatogene associations were reduced to small areas (marginal areas with regard to the human habitat) and associations with high diversity occupied most of the Mediterranean Basin: this was the reverse situation of the previous time (see Fig. 7).

The social system too can be represented by an inverse pyramid. The Mediterranean culture showed a high diversity. The atomistic unit of most cultures in this area was the fortified town, built on the top of a small hill and surrounded by a stone wall. This model can be seen from Mykenes to Athens, Carthago and Rome, from the towns of Phoenicians and Etrusks to the *polis* of the Greeks, the *civitas* of the Latins and even the *commune* in the Middle Ages. The centre was represented by the temple (in more recent times by the church) in which local gods and tradi- tions were worshipped; consequently each polis or civitas or commune was a small state, independent of others and with its own history and culture. Continuously changing alliances and wars hindered the dominance of one culture over the others; even when most of the Mediterranean Basin was unified in a single political unity by the Romans or Arabs, the towns conserved a broad autonomy.

The social system was based on the uptake of organic matter from the natural ecosystems, and even a flow of information from the ecological to the social system can be recognized. Local climate, geomorphology, vegetation, plant products and even fauna had an influence on the local culture, but this influence was not sufficient to account for the extreme diversity of the ancient social systems in the Mediterranean Basin. The vegetational diversity seems not to be the cause of the cultural diversity and evidently the contrary was also impossible: both diver- sites have developed independently an interesting iso- morphism, which may be caused by the same environ- mental factors.

It is important to observe a further difference: the natural ecosystems of the Pliocene and Pleistocene pos- sessed a feed-back, i.e. they were functioning more or less in a cycle. The urban ecosystems phase shows a transfer of energy from the solar radiation to photosynthetic structures, vegetation and finally to the social system: therefore it worked linearly. The flow of organic matter, minerals, energy, information, etc. was always from the environment to the social system; it was uni-directional. This caused in many cases the exhaustion of local resources and the social systems collapsed: examples of regions with an ancient culture which disappeared because of deserti- fication are Cyrenaica, Syria, Numidia, Andalusia, Sicily, Peloponnesos, etc.

Also Central Italy, strongly exploited during the Roman Empire (I–IV century), impoverished and its population (and culture) sank to a very low level: this gave rise to a widespread regeneration of the plant cover (in the V–X centuries) and later the ecosystem was again adequate to sustain a second cycle of social development. Many similar cases can be quoted. Consequently, it is possible to recognize a super-system, including both natural and social ecosystems, which depend on one another, interact- ing as it were in a 'predator-prey' relationship described by Volterra (1931). This super-system is working again in a cycle: the equilibrium is reached at a high hierarchic level.

The technological impact (phase IV)

The latest phase in the Mediterranean ecosystem begins more or less after World War II: the social system based on technology expands progressively over the whole area and this process is still going on. Optimalisation of the methods of work and nearly unlimited sources of energy (both reached by the application of scientific knowledge)

163

leads the social system to a high efficiency. Its dependence on the ecosystem is nearly overcome. Even the need for wood or edible substance is satisfied by imports from alien lands. Traditional agriculture, sheep-farming and fishing collapse and are substituted by monocultures producing for export; in most cases large surfaces of agricultural land are abandoned. A secondary cycle with regrowth of vegetation is hampered by the destructive influences of the industrial system, mainly through discharge and pollution. Further diffusion of the socio-technological structures into the natural with serious consequences for the ecosystem include: touristic exploitation of the coasts, the creation of industrial landscapes, urbanisation, reforestation with exotic woods, chemical fertilizers, errors in agricultural planning.

The natural ecosystems cannot properly respond to the new conditions imposed by the technological system. The input of information derived from geomorphology, climate and biotic mechanisms is overruled by the information derived from man (*Homo faber*, i.e. the source of technological information). Even the fixation of solar energy falls down: coastal vegetation with a high energetical input, but disturbed by pollution and human management, has a smaller production than the undisturbed mountain vegetation with low energetical input.

The production of humus and in general all natural feedback systems are strongly reduced or become completely inefficient. Consequently, the ecosystem operates linearly under the predominance of technological information. This condition can be described as the *implementation of man's mental scheme onto the environment* (Fig. 6).

It must be stressed that technological information and industrial methods of work are based on standardisation, uniformity and efficiency; they are completely independent of the local ecosystem and even of the local culture. The social system of the technological phase reproduces the pyramid which was typical of the undisturbed natural ecosystems in the Pliocene and in the ice age: maximum of productive structures, minimum of diversity. Therefore

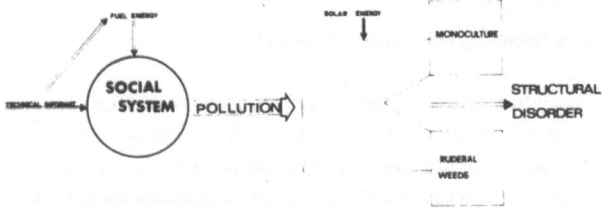

Fig. 6 The Mediterranean vegetational system during the present time (technological phase).

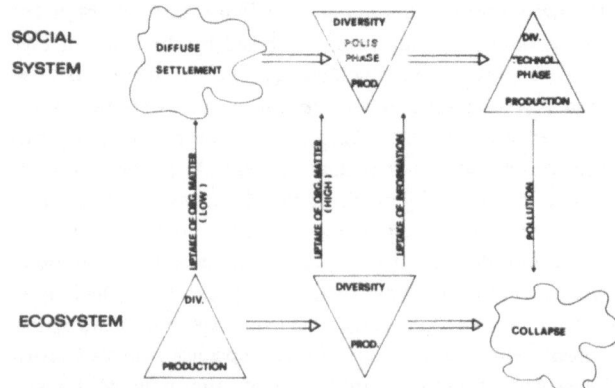

Fig. 7 Relationships between vegetational system and social system in the Mediterranean Basin.

it can be regarded as the imposition of an alien model over the Mediterranean cultures, in strong contradiction with their diversity. The synergism between the ecological and social system, as observed in the polis phase is impossible and by the steady action of the disturbing factors, the natural ecosystems are condemned to collapse (Fig. 7). The desertification of large areas caused by the disturbance of the water balance (e.g. in lower Egypt) or by the penetration of sea water (e.g. in the Lagoon of Venice) or by plantation of *Eucalyptus* and *Cupressus* (e.g. in Algeria) or by air pollution (e.g. Etang de Berre) are forebodings of a new environment in the Mediterranean Basin. The dying pines of the Pineta di Ravenna (because of the pollution of ground water) and the crumbling sculptures of the Parthenon are the proof that both Mediterranean vegetation and mediterranean culture are unable to bear the new conditions.

Polyploidy as additional evidence for human impact on Mediterranean vegetation

The development of the Mediterranean vegetation as explained above is well supported by historical evidence only for phase IV and the last centuries of phase III. For the reconstruction of the more ancient phases only little evidence from palynological and paleontological data or from the interpretation of chorological patterns are available. The explanation in terms of systems is therefore highly hypothetical, at least for phases I and II. Therefore, the problem of correspondence between our reconstruction and reality must be put forward here. A direct proof of our inferences is impossible at present and an adequate method of investigation for these problems does not exist. An indirect indication can be derived from the analysis of polyploidy.

The appearance of polyploid species is directly related to their evolutionary chances (Pignatti 1960). Therefore, the frequency of polyploids is high in open, unstable or recently developed vegetation types, whereas it is low in associations with relict character which are well integrated and stable. For a very large number of association tables the index of diploidy has been calculated following the method already proposed by us (Pignatti 1960). Indexes under 1.0 correspond to the predominance of polyploids, higher values to the predominance of diploids. Mean values for each vegetation class (in the syntaxonomical sense) have been obtained by averaging the values of the associations belonging to it. The values for the vegetation classes here discussed are:

	No of tab.	Index of diploidy
Ononido-Rosmarinetea	12	0.471
Thero-Brachypodietea	10	1.109
Stellarietea mediae	7	1.112
Querco-Fagetea	43	1.161
Artemisietea vulgaris	5	1.374
Cisto-Lavanduletea	4	1.500
Asplenietea rupestria	5	1.579
Crithmo-Staticetea	5	1.628
Quercetea ilicis	11	1.880

The vegetation classes included in Fig. 8 are ordered according to the diploids/polyploids ratio (ordinate) and the features of the vegetation (abscissa) from evergreen woody types, to deciduous, herbaceous and annual vegetation. The frequency of polyploids is in accordance with our reconstruction of vegetational history. In the *Quercetea ilicis* polyploids have the lowest values, and this suggests that this vegetation remained nearly stable for a very large time: consequently the *Quercetea ilicis* may be regarded as a Tertiary relict as supposed by our ecosystemic theory. The same can be stated for *Crithmo-Staticetea*, which contains associations of marginal habitats. Compared with the *Quercetea ilicis*, the deciduous communities (*Querco-Fagetea*) seem to be relatively young. As expected, the highest values of polyploids occur in *Ononido-Rosmarinetea* and in weed vegetation (*Artemisietea, Stellarietea*).

Some general features of ecosystems

In the ecosystems (as operative units) it is possible to distinguish some features:

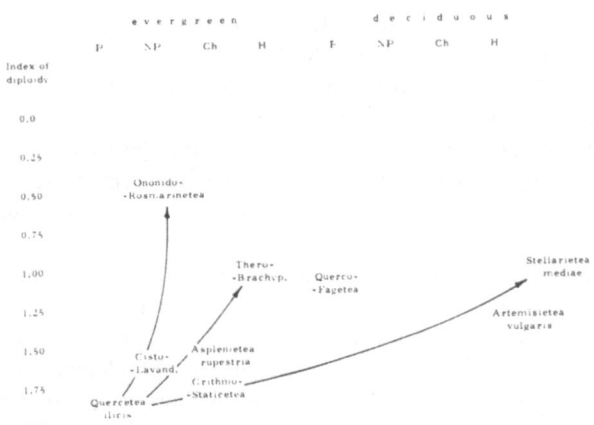

Fig. 8 Structural features of Mediterranean vegetation in relation to their index of diploidy.

cyclic – systems operating in cycles, i.e. when the organic matter produced remains in the system and is utilized by the components of the system itself.

linear – systems operating linearly, i.e. when the organic matter is transferred to alien systems.

autonomous (selfregulating) – the turnover of the system is regulated by the feed-back effects.

allonomous – the system works as a function of external factors.

With regard to the order content, the systems can be considered as:

accumulating order, in form of spatial structures or temporal organisation, diversity, etc.

transferring order to an alien system or

dissipating order, when the structures and diversity are destroyed.

An important difference exists between those systems (both ecological and social systems) which are dependent on the features of the environment and others which operate in an independent way. I propose the following nomenclature:

Ecogenous systems: solar and geothermal energy are the principal energetical sources; material and informational input are supported by the environment. The activity of the living component can be dominated by the primary producer (phytonomous) or by man's activity (anthroponomous).

Technogenous systems: solar and geothermal energy support most of the energetical input, but fossil (or atomic) energy applied on crucial points becomes dominating; material and informational input are imported from alien environments. Such systems are always anthroponomous.

The ecosystems of the Mediterranean Basin (specialized during the Pliocene and the Pleistocene) are ecogenous and of the phytonome type; they are cyclic, autonomous, and accumulate order in form of spatial structures and temporal organisation; they have the possibility of existing interminably. During the last millennia man settled in this area and built a splendid culture: the system shifted to the anthroponomous type but remained cyclic (at least as super-system) and autonomous; the order was further on accumulated as diversity: this system too has a possibility of indefinite existence. The recent evolution leads to a technogenous system, linear and allonomous, which dissipates order. Because of the linearity it cannot regenerate order and consequently it has a finite possibility of existence.

Summary

The vegetation of the Mediterranean Basin was originally composed of evergreen forests; during the Pleistocene deciduous forests expanded, chiefly in the mountains. In historical time the forest belt was strongly reduced by human activity and substituted by anthropogenous vegetation types (macchia, garigue, weed-communities). The frequency of polyploids in the present vegetation types support this interpretation. Reciprocal relationships between the vegetational system and social system are discussed and a terminology is proposed. During ancient times and the middle ages a reciprocal control of vegetation and human activity was possible (cyclic system), stabilizing the vegetation in a steady state; the technological impact modified these conditions in a linear sense, and now the vegetation is menaced by irreversible changes.

Zusammenfassung

Die Vegetation der Mittelmeerländer bestand ursprünglich aus immergrünen Hartlaubwäldern; während des Pleistozäns breiteten sich sommergrüne Laubwälder aus, besonders in den gebirgigen Teilen. Die dichtere menschliche Besiedlung nach der Steinzeit bewirkte eine Einschränkung des Waldgürtels der zum Grossteil durch anthropogene Vegetationstypen (Macchie, Garigue, Unkrautgemeinschaften) ersetzt wurde. Die Spuren dieses Prozesses spiegeln sich in den Polyploidieverhältnissen. Bis zu diesem Punkt entwickelte sich das vegetationelle System autonom. Mit dem Auftreten des Menschen beginnt eine rege Dialektik zwischen dem vegetationellen und dem sozialen System. Einige Gesetzmässigkeiten werden dargestellt und eine geeignete Terminologie wird vorgeschlagen. Während des Altertums und des Mittelalters war eine gegenseitige Kontrolle zwischen Vegetátion und menschliche Einwirkung möglich (zyklisches System), und die Vegetation wurde dadurch in einem Zustand von Fliessgleichgewicht stabilisiert; durch die Technologie wurden diese Verhältnisse verändert und es entstand ein lineares System, sodass nun die Vegetation unter der Drohung einer irreversiblen Aenderung steht.

References

(Floristic papers cited in the text are not listed)

Bharucha, F.R. 1933. Etude écologique et phytosociologique de l'association à Brachypodium ramosum et Phlomis lychnitis des Garigues languedociennes. Beih. Bot. Centralbl. II, 50: 247–259.

Bolos, O. & R. Molinier. 1958. Recherches phytosociologiques dans l'Ille de Majorque. Collect. Bot. 5: 699–865.

Braun-Blanquet, J. 1923. L'origine et le développement des Flores dans le Massif Central de France. ed. Lhomme-Beer, Paris, Zürich.

Braun-Blanquet, J. 1933. L'association végétale climatique et le climax du sol dans le Midi méditerranéen. Bull. Soc. Bot. Fr. 80: 715–722.

Braun-Blanquet, J. 1936. La forêt d'Yeuse languedocienne (Quercion ilicis). Monographie phytosociologique. Mem. Soc. Etude Sc. Nat. Nimes 5: 1–147.

Davis, P.H., Harper, P.C. & I.C. Hedge. 1971. Plant life of South-West Asia. Oliver & Boyd, Edinburgh. 335 pp.

Di Castri, F. & H.A. Mooney. 1973. Mediterranean type ecosystems. Springer, Berlin, Heidelberg, New York. 405 pp.

Dickinson, O. 1934. Les espèces survivantes tertiaires du Bas-Languedoc. Comm. SIGMA no 34. Toulouse.

Di Martino, A., Marcenò C. & F.M. Raimondo (in press). Sintesi degli studi condotti sulla vegetazione delle Madonie.

Emiliani, C. 1955. Pleistocene temperatures. J. Geol. 63: 538–578.

Feoli, E., Lausi, D. & S. Pignatti. 1975. Grundsätze einer Kausalen Erforschung der Vegetationsdynamik. In: W. Schmidt: Sukzessionsforschung, pp. 1–12. J. Cramer, Vaduz.

Frobenius, L. 1933. Kulturgeschichte Afrikas Phaidon, Zürich. 652 pp.

Gamisans J. 1975. La végétation des montagnes Corses. Thèse Univ. Marseille. 295 pp.

Lang, G., 1970. Florengeschichte und mediterran-mitteleuropäisch Florenbeziehungen. Feddes Rep. 81: 315–335.

Mai, D.H. 1965. Der Florenwechsel im jüngeren Tertiär Mitteleuropas. Feddes Rep. 70: 157–169.

Meusel, H. 1970. Wuchsformenreihen mediterran-mitteleuropäischer Angiospermen-Taxa. Feddes Rep. 81: 315–335.

Pasa, A. 1953. Appunti geologici per la paleogeografia della Puglia. Mem. Biogeogr. Adriat. 2: 175–286.

Pignatti, S. 1960. Il significato delle specie poliploidi nelle associazioni vegetali. Atti Ist. Ven. Sc. Lett. Arti 118: 75–98.

Pignatti, S. 1978. Flora d'Italia. Cramer, Vaduz. C. 1300 pp.

Pignatti, S. In press. Plant geographical and morphological evidences in the evolution of the mediterranean flora. Webbia.

Prigogine, I. 1967. Thermodynamics of irreversible processes. Wiley, New York, London, Sydney. 147 pp.

Raven, P.H. 1973. The evolution of mediterranean Floras in: Di Castri F. & H.A. Mooney (cit.) p. 213–224.

Rikli, M. 1942–1948. Das Pflanzenkleid der Mittelmeerländer, 3 vol. H. Huber, Bern, 1418 pp.

Termier, H. & G. Fermier 1960. Atlas de Paleogeographie. Masson, Paris, 100 pp.

Volterra, V. 1931. Leçons sur la théorie mathématique de la lutte pour la vie. Gauthier-Villars, 146 pp. Paris.

PLANT SPECIES AND PLANT COMMUNITIES: SOME CONCLUSIONS

M.J.A. WERGER & E. VAN DER MAAREL

Division of Geobotany, Toernooiveld, Nijmegen, The Netherlands

In this attempt to integrate the headlines and main tracks of the papers presented here we can only touch upon those of the many interesting considerations that fit directly into the general theme of this symposium. Let us start these remarks with some basic considerations on species ecology and vegetation structure.

Every plant species, or rather ecotype, has its physiological amplitude that allows it or prevents it to occur under certain environmental conditions. Since the experiments of Ellenberg (1953) and later of Van den Bergh (1968, 1973, van den Bergh & Elberse 1970; cf. Knapp 1967, Larcher 1976, Stöcker 1977) and others, it has become clear that not only the physiological amplitude but also mycorrhizal and bacterial activity in the soil and interaction between species, notably competition, are major factors determining the occurrence and performance of a species. Consequently, the ecological amplitude of a species (in which all biotic factors are taken into account), may be quite different from its physiological amplitude.

Apparently ecological amplitudes of species overlap, and frequently their optima or their lower or upper limits may coincide. This is a major cause in the formation of pattern in vegetation (sociological pattern sensu Kershaw 1973).

Another important source of variation in the species composition of phytocoenoses is brought about by historical factors. Dale (this symposium, Dale & Clifford 1976) and Pignatti (this symposium) have elucidated various of these factors (in completely different approaches). Dale showed how vegetational variation in a large area can be interpreted in an evolutionary perspective by analysing floristic relations between stands on higher idiotaxonomical levels (cf. van der Maarel 1972a, del Moral & Benton 1976). Thus, on the basis of taxonomic (or genetic) relationships a simplified picture might be obtained which enables an ecological interpretation of the vegetation of a given area.

Pignatti emphasized the importance of historical and genetical factors in the evolution of the vegetation of the western Mediterranean Basin. Certain immigrant or residual, possibly relict, plant taxa quickly evolved and adapted to the available niches as the climate changed towards the end of the ice ages, and made up new plant communities. This process is in effect rather common when new peripheral habitats are invaded (cf. Terborgh 1973). The vegetation types thus evolved should be studied in the light of the evolution of taxa in correlation with the development of their habitat.

Some practical implications of this niche occupation process during the development of woodland communities were discussed by Sissingh (this symposium): man selects and introduces suitable races of allochthonous plant species and by environmental management maintains a favourable environment for these exotics. Such management measures also maintain a diversified structure of the newly generated plant community (cf. also Brunig 1976) developing from the interactions of the spontaneously occurring local flora and the spontaneous regeneration of originally introduced exotics. These newly developing plant communities induce their own habitat in terms of litter development and turn-over and soil meso- and microbiological features.

Each species, by virtue of its ecological amplitude, is an environmental indicator. Hence, the phytocoenosis, the relatively homogeneous stand of vegetation at a particular site consisting of interacting plant populations (Whittaker 1970, Westhoff & van der Maarel 1973), is the best indicator of the species-environment relationships at that site. Consequently, the analysis of the phytocoenosis, and the subsequent identification of the plant community which is an abstraction or typification process related to the former (cf. von Glahn 1968, Westhoff & van der Maarel 1973) is the best tool to recognize and delineate the functional plant-habitat relationships at any site, and therefore, the best way towards identifying the ecosystem (cf. Major 1969, Ellenberg 1973, Werger 1977a).

The indicator value of plant communities can be used in landscape analysis as has been demonstrated by Lebrun (this symposium) and Stumpel & Kalkhoven (this symposium, also Kalkhoven, Stumpel & Stumpel-Rienks 1976). Such applications make use of plant community complexes in relation to the potential natural vegetation of landscape units (Seibert 1968, Trautmann 1972, Tüxen 1973, Werkgroep GRIM 1974).

Géhu (1974, this symposium) elaborated a formal classification system of so-called sigma-associations, i.e. association-complexes, which can be arranged in higher complex units. The value of this 'sigma-synsystematical' system for landscape ecology has still to be proven. In any case numerical methods like the ones Dale (this symposium, cf. van der Maarel, Orlóci & Pignatti 1976) presented will be an essential tool in the analysis and synthesis of such highly complex data.

The first attempts to understand vegetation patterns in the landscape were concerned with aspects of structure: physiognomy of the plant cover of the landscape was used to establish a classification of plant communities. Fairly soon it was realized, however, that such a classification did not allow a detailed interpretation of the landscape. For such a more detailed analysis the floristic criterium appeared to be indispensable. This idea was particularly developed in the Braun-Blanquet approach (Whittaker 1962, Westhoff & van der Maarel 1973, Werger 1974, van der Maarel 1975). The plant cover of an area was classified according to the distribution pattern shown by the species in that area in such a way that community types could be defined on co-occurrences as well as joint absences of species; the marked shift in species composition can be used to indicate vegetational boundaries which are supposed to be ecologically significant (cf. Van der Maarel 1976, Louppen, Werger & Eppink 1978). As these classificatory principles proved to be applicable at various levels of abstraction, an hierarchial system became obvious. An illustrative recent example of the difficulties involved in this classificatory approach is presented by Meyer (1976, this symposium). Thus, the basic unit of the floristic classification system, the association, was established. It was realized, from the beginning, that in most cases, an association being typified by a characteristic species composition and a specific constellation of environmental factors, also possesses a specific structure. Therefore, this property was included in the first definition of the association as approved of at the International Botanical Congress at Brussels in 1910.

Nevertheless the development of the Braun-Blanquet system was mainly based on the floristic analysis of phytocoenoses, in some cases at the expense of the structural homogeneity of plant community types; i.e. vegetation units with a weak floristic affinity to a syntaxon and a deviant structure were still assigned to that syntaxon (e.g. the subassociation *juniperetosum phoenicaeae* of the *Rosmarino-Lithospermetum* (Braun-Blanquet, Roussine & Nègre 1952) or the *Cichorietum intybus* of the alliance *Lolio-Plantaginion* (Sissingh 1969). In other cases, however, floristically rather similar plant communities differ strongly in structure, particularly in the vertical arrangement of plant biomass. It was especially Westhoff (1967, 1968) who drew attention to this situation and discussed a number of examples, e.g. the notorious, but according to Westhoff nevertheless real *Blechno-Quercetum extrasilvaticum*, a herb community in West Ireland. A similar feature was briefly mentioned by Moore (this symposium). Westhoff explained that this discrepancy was especially the case in areas where vegetation types met with a predominant, extreme, ecological factor. Such a factor could be a strong seasonal fluctuation between waterlogged and dry soil conditions, or extreme temperatures, extreme exposure, or frequent fires. Severe and prolonged overgrazing may cause similar effects (cf. Werger 1973, 1977b). In such cases structurally different but floristically similar vegetation types can occur next to one another, and Westhoff used Katz's term twin communities for this phenomenon.

Thus, important structural differences in the vegetation of an area which do not necessarily coincide with equally important differences in floristics can still be correlated with environmental differences. This has been elaborated in ecophysiological and ecomorphological research by Mooney and others, who worked on the ecological meaning of vegetation structure. The total structural variety within a vegetation type is analysed and the relationships between plant form and plant function are studied. This research first focused on the remarkable structural convergence in some geographically separated and floristically diverse vegetation types in comparable climatic regions. Some aspects of such a structural convergence were already noted at the beginning of this century (e.g. Bailey & Sinnott 1916), but particularly in the framework of the American IBP programme this structural convergence was systematically and thoroughly investigated (various papers in Di Castri & Mooney 1973, Mooney et al. 1970, Solbrig 1972, Mooney 1974, Parsons & Moldenke 1975, Parsons 1976, Gulmon 1977). In this branch of research the emphasis is not so much on the plant species, but rather on its functional form as part of the plant community. Mooney (1974) stated that for any given environmental condition there apparently is a plant form which represents the optimal form-behavioural strategy for carbon gain. Plant forms of this particular type will be dominant. This implies that structural research in vegetation should no longer emphasize the general physiognomy of the various phytocoenoses, but should concentrate on the structural-morphological and structural-functional characteristics of the plant species making up the phytocoenosis.

Not only the actual plant form occurring in a certain vegetation type is relevant for an understanding of the vegetation structure. Knowledge of the maximal performance capacity of a plant species in utilizing the habitat resources (cf. Grime & Hunt 1975) also is important and indicates the significance of distinguishing functional-ecological groups of plants (cf. Reed 1977) in a phytocoenosis. In old-field succession series (e.g. Parrish & Bazzaz 1976) such functional-ecological groups occur in a sequence leading to a higher degree of niche separation between the participating populations. Also the seasonal and yearly fluctuations in the annual weed communities discussed by Holzner (this symposium) and the vigorous responses of the annual *Suaeda maritima* and the biennial *Aster tripolium* following disturbances in the *Halimionetum portulacoidis* stands described by Beeftink et al. (this symposium) can be interpreted in this light, though Holzner pointed out that allelopathic effects also play a role in these dominant plant form patterns. The general strategical aspects of such sequences of annuals, biennials and perennials have been discussed recently by Hart (1977).

Mooney & Dunn (1970) suggested, that where the vegetation is open and thus competition for light will not be strong, a greater variety of morphological types may develop, all specializing in a different way on the efficient use of other resources for which there might be joint demands. A very clear example was demonstrated by Gimingham (e.g. 1972 and this symposium) who showed that the horizontal structure of a *Calluna*-heath (that is: the pattern in a *Calluna*-stand resulting from the age-differentiation of the *Calluna*-individuals) can be correlated directly with the types of companion species of *Calluna* according to their vertical structure or to their survival strategy. Gimingham distinguished six such strategy types of which only one is characterized as a competitor type and five are considered to be of a 'complementary strategy' type. A comparable example was presented by Barkman et al. (1977, this symposium)

who showed how the horizontal patterns in mosses (and some higher plants) directly correlate to the rather strong microclimatological pattern in the phytocoenosis, caused by the taller plants in the stand.

Another example is provided by the semi-arid savanna areas, where two basic growth forms, namely grasses with their shallow, but intensive root systems, and woody plants with their deeper but extensive root systems, together form an equilibrium and compete for the available moisture. Apart from these differences in growth forms that are relevant in the competition for the available moisture in these habitats, there is also a specialization in seasonal rhythm: the grasses are physiologically active from the first rains and rather quickly complete their seasonal cycle, while the woody plants profit more from the rains falling later in the season (Roux 1966). If the equilibrium in competition for the available moisture is disturbed, the growth possibilities of the dominant life forms are affected and the structure of the vegetation changes. This does not necessarily mean a change in qualitative floristic composition, though a quantitative change often is involved. If, for example, a severe and prolonged overgrazing in the semi-arid savanna ecosystem occurs the grass component is severely restricted in growth, or better, in moister usage. Thus, more moisture remains available in the soil to be used by the woody plants, and the result is a bush encroachment, a structural change towards a strongly more woody vegetation (cf. Walter 1954, 1973, Werger 1977b).

Different plant forms may also compete in the same way for a restricted resource. In the arid, sandy Kalahari, for example, the two important life forms are therophytes and hemicryptophytes, both characterized by a relatively shallow and intensive root system. Consequently these life forms are strong competitors for the strongly limited moisture supply. In several vegetation types in the Kalahari this competition is to some extent avoided as they possess relatively low percentages of these life forms amongst the constituent plant species. In only one community the percentages of both life forms are relatively high and thus competition can be expected to be relatively severe in this community. Here competition is largely avoided, however, because the community only occurs on the compact soils in dry riverbeds where the constituent species are physiologically active solely during the flooding periods when there is no shortage of moisture (Werger 1978).

Conclusions by Beeftink et al. on the changes in life strategies in life time and form of the species in stabilizing salt marsh vegetation under different habitat conditions, illustrate the complicated interactions between structure and floristic composition of vegetation types. They also show clearly how under certain habitat conditions, with a shorter or longer time-lag response, some plant strategy types may temporarily reach dominance, while gradually a more stable situation, usually with other dominants, evolves. Also the patchy distribution of locally abundant minerotrophic species in the ombrotrophic bogs on sites which are locally richer in nutrients, as discussed by Damman (this symposium), are expressions of the same phenomenon.

Usually the dominants of any vegetation type only use a proportion of the resources available: a generally adapted type (form) cannot compete with the specialized ones which are adapted to efficient utilization of only a certain segment of the available resources (Mooney 1974). Therefore, there is a variety of growth forms and functional-morphologic adaptations in almost any environment and together these forms indicate the response of the plant component to its environment both in stable and in evolving ecosystems. Malmer, Lindgren & Persson (this symposium) have shown how species of different growth form and height interact in a complicated, shifting picture of change in structure and floristic composition in a dense type of vegetation like the older growing forests of middle latitudes; due to the disruption of horse grazing and cutting an egalization of vegetation structure occurred in these forests, as open patches closed and the vertical arrangement of the phytomass became less patchy. This egalization certainly implied a decrease in available niches and thus in plant species and possibly also in plant forms. In natural communities this egalization will never reach the level of the even-aged monocultures of the old-fashioned foresters discussed by Sissingh. On a different scale and along another dimension Damman showed how patterns in floristically and structurally clearly different bog communities change markedly in a climate-related gradient from the coast inland, and how this changing pattern results in a different landscape altogether. It is clear that some promising possibilities for research lie in the contact zones between structurally differing vegetation types, like the tree-line at high altitudes or latitudes, the savanna-forest ecocline, the shrubland-dry grassland ecocline, the heathland-shrubland transition, the salt marsh patterns, etc.

There is often a clear morphological range in the populations of a certain species over its entire distribution area (cf. Meusel et al. 1965, Meusel 1976) and this can be explained ecologically.

A striking demonstration of this is presented by Landolt (this symposium), who sketched the series of 'Kleinarten' or ecological or morphological types within the sensu lato form of several species, each with their clearly definable habitat preferences. Such series can comprise a considerable number of taxa either gradually or distinctly differing in form, ecological preferences, or both. Some morphological differences, also if they indicate habitat differences, are modifications and are not genetically fixed (compare the findings of Beeftink et al.). It should be established experimentally whether one has to do with modifications or with morphologically and genetically distinct subspecific taxa. Usually only these latter, if they are rather sharply distinct, are attractive for use as differential taxa in phytosociological classifications.

There is, however, also a shift in ecological indicator value for a given species, even for a given morphological form of a species. As recently discussed by Werger & Van Gils (1976) and also indicated by Willems (this symposium) near the margins of their distribution area species are more specific environmental indicators than in the more central parts of their distribution area. This does not mean that a certain species indicates the same environmental characteristic, however coarsely defined, at all margins of its distribution area (cf. Meusel 1969). This was already found by Walter (1953) and formulated as the 'law of relative habitat constancy' (Gesetz der relativen Standortskonstanz). Near one marginal area a species is for instance an indicator of well-drained sites, while at another marginal area it rather serves as an indicator of lime, or of shady sites. This phenomenon is also known as ecological compensation (Vinogradov 1965, see Holland & Hove 1975)

Holzner (this symposium) explained that depending on the area of origin of weedy species their competitive power and their indicator value in the present distribution area change. Species occurring at segetal and ruderal sites towards the centre of their distribution areas, only occur ruderal at a more marginal part. Holzner cited similar examples of calciphilous weeds. The explanation for this so-called calciphily are clear: near the centre of a species distribution area its ecological amplitude allows it to meet the kinds, and intensities of the governing environmental factors and thus it is strongly competitive to other species, whilst towards the margins the conditions become quite different. One by one the various environmental factors (e.g. climatical factors) change and the tolerance limits of the species or ecotype for these factors are reached until one factor (in this case lime content) becomes critical and determines the occurrence or absence of the species. Therefore, species characterize the plant community, and hence the site and the ecosystem more distinctly near the margin of their distribution area. This was demonstrated by Hengeveld & Haeck (1978) who showed that a relatively large proportion of Dutch character-taxa for lower synsystematical units (Westhoff & Den Held 1969), i.e. associations and alliances, find a limit of their distribution area in or near the Netherlands, whilst a significantly lower proportion of character-taxa for orders and classes do so. Such observations confirm the more general statement of Adriani & Van der Maarel (this symposium) that some species are more sensitive than others.

Hundt (this symposium) described how in the youngest stages of the planted woods, species with the widest distribution range were most common, while their importance decreased in the older woodland stands and they were replaced by species with an increasingly smaller distribution area. This is probably another expression of the phenomenon described.

Turning back to synsystematics we may remark here that complex distribution patterns of characteristic species of certain syntaxa may complicate the syntaxonomy. Especially Oberdorfer (1957, 1968) paid attention to this problem. He sharply distinguished between 'association' 'regional association' and 'geographical race'. A regional association ('Gebietsassoziation') has a limited distribution area; it is characterized by a combination of regional (or local) character taxa. A geographical race is a regional expression of a larger regional or general association, without such a typical combination of alliance character taxa and regional association character taxa.

To understand these phenomena in a more quantitative way and to allow an ecological interpretation of the species-habitat interrelationships, and hence of the ecosystem, experimental ecological research is inevitable. Both autecological and interaction (particularly competition) experiments have to be carried out, wherever possible in relation to real field situations. The contributions by Van den Bergh, Oomes & Braakhekke (this symposium, see also van den Bergh 1968, 1973) and Blom (this symposium, see also 1973, 1976, 1977) present examples of such field-orientated experiments. In this connection, the type of work carried out by Westhoff and some of his collaborators and emphasized by Adriani & Van der Maarel and by Barkman in their papers, in exactly describing the phytosociological position of individual species and deriving general ecological 'clues' from these descriptions, remains a valuable ecological tool. It is also for that reason

that we still need the species, or equivalent taxonomic units like ecotypes for our descriptive and experimental plant ecological studies, as long as the plant-habitat relationships of the various species or ecotypes have not been properly ascertained and as long as the study of vegetation structure is still in its infancy.

It will take still many years to advance along these two lines of research and with Pignatti and others we may doubt whether we will have any well-developed ecosystems left by the time we are really able to understand them. It is not surprising therefore that so many distinguished phytosociologists are devoted to nature conservation and Westhoff (e.g. 1970, 1971) is both an outstanding and inspiring example!

References

Bailey, I.W. & Sinnott, E.W. 1916. The climatic distribution of certain types of Angiosperm leaves. Amer. J. Bot. 3: 24–39.
Barkman, J.J., Masselink, A.K. & Vries, B.W.L. de. 1977. Über das Mikroklima in Wacholderfluren. In: H. Dierschke (ed.), Vegetation und Klima. Ber. Int. Symp. Rinteln 1975. pp. 35–81. Cramer, Vaduz.
Bergh, J.P. van den. 1968. An analysis of yields of grasses in mixed and pure stands. Agric. Res. Rep. Wageningen 714: 1–71.
Bergh, J.P. van den. 1973. Competitive studies based on mono- and bicultures. Acta Bot. Neerl. 22: 259–260.
Bergh, J.P. van den & Elberse, W. Th. 1970. Yields of monocultures and mixtures of two grass species differing in growth habit. J. App. Ecol. 7: 311–320.
Blom, C.W.P.M. 1973. The influence of trampling and soil compactness on the distribution of some Plantago species. In: Inst. of Ecol. Res. Progress Rep. 1972: 106–112.
Blom, C.W.P.M. 1976. Effects of trampling and soil compaction on the occurrence of some Plantago species in coastal sand dunes. I. Soil compaction, soil moisture and seedling emergence. Oecol. Plant. 11: 225–241.
Blom, C.W.P.M. 1977. Effects of trampling and soil compaction on the occurrence of some Plantago species in coastal sand dunes. II. Trampling and seedling establishment. Oecol. Plant. 12: 363–381.
Braun-Blanquet, J., Roussine, N. & Nègre, R. 1952. Les groupements végétaux de la France méditerranéenne. C.N.R.S., Montpellier.
Brunig, E.F. 1976. Ökologische Stabilität von forstlichen Monokulturen als Problem der Bestandesstruktur. Verh. Ges. Ökologie Göttingen 1976: 189–204.
Castri, F. di & Mooney, H.A. (eds.) 1973. Mediterranean type ecosystems. Ecol. Stud. 7. Springer, Berlin-New York.
Dale, M.B. & Clifford, H.T. 1976. On the effectiveness of higher taxonomic ranks for vegetation analysis. Austr. J. Ecol. 1: 37–62.
Ellenberg, H. 1953. Physiologisches und ökologisches Verhalten der selben Pflanzenarten. Ber. Dtsch. Bot. Ges. 65: 350–361.

Ellenberg, H. 1973. Ökosystemforschung. Springer, Berlin.

Géhu, J.-M. 1974. Sur l'emploi de la méthode phytosociologique sigmatiste dans l'analyse, la définition et la cartographie des paysages. C.R. Acad. Sc. 279: 1167–1170.

Gimingham, C.H. 1972. Ecology of heathlands. Chapman & Hill, Edinburgh.

Glahn, H. von. 1968. Der Begriff des Vegetationstyps im Rahmen eines allgemeinen naturwissenschaftlichen Typenbegriffes. In: R. Tüxen (ed.), Pflanzensoziologische Systematik. Ber. Int. Symp. Stolzenau 1964. pp. 1–13. Junk, The Hague. Published earlier in Ber. Geobot. Inst. Rübel 36: 14–27.

Grime, J.P. & Hunt, R. 1975. Relative growth-rate: its range and adaptive significance in a local flora. J. Ecol. 63: 393–422.

Gulmon, S.L. 1977. A comparative study of the grassland of California and Chile. Flora 166: 261–278.

Hart, R. 1977. Why are biennials so few? The Amer. Nat. 111: 792–799.

Hengeveld, R. & Haeck, J. 1978. The distribution of abundance. J. Biogeogr.: in press.

Holland, P.G. & Hove, A.R.T. 1975. The distribution of Euphorbia candelabrum in the Southern Rift Valley, Kenya. Vegetatio 30: 49–54.

Kalkhoven, J.T.R., Stumpel, A.H.P. & Stumpel-Rienks, S.E. 1976. Landelijke Milieu-kartering. Environmental Survey of the Netherlands, a landscape ecological survey of the natural environment in The Netherlands for physical planning on national level. Staatsuitgeverij, The Hague.

Kershaw, K.A. 1973. Quantitative and dynamic ecology. 2nd. ed. Elsevier, New York.

Knapp, R. 1967. Experimentelle Soziologie. Ulmer, Stuttgart.

Larcher, W. 1973. Ökologie der Pflanzen. UTB 232. Ulmer, Stuttgart.

Louppen, J.M.W., Werger, M.J.A. & Eppink, J.H.M. 1978. Vegetation patterns and species performances along an environmental gradient. 2nd Int. Congr. Ecology, Jerusalem.

Maarel, E. van der. 1972a. Ordination of plant communities on the basis of their plant genus, family and order relationships. In: E. van der Maarel & R. Tüxen (eds.), Basic problems and methods in phytosociology. Ber. Int. Symp. Ver. Vegetationskunde, Rinteln 1970: 183–190. Junk, The Hague.

Maarel, E. van der. 1972b. On the transformation of cover-abundance values in phytosociology. Report Bot. Lab., Nijmegen.

Maarel, E. van der. 1975. The Braun-Blanquet approach in perspective. Vegetatio 30: 213–219.

Maarel, E. van der. 1976. On the establishment of plant community boundaries. Ber. Dtsch. Bot. Ges. 89: 415–443.

Maarel, E. van der, Orlóci, L. & Pignatti, S. 1976. Data-processing in phytosociology, retrospect and anticipation. Vegetatio 32: 65–72.

Major, J. 1969. The historical development of the ecosystem concept. In: G.M. van Dyne (ed.), The ecosystem concept in in natural resource management. pp. 9–22. Academic Press, London.

Meusel, H. 1969. Chorologische Artengruppen der mitteleuropäischen Eichen-Hainbuchenwälder. Feddes Repert. 80: 113–132.

Meusel, H. 1976. Die Evolution der Pflanzen in pflanzengeographisch-ökologischer Sicht. In: H. Boehme, R. Hagemann & R. Loether (eds.), Beiträge zur Geschichte und Abstammungslehre. pp. 521–555. Berlin.

Meusel, H., Jäger, E. & Weinert, E. 1965. Vergleichende Chorologie der zentraleuropäischen Flora. Fischer, Jena.

Meyer, M. 1976. Pflanzensoziologische und ökologische Untersuchungen an insubrischen Trockenwiesen karbonathaltiger Standorte. Veröff. Geobot. Inst. ETH, Stift. Rübel, 57: 1–145.

Mooney, H.A. 1974. Plant forms in relation to environment. In: B.R. Strain & W.D. Billings (eds.), Vegetation and environment. Handbook of vegetation science 6: 111–122. Junk, The Hague.

Mooney, H.A. & Dunn, E.L. 1970. Photosynthetic systems of mediterranean climate shrubs and trees of California and Chile. The Amer. Nat. 104: 447–453.

Mooney, H.A., Dunn, E.L., Shropshire, F. & Song, L. 1970. Vegetation comparisons between the mediterranean climatic areas of California and Chile. Flora 159: 480–496.

Moral, R. del & Benton, M. F. 1977. Analysis and classification of vegetation based on family composition. Vegetatio 34: 155–165.

Oberdorfer, E. 1957. Süddeutsche Pflanzengesellschaften. Fischer, Jena.

Obendorfer, E. 1968. Assoziation, Gebietsassoziation, Geographische Rasse. In: R. Tüxen (ed.), Pflanzensoziologische Systematik. Ber. Symp. Int. Ver. Vegetationskunde, Stolzenau 1964: 124–131. Junk, The Hague.

Parrish, J.A.D. & Bazzaz, F.A. 1976. Underground niche separation in successional plants. Ecology 57: 1281–1288.

Parsons, D.J. 1976. Vegetation structure in the Mediterranean scrub communities of California and Chile. J. Ecol. 64: 435–447.

Parsons, D.J. & Moldenke, A.R. 1975. Convergence of vegetation structure along analogous climatic gradients in California and Chile. Ecology 56: 950–957.

Reed, F.C.P. 1977. Plant species number, biomass accumulation and productivity of a differentially fertilized Michigan old-field. Oecologia 30: 43–53.

Roux, P.W. 1966. Die uitwerking van seisoensreënval en beweiding op gemengde Karooveld. Proc. Grassl. Soc. Sth. Afr. 1: 103–110.

Seibert, P. 1968. Gesellschaftsring und Gesellschaftskomplex in der Landschaftsgliederung. In: R. Tüxen (ed.), Pflanzensoziologie und Landschaftsökologie. Ber. Symp. Int. Ver. Vegetationskunde, Stolzenau 1963: 48–60. Junk, The Hague.

Sissingh, G. 1969. Über die systematische Gliederung von Trittpflanzen-Gesellschaften. Mitt. Flor. Soz. Arbeitsgem. NF 14: 179–192.

Solbrig, O.T. 1972. New approaches to the study of disjunctions with special emphasis on the American amphitropical desert disjunctions. In: Valentine, D.H. (ed.), Taxonomy, phytogeography and evolution. pp. 85–100. London, Acad. Press.

Stöcker, G. 1977. Ein Modell der Dominanzstruktur und seine Anwendung. 2. Bioindikation, allgemeine Ergebnisse. Arch. Naturschutz u. Landschaftsforsch. 17: 89–118.

Terborgh, J. 1973. On the notion of favorableness in plant ecology Amer. Nat. 107: 481–501.

Trautmann, W. 1972. Erläuterungen zur Karte 'Vegetation' (potentielle natürliche Vegetation). Deutscher Planungsatlas Band I: Nordrhein-Westfalen. Lieferung 3: Gebrüder Jänecke Verlag, Hannover.

Tüxen, R. 1973. Vorschlag zur Aufnahme von Gesellschaftskomplexen in potentiell natürlichen Vegetationsgebieten. Acta Bot. Acad. Scient. Hung. 19: 379–384.

Vinogradov, B.V. 1965. Ecological compensation and replaceability, and the extrapolation of plant indicators. In: A.G. Chikishev (ed.), Plant indicators of soils, rocks and subsurface waters. pp. 180–187. Consultants Bureau, New York.

Walter, H. 1954. Die Verbuschung, eine Erscheinung der subtropischen Savannengebiete, und ihre ökologischen Ursachen. Vegetatio 5–6: 6–10.

Walter, H. 1973. Die Vegetation der Erde. Bd. I. 3. Aufl. Fischer, Stuttgart.

Walter, H. & E. 1953. Einige allgemeine Ergebnisse unserer Forschungsreise nach Südwestafrika 1952/53: Das Gesetz der relativen Standortskonstanz; das Wesen der Pflanzengesellschaften. Ber. Dtsch. Bot. Ges. 66: 228–236.

Werger, M.J.A. 1973. Phytosociology of the upper Orange River Valley, South Africa. A syntaxonomical and synecological study. Thesis Nijmegen. V & R, Pretoria.

Werger, M.J.A. 1974. The place of the Zürich-Montpellier method in vegetation science. Folia Geobot. Phytotax. 9: 99–109.

Werger, M.J.A. 1977a. Applicability of Zürich-Montpellier methods in tropical and subtropical range vegetation in Africa. In: W. Krause (ed), Handbook of Vegetation Science. Vol. 13: 123–145: Application of vegetation science to grassland husbandry. Junk, The Hague.

Werger, M.J.A. 1977b. Environmental destruction in southern Africa: the role of overgrazing and trampling. In: A. Miyawaki & R. Tüxen (eds.), Vegetation Science and Environmental Protection. Proc. Int. Symp. Tokyo 1974. pp. 301–305. Maruzen & Co, Tokyo.

Werger, M.J.A. 1978. Vegetation structure in the southern Kalahari. J. Ecol. In press.

Werger, M.J.A. & Gils, H. van. 1976. Phytosociological classification problems in chorological border line areas. J. Biogeogr. 3: 49–54.

Werkgroep G.R.I.M. 1974. Landschapsecologische basisstudie voor het streekplangebied Midden-Gelderland. Provinciale Planologische Dienst, Arnhem.

Westhoff, V. 1967. Problems and use of structure in the classification of vegetation. The diagnostic evaluation of structure in the Braun-Blanquet system. Acta Bot. Neerl. 15: 495–511.

Westhoff, V. 1968. Einige Bemerkungen zur syntaxonomischen Terminologie und Methodik, insbesondere zu der Struktur als diagnostischen Merkmale. In: R. Tüxen (ed.), Pflanzensoziologische Systematik. Ber. Int. Symp. Stolzenau 1964. pp. 54–70. Junk, The Hague.

Westhoff, V. 1970. New criteria for nature reserves. New Scientist 46: 108–113.

Westhoff, V. 1971. Choice and management of nature reserves in the Netherlands. Bull. Jard. Bot. Nat. Belg. 41: 231–245.

Westhoff, V. & Den Held, A.J. 1969. Plantengemeenschappen in Nederland. Thieme, Zutphen.

Westhoff, V. & Maarel, E. van der. 1973. The Braun-Blanquet Approach. In: R.H. Whittaker (ed.), Handbook of Vegetation Science Vol. 5: 617–726. Junk, The Hague.

Whittaker, R.H. 1962. Classification of natural communities. Bot. Rev. 28: 1–239.

Whittaker, R.H. 1970. The population structure of vegetation. In: R. Tüxen (ed.), Gesellschaftsmorphologie. Ber. Int. Symp. Rinteln 1966. pp. 39–62. Junk, The Hague.

Reference to original publications in VEGETATIO

W. G. Beeftink et al.	1978. Vegetatio 36: 31– 43
J. Braun-Blanquet	1978. Vegetatio 36: 115–117
M. B. Dale	1977. Vegetatio 35: 131–136
A. W. H. Damman	1977. Vegetatio 35: 137–151
J.-M. Géhu	1977. Vegetatio 34: 117–125
C. H. Gimingham	1978. Vegetatio 36: 179–186
W. Holzner	1978. Vegetatio 38: 13– 20
A. O. Horvath	1978. Vegetatio 37: 119–122
R. Hundt	1978. Vegetatio 38: 1– 12
E. Landolt	1977. Vegetatio 34: 179–189
J. Lebrun	1977. Vegetatio 35: 123–129
N. Malmer et al.	1978. Vegetatio 36: 17– 29
M. Meyer	1977. Vegetatio 35: 107–114
S. Pignatti	1978. Vegetatio 37: 175–185
G. Sissingh	1977. Vegetatio 35: 187–191
A. H. P. Stumpel & J. T. R. Kalkhoven	1978. Vegetatio 37: 163–173
J. H. Willems	1978. Vegetatio 37: 141–150

List of unpublished contributions to the symposium

J. J. Barkman: Ecological aspects of microvariation in Juniper scrub

J. P. van den Bergh, W. G. Braakhekke & M. J. M. Oomes: Species diversity in grasslands

C. Blom: Effects of trampling and soil compaction on the occurrence of some Plantago species in coastal sand dunes

L. F. M. Fresco: An autecological model for the analysis of vegetation

J. J. Moore: Heathland vegetation of Ireland